Textbooks in Telecommunication Engineering

Series Editor
Tarek S. El-Bawab, PhD
Jackson State University
Jackson, MS, USA

Telecommunications have evolved to embrace almost all aspects of our everyday life, including education, research, health care, business, banking, entertainment, space, remote sensing, meteorology, defense, homeland security, and social media, among others. With such progress in Telecom, it became evident that specialized telecommunication engineering education programs are necessary to accelerate the pace of advancement in this field. These programs will focus on network science and engineering; have curricula, labs, and textbooks of their own; and should prepare future engineers and researchers for several emerging challenges. The IEEE Communications Society's Telecommunication Engineering Education (TEE) movement, led by Tarek S. El-Bawab, resulted in recognition of this field by the Accreditation Board for Engineering and Technology (ABET), November 1, 2014. The Springer's Series Textbooks in Telecommunication Engineering capitalizes on this milestone, and aims at designing, developing, and promoting high-quality textbooks to fulfill the teaching and research needs of this discipline, and those of related university curricula. The goal is to do so at both the undergraduate and graduate levels, and globally. The new series will supplement today's literature with modern and innovative telecommunication engineering textbooks and will make inroads in areas of network science and engineering where textbooks have been largely missing. The series aims at producing high-quality volumes featuring interactive content; innovative presentation media; classroom materials for students and professors; and dedicated websites. Book proposals are solicited in all topics of telecommunication engineering including, but not limited to: network architecture and protocols; traffic engineering; telecommunication signaling and control; network availability, reliability, protection, and restoration; network management; network security; network design, measurements, and modeling; broadband access; MSO/cable networks; VoIP and IPTV; transmission media and systems; switching and routing (from legacy to next-generation paradigms); telecommunication software; wireless communication systems; wireless, cellular and personal networks; satellite and space communications and networks; optical communications and networks; free-space optical communications; cognitive communications and networks; green communications and networks; heterogeneous networks; dynamic networks; storage networks; ad hoc and sensor networks; social networks; software defined networks; interactive and multimedia communications and networks; network applications and services; e-health; e-business; big data; Internet of things; telecom economics and business; telecom regulation and standardization; and telecommunication labs of all kinds. Proposals of interest should suggest textbooks that can be used to design university courses, either in full or in part. They should focus on recent advances in the field while capturing legacy principles that are necessary for students to understand the bases of the discipline and appreciate its evolution trends. Books in this series will provide high-quality illustrations, examples, problems and case studies. For further information, please contact: Dr. Tarek S. El-Bawab, Series Editor, Department of Electrical and Computer Engineering, Jackson State University, telbawab@ieee.org; or Mary James, Senior Editor, Springer, mary.james@springer.com

More information about this series at http://www.springer.com/series/13835

Eugene I. Nefyodov • Sergey M. Smolskiy

Electromagnetic Fields and Waves

Microwave and mmWave Engineering with Generalized Macroscopic Electrodynamics

 Springer

Eugene I. Nefyodov
Durban Technological University (RSA)
Durban, South Africa

Sergey M. Smolskiy
Durban Technological University (RSA)
Durban, South Africa

Additional material to this book can be downloaded from http://extras.springer.com.

ISSN 2524-4345   ISSN 2524-4353   (electronic)
Textbooks in Telecommunication Engineering
ISBN 978-3-030-08114-0   ISBN 978-3-319-90847-2   (eBook)
https://doi.org/10.1007/978-3-319-90847-2

This Springer imprint is published by the registered company Springer Nature Switzerland AG
The registered company address is: Gewerbestrasse 11, 6330 Cham, Switzerland

This book is devoted to:
Kotelnikov Institute of Radio Engineering and Electronics
of the Russian Academy of Sciences and

National Research University
"Moscow Power Engineering Institute"

Foreword

University teachers (professors, associate-professors, senior lecturers) perfectly understand that a clearly dedicated lecture course, which is supported by a textbook, practical laboratory training, exercises, and computer modeling, is the one of the main common properties of the modern university. Such a lecture course (on the definite dedicated educational direction) has been conceived, formed, and developed as a result of the longstanding experience of leading professors in their work with students and in their research projects in the interests of industry, in which these professors and active students participate. These lecture courses are permanently "polished" on the basis of new, developing technologies in science, industry, and in education, which introduce the constant need for renewal of student courses, not only in particular applied programs, but also in fundamental subjects.

The educational course in electromagnetics, which can have different names in different programs (for instance, electrodynamics, electromagnetic fields and waves, excitation of electromagnetic waves, etc.), is, in essence, a fundamental lecture course, which may be necessary for students of technology at classical universities, who are trained in electronics, telecommunications, radio physics, radio biology, medicine, and many other specialties. Of course, the specialization (dedication) of such a fundamental lecture course is different for future engineers, specialists in the excitation and propagation of electromagnetic waves, physicists, biologists, medical students, etc. Moreover, students of various fields have a different preliminary knowledge in mathematics, physics, and computers; therefore, the fundamental lecture course in electromagnetics should be oriented toward the future specialty of the student trained at university.

I have known the authors of this university lecture course for many decades as the much-experienced professors, who have delivered and are delivering lectures to students at many Russian and foreign universities. The authors permanently participate in research projects in the various fields of radio-engineering and electronics, which serve as the basis for the modernization of students' lecture courses.

This textbook, on the one hand, supports the fundamental lecture course in electrodynamics, but it also represents new important opportunities.

1. This lecture course not only relies on the strict mathematical description of electromagnetic phenomena (usually, similar courses are introduced into the program of elder academic years of the university, after studying the complex mathematical issues, for instance, equations of mathematical physics). On the contrary, authors rely on the more exact physical description, which (as is well-known for university teachers) is much more easily remembered by the students than the mathematical transformations. Evidently, the authors cannot manage without mathematics in this lecture course: classic Maxwell's equations, the laws of Coulomb, Ampere, Faraday, and many other classical laws of electromagnetism, but mathematics in this educational course is not the "main" instrument for study; it is used together with physical explanation. This is especially important, for example, not for the explanation, say, of classic metallic waveguides, but the waveguide structures of the new type: strip, microstrip, slot, corrugated, dielectric, etc., for which the strict mathematics is at

the developmental stage. A similar approach is not usual for students' courses of electromagnetics, but it makes this textbook available for students of different specialties, who have only an initial knowledge of higher mathematics.

2. Another distinctive feature of the offered textbook is that the authors, who know the classic Maxwell's theory of electromagnetism in depth, clearly see some of its drawbacks (by the way, openly mentioned by Maxwell and his fellow-fighters), which lead to the so-called paradoxes of electrodynamics. In very recent years, many experts in electromagnetics describe many of these paradoxes, which cannot be explained by Maxwell's theory. We know perfectly well that many parts of physics had a hard time in their development, which was considered a crisis phenomenon, the outlets from which opened up new doors in physics, yielding to new theories, new phenomena, new technologies, and devices. This textbook is probably the first students' textbook, in which authors (having discovered the new serious stratum of results of many modern authors) describe at student level the paradoxes of Maxwell's theory discussed in the scientific publications. An outlet is offered by many experts in the form of the introduction of the so-called "scalar magnetic field," which provides symmetry to classic equations by introducing new terms. This approach is often criticized by many experts, who stick to the classic Maxwell's equations: strictly speaking, the scalar magnetic field in the explicit form has not yet observed by anybody, but in Maxwell's times, for many years, nobody observed the electromagnetic Maxwellian field! In this text book, probably for the first time, these contradictions of classical electromagnetism are described for students. Let us remember that Heinrich Hertz, who factually proved the existence of the electromagnetic field, performed his famous investigation at a very young age (less than 30, like all modern students).

3. The third feature of this textbook is that the authors insist on the following position: in the absence of a strict theory on existing waveguide structure operation, special significance should be given to methods of computer modeling of processes in such structures. Thus, the software package for such modeling should be simple and must provide very fast independent modeling with variation of many parameters of the task, which is necessary for engineering practice. Authors have chosen the software packet MathCAD for this purpose, with which modern students, as a rule, are acquainted. For the various chapters of this textbook, the authors developed almost 50 programs for the modeling of processes in the different devices. Students and engineers who design waveguide transmission lines and resonance structures based on it, have the possibility of independently changing parameters and better understanding its operation. This is doubtless the advantage of this textbook.

4. Of course, any student textbook, besides the main subject, should teach the succession of science. Students should know their ancestors. Therefore, the authors include a list of leading world-wide scientists who made a serious contribution to the foundations of electromagnetism. Of course, it is impossible to enumerate all experts, and this list includes the very main contributors. Thus, taking into consideration that in western scientific and especially educational literature, there is rather poor information about many scientific schools of electromagnetism acting in Russia, the authors include in this list many Russian experts in electromagnetism.

The aforementioned distinctive features of the present textbook, to my mind, will be of interest to various universities worldwide and it will be used by students of many specialties. Obviously, PhD candidates and researchers from industry will have an interest in this book.

William I. Kaganov
Doctor of Engineering Sciences,
Professor of Moscow State University of
Radio Engineering, Electronics and Automatics
Moscow, Russia

Preface

The educational lecture course offered in this book has been composed as the corresponding systematic lecture course according to the Russian Educational Standard for the dedicated discipline "*Electromagnetic Fields and Waves*" in the educational direction 210700 – "Information communication technologies and communication systems" (Bachelor's course), and also 210400 "Radio Engineering" (Bachelor's course) for university students on specialty 200800 – "*Design and technology of radio electronic means.*" It can be also used for Bachelor's, engineers and Master's in engineering in directions 200700 ("*Radio Engineering*"), 201100 ("*Radio communications, broadcasting, and television*"), 201000 ("*Multi-channel telecommunication systems*"), 200900 ("*Communication systems for mobile objects*"), 071700 ("*Physics and technique of optical communication*"), and some other directions of professional education, for which electrodynamics is the fundamental discipline. We can list here the following disciplines: "*Engineering electrodynamics,*" "*Electromagnetic fields and waves,*" "*Antenna-waveguide devices,*" "*HF devices and antennas,*" "*Electrodynamics and radio wave propagation,*" "*HF devices,*" "*Satellite and relay systems for information transmission,*" etc.

In spite of the orientation toward the Russian educational system in this discipline, the authors hope that the educational programs of leading western and eastern universities for include educational disciplines of similar directions. In this case, the authors' multi-decade experience of teaching and the constant modifications of issues, which are considered in this book, will be interesting for undergraduate, graduate, and PhD students and teachers of various universities worldwide. In addition, authors perform the long-term scientific and technological investigations in the field of modern electrodynamics, radio wave propagation, microwave and millimeter-wave technologies, and, as always occurs in universities, new modern knowledge is introduced into educational courses, which serves as training modifications for specialists in higher education.

In this lecture course, the main laws of classic and generalized electrodynamics are briefly described. The deductive approach (from general to specific) is taken for presentation and explanation, when, at first, the *general electrodynamics regulations* are postulated. Mainly, attention is paid to high-frequency electrodynamics, and then different specific cases are analyzed. At the same time, frequently, the static problem solution, for example, is used as the zero approximation for obtaining a solution for a high-frequency region. In some cases, this approach allows, first, the required solution to be obtained by the shortest path (and sometimes, with its error estimation) and, second, to the features of the physical content of the problem to be discovered most clearly, which is especially important in pedagogical activity. Precisely the latter circumstance seems to the authors to be the main and crucial one in the training system of the modern engineer, the specialist with a complete higher education, the PhD student, and the researcher of scientific and technological institutions. This is especially significant in the period of the general enthusiasm for personal computers, when many persons seem to think that by having a powerful enough computer (personal or network), any problem can be solved. Nevertheless, it is our deep and long-term pedagogical and research belief that this occurs far from always, and having obtained (independently, from books, or from a consultant-professor) a clear enough insight into the *task of physics*, one can proceed to its *analysis* and/or *synthesis* under properties (characteristics, parameters) given in advance.

The present book was written during a remarkable period for electrodynamics and physics as a whole (this was a motivation for the authors), when in the worldwide scientific community in electrodynamics problems, the new knowledge and a large number of results (sometimes as paradoxes) were accumulated, for which the "usual traditional" electrodynamics cannot give a reasonable explanation. In essence, this always takes place: new facts and experimental results do not remain within the Procrustean bed of existing theory, and new approaches are required, which, at times, change the classic interpretations strongly enough. Hence, we need new approaches to the fundamentals of physics and to electrodynamics in particular. There are many such theories; they have different areas of application, different restrictions, different methods, etc.

It is already good that the "spring ice drift begins in rivers" and, although this is not always supported by official scientific and technological persons, this always occurs and this will always occur in the future: new ideas always have a difficult time breaking through.

The authors are glad that toward the end of their pedagogical life, we are able to dip into an avalanche of new ideas and to try to explain our insights into much new knowledge, to compare this with classic ideas, which still "work" in many problems and will work within the new limits for a long time, and to propose how to pass new knowledge onto a large group of university students of different fields.

The offered lecture course entitled "*Electromagnetic fields and waves*" has at least two features that make it different from the usual traditional (or associated) lecture courses of the same name, and from the others in which electrodynamics plays an important, decisive role. First of all, this course is partially computerized, i.e., most processes and devices are accompanied by software in the MathCAD medium, which allow readers to independently perform calculations and experiments for extension, deepening, and specification of the described theoretical materials. More completely, this issue is described in the Russian book by S.B. Kliuev and one of the authors of this book [1], and also in the book [2]. In essence, the idea of the computerized lecture course in physics and electronics is not new. Many interesting methodological and scientific results have been obtained by Russian professor V.I. Kaganov in his books [3]. To date, evidently, we understand that only via the *digit* can we become acquainted to the surrounding world and, in particular, the world of physics, engineering, and, of course, the infinite world of radio electronics. Because of the above, great attention in this book is paid to the utilization of a unique and sufficiently simple educational approach for not only future engineers, but for specialists from industry, especially young persons. The problem is that reforms of the educational system within the limits of the modern development of Russia adversely affected the state financing of both higher education and of science. Russian students and engineers from industry actually lost the possibility of performing natural experiments, without which (as the famous Russian scientist *M.V. Lomonosov* wrote) there is no education or science. It is important that this situation is typical not only for Russia and countries of the former USSR. Western professors and researchers also complain about financing limitations and, obviously, all of us cannot be sure to think about improving this situation. The same "loophole" is the wide simulation of experiences, processes, technology, and theoretical analysis using modern computers and computing systems. This is exactly why, therefore, the authors of this book emphasize the application of the distributed promising software package MathCAD (or other modern packages) in this lecture course. Great attention is paid to this issue in our course, although we do not want to move all education to computers: we cannot rear a good specialist without a clear understanding of the essence of physical processes.

Numerous examples are represented in our lecture course, in which the final results are given not only in the form of an algorithm of a boundary problem solution or its analytical solution (which is not always available to the reader), but in the form of programs in the MathCAD medium. This represents for the reader an excellent opportunity to performing the calculation personally and independently together with a physical analysis of the particular process of the device and to reveal its physical content. In most practical cases, it is difficult or even often impossible to carry out this analysis. Of course, the classic method of equivalent electromagnetic circuits still remains, but it is extremely not simple, not always understandable, requiring large efforts and long-term accumulated intuition from a designer (see, for instance, Stratton, and Hoffmann [29, 113]).

Together with the "usual" system of description of Maxwell's equations – classic macroscopic electrodynamics – another new feature of our lecture course is the discussion of generalized macroscopic electrodynamics, which is not considered in students' courses. The problem is that since the time (1873), when *James Clerk Maxwell* suggested his famous equations, much time had gone by and a large number of cases were revealed (they are usually called "paradoxes"), when these equations of classic macroscopic electrodynamics do not permit explanation of some experimental results. Many scientists and engineers from different countries tried to generalize Maxwell's equations and give them another form. Apparently, the most important results in this direction were obtained for the first time by Russian scientist *N.P. Khvorostenko* [5], who was a messmate of one of the authors of this book (they formed together a mutual scientific world view more than half a century ago). The results of Russian scientists *N.P. Khvorostenko* and *A.A. Protopopov* with an account of the considerations of other authors (including our results in Khvorostenko, and Nefyodov et al. [5, 6)]) are described very briefly in Nefyodov [38]. In these publications, the quantum effects were touched upon.

Owing to ideas of generalized electrodynamics, we can consider a problem of longitudinal electromagnetic waves, which has led to a strong and controversial discussion in the worldwide community [7, 10]. We would like to note here the old problem of unity of electromagnetic and mechanic phenomena, which arose as far back as 1903 [8] and continues today [43, 44].

We shall mainly describe in this book the generalization of classic macroscopic electrodynamics based on the results of *A.K. Tomilin*, *G.V. Nikolaev*, and *V.V. Erokhin* [7, 10]. Using this approach, we can clarify the series of complexities, paradoxes of classic electrodynamics, and give them a rather full physical treatment. It is clear that generalized classic macroscopic electrodynamics is not "a truth in the last instance"; however, its appearance and some new results constitute an undoubted step forward in the construction of perfect electrodynamics.

Together with traditional materials, which are usually encountered in lecture courses on "*Electromagnetic fields and waves*," we include in this book a series of new issues in particular related to the theory and calculation of *integrated circuit* (IC) elements in microwave and millimeter-wave ranges and their highest stage of development: *three-dimension integrated circuit* (3D-IC) of microwave and millimeter waves [13–17, 123], based on the fundamental principles of 3D-IC: the *principle of basing element optimality* and the *principle of constructive conformity*. Naturally, increased attention is paid in the book to this circle of themes, not only because one of the authors had an immediate relationship with the introduction and substantiation of many 3D-IC applications, but also because 3D-IC application leads to tremendous successes in the creation of *ultra-fast information processing systems* (UFIPS) operating directly in radio frequency at the highest frequencies. Thus, the operation speed of UFIPS increases by 1–3 orders and mass-dimension parameters of the high-frequency module can be decreased in such a manner. The advantages of 3D-IC are especially manifested in the designing, for example, of beam-forming matrices for *phased antenna arrays* (PAAs), active PAAs, adaptive PAAs, filters for spatial–temporal signal processing, and many other devices and systems. Great successes in the filter development for spatial–temporal signal processing are attributed to *H.F. Harmuth* [22]. Nevertheless, their implementation in acceptable (for aerospace practice) sizes was realizable only with 3D-IC application.

Frequently, it happens that, as it would seem, the well-studied and widely used high-frequency guiding structure has a series of previously unnoticed remarkable properties. Thus, for instance, the coaxial transmission line (a waveguide) is used on the main transverse, so-called, T (or TEM) mode. At the same time, utilization of one class of higher types of *magnetic waves* allows the creation of open cavities with a *uniquely rare spectrum of fundamental oscillations* based on the coaxial line [23]. Similar opportunities, as could be shown, are typical of a series of other guiding structures (see, for instance, Neganov et al. [18, 19]).

The active familiarization of the described materials allows the reader to proceed to original publications, mentioned in the list of recommended literature, in addition to deep assimilation with other lecture courses in the chosen direction, and prepares graduate and PhD students for performing different projects and theses. Under the term "active familiarization" we understand not only (and may be not so much) assimilation of the material itself, but also the necessary solution of tasks and exercises suggested in our books, a creative attitude to project fulfillment, the necessary research activity at a university department in one of the research directions. This interaction with *electrodynamics* should necessarily be the continuous everyday work, and will pave the way to success.

We would like to add one consideration to this, based on our own experience. This can be reduced to a simple rule: *we must be prepared in advance for each new lecture* and then the effectiveness of the assimilation of new material will be repeatedly increased, a deep insight will be achieved, and, therefore, the retention and possibility of using it in practice will come.

As the basis of this book, we use the lectures that were delivered over more than 40 years in different Russian universities: Taganrog Radio Engineering Institute, Moscow Power Engineering Institute (MPEI), Friazino branch of MPEI, and Moscow Institute of Radio Engineering, Electronics, and Automatics. Some of the materials were delivered at Bauman Moscow State Technical University, in lectures for engineers and PhD students in Central R&D Institute of Communication, R&D Institute of Precision Machinery. The authors delivered lectures in Germany, Kazakhstan, Poland, USA, France, Czech Republic, Argentina, Brazil, and China.

The smooth and natural entrance into the circle of considered problems, ideas, and methods can be used by readers after assimilation with a small book [25]. The role of electrodynamics is the basis for both radio engineering and other disciplines necessary for a radio electronic engineer, but not only for them.

We would like to note that Soviet and Russian schools of thoughts on electrodynamics differed at all times by the highest class of results, which were undoubtedly accepted worldwide, and had a great influence on electrodynamics as a science and as applications in practice. It is impossible to enumerate all the famous Russian scientists whose contribution was notable. Here are some of them: *G.I. Veselov* (1935–1987), *G.V. Kisun'ko* (1918–1998), *M.I. Kontorovich, G.D. Maliuzhinets, L.I. Mandelstam* (1879–1944), *V.A. Fock* (1898–1974), *A. N. Schukin* (1900–1990), and many others.

As already mentioned, this book mainly addresses university students. However, some sections will also be useful, in our opinion, for PhD students, university teachers, and engineers from industry.

The authors believe that if the reader does not assimilate materials on a computer according to our computer programs, the aim of this course will be achieved in no more than 10–15%. It is not quite enough for a transition to an assimilation of new original scientific papers.

Many references are to be found in this book, which is untypical for educational literature. Evidently, at first reading and for general study these references may be ignored. However, when studying in detail, these references will be useful and, in some cases, they are necessary.

In addition, the authors would like to note some additional circumstances.

The authors intended this book as an autonomous textbook for beginning students. Therefore, we stepped aside from the standard structure of the textbook on electrodynamics, which involves advanced mathematics, which is not yet familiar to beginning students. Instead, we pay attention to the physical side of studying the fundamentals of modern electrodynamics. Therefore, we do not draw the serious attention of this (main) group of readers to references on the modern textbooks in electrodynamics: without false modesty, we think that our book will be sufficient to achieve our goals. We hope that our interested readers, on further study of this most interesting and important scientific field, will be capable of easily finding the suitable and remarkable educational and scientific literature to improve the initial education.

Nevertheless, references to literature, of course, are necessary for readers at another level (trained specialists, PhD students, researchers from scientific and R&D institutions, consultants, teachers of colleges and universities, especially for those who deal with the history of the origin and development of modern electrodynamics (including nontraditional generalized electrodynamics). Precisely for this category of readers, we provide a complete reference list. Thus, we take into account the following circumstances.

In the world literature on electrodynamics, some excellent books have been published for the very different directions and orientations. Unfortunately, it seems to us that among these worldwide treasures, the scientific and engineering information on Soviet and Russian schools of thought in electrodynamics is practically completely absent. We are trying in our book (in different chapters) to partially bridge this gap and to give western readers the possibility of estimating scientific, physical, engineering, practical, and experimental results of the Russian school of electrodynamics, antenna systems, radio wave propagation, microwave and millimeter-wave engineering, volumetric microwave integrated circuits, and other modern problems. Connected with this, we must refer to the principal publications of Russian classics, which were published in most cases in the Russian language. We perfectly understand that these publications may not always be suitable for English-speaking readers, but, nevertheless, we hope that strongly motivated readers will find ways of obtaining the necessary publications (for instance, in the excellent U.S. Congress Library or in libraries of leading western universities) and be able to assimilate these essential works. Therefore, we consider it important to include in the textbook brief biographical information about famous world scientists and representatives of the Russian electrodynamics school (Appendix A).

One more additional thing. As we wrote above, it is insufficient for engineers entering this field to study purely theoretical electrodynamics (however excellently and strictly it would be described). A physical insight is extremely important, but it is also insufficient. It is very important to have skills in the statement and quick solution of various practical tasks, including computer modeling with the help of modern widespread computer packages. With this goal, we discuss in our book a number of examples of calculations and modeling of many (more than 50) practical tasks distributed over appropriate chapters and briefly described at the end of each chapter. We use for this the MathCAD package, which is widely distributed and well-known to students and engineers, and allows easy and independent change of the initial conditions, parameters, and aims of calculation and modeling, giving students and engineers not only an additional deep understanding, but certain skills of independent scientific and engineering investigations in this promising and attractive area.

This textbook is organized into 11 chapters. The contents of the chapters are listed in the Table of Contents.

The authors are not native speakers of the English language and we beg forgiveness for our imperfect language in this course. Finally, we would like to remember the comment of the famous Georg Christoph Lichtenberg: (1742–1799): *"The preface can be called the lightning-rod."* Probably, it is so. The reader will judge.

Moscow, Russia
2017

Eugene I. Nefyodov
Sergey M. Smolskiy

Acknowledgements

The authors remember with sincere gratitude their great predecessors – teachers and colleagues: *L.A. Weinstein (1920–1989), B.Z. Katsenelenbaum (1919–2014), M.I. Kantorovich (1906–1987), G.D. Maliuzhinets (1910–1969), G.T. Markov (1909–1981), A.G. Sveshnikov (1919), Y.N. Feld (1912–1995), A.F. Chaplin (1931–1993), Y.A. Dubinskii, G.M. Utkin (1925–1989), V.M. Bogachev (1938)*, who did so much for electrodynamics in our country.

It is a pleasure for us to thank Russian professors *S.A. Uvaisov* and *A.S. Petrov* for valuable recommendations, Professors *V.I. Kaganov, I.P. Korshunov, A.D. Shatrov* for their help in the discussion of this book's materials and notes.

We are grateful to *S. Grigo*, PhD (Radeberg, Germany) and *Wusheng Ji*, PhD (Langzhou, China) for their help with MathCAD, and to Professor *A.A. Trubin* (Ukraine) for programs in the MathCAD medium.

The authors thank Professor *N.P. Khvorostenko* for the possibility of becoming acquainted with his unpublished report.

Many recommendations and remarks were made by Professors *V.A. Malyshev* and *Yu V. Pimenov*. Useful remarks were made by Professors *B.M. Petrov, G.S. Makeeva, V.A. Neganov,* and *S.V. Raevskiy*. These recommendations were taken into account with gratitude.

We are grateful to Professor *A.M. Tomilin*, who assessed the Russian manuscript and gave us a series of useful suggestions.

S.B. Kliuev, S.V. Lebedev, and *E.V. Mezhekova* were our assistants and we thank them.

Special thanks to Russian Publisher KURS (General Director *Alla V. Nikitina* and her advisor Dr. *Yuri V. Korovin*) for their permission to use figures from our Russian book. We also thank Md. *Nadezhda G. Sergeiko* for help in the preparation of the figures.

We thank Professor *Dimov Stoice Ilchev* from Durban University of Technology and the senior editor of Springer, *Mary E. James,* for valuable advice on improving this book.

And last but not least: this textbook could not have seen the light of the day without permanent help and the boundless patience of our dear wives Valentina F. Nefyodova (1933–2014) and Natalia N. Smolskaya. We offer them our sincere thanks for their understanding and support of all kinds. The specific motivation to work was our grandsons: Yaroslav Nefyodov and Ivan Smolskiy. God grant them all good in their life of mystery (which is good) and of catastrophe (which is bad but exists).

Introduction

Radio Electronics, Radio Physics, and Our Lecture Course

The great Russian scientist, the Nobel laureate in physics, *Piotr L. Kapitsa* (1894–1984), wrote many years ago: *"There are still a few scientific areas in which we* (the USSR – *authors' note) rank as leaders; ... a number of areas, where we do not have leadership, now increases»*. These words may serve as an epigraph, in essence, to our whole book and to all that has happened, to our great sorrow, and what is happening with acceleration just now with Russian science and technologies. During the past 30 years, the situation has essentially degraded and continues to degrade with increasing speed. This is a real fact and we cannot contradict this.

Our lecture course is devoted to examination of the main laws and regulations of *classic macroscopic electrodynamics* and to some of its generalization that essentially widen our knowledge and opportunities in our understanding of the processes and devices of the radio electronics and radio physics, in addition to an analysis of various electrodynamics structures and their physical and mathematical models. These models are used in the *ultra-fast systems for information processing*, which operate (function) exclusively in the microwave and millimeter-wave ranges. Thus, we implicitly assume that the microwave and millimeter-wave devices (block, module) of ultra-fast systems can be constructed on the basis of the predominantly *plane* (planar) IC and/or the high-frequency 3D-IC. Sometimes, these ICs are called three dimensional ICs (3D-IC) [13–15].

According to recommendations of the International Consultative Committee on Radio (ICCR), electromagnetic wave ranges are defined as follows: $(0.3-3) \cdot 10^N$, (Hz), where N is the range number. For radio communication, as a rule, the ranges with $N = 4-12$ are used.

According to the ICCR recommendation, the super-high-frequency (SHF) range includes the range with $N = 10$ only. However, owing to the great generality of principles lying on the basis of devices and equipment constructions within 9–11 ranges, they are often considered to be the unified SHF range, both in native and foreign literature. Therefore, in this book, we understand under SHF range the totality of the 9–11th electromagnetic wave ranges.

We can nominate the following features of the SHF range.

1. The value of the SHF range is its broadbandness. In three ranges ($N = 9-11$), with the bandwidth $\Delta f = 300$ GHz, we can transmit 10^4 times more information during the time period than in the five ranges together ($N = 4-8$). Broadbandness allows utilization of the jam-resistive frequency and phase modulations for which the signal level at the receiver output does not depend (within definite limits) on the input signal level varying because of fading. These types of modulations allow provision of high-quality telephone and television communication, to transmit digital data with high speed without information losses in different networks.

2. In this range, it is rather simple to create antennas with dimensions, which are much more than the wavelength and havepencil-beam radiation. This results in an increase in the noise immunity of radio communication and ensures the possibility of operating in the given

geographic region of several radio links with the same frequencies without disturbance of the electromagnetic compatibility conditions.

3. SFH waves pass without obstruction through the layers of the ionosphere, which allows communication of the ground-based stations with artificial Earth satellites and space apparatus. When SFH waves propagate near the Earth's surface, their diffraction and refraction are small. Therefore, for communication between objects that are located outside of straight vision, relay stations are necessary.

4. In the SHF range, the level of atmosphere and industrial interference is low and the time of day and the season of the year do not affect the propagation conditions. Nevertheless, for frequency growth, the attenuation of these waves due to rain and the resonant absorption in gases increases. This is especially demonstrated in millimeter-waves, which are intensively used for communications. In the millimeter range, the atmosphere has some transparent windows and absorption peaks (for details, see Harmuth, and Nefyodov [22, 38]).

5. Radio communication, as a rule, is provided in the transparent windows. The greater absorption of millimeter-waves in hydrometeors compared with microwaves leads to degradation of the communication distance, which requires growth of the energy potential of the radio link to compensate for attenuation. The millimeter range is not yet overloaded, and their communication equipment has good electromagnetic compatibility with the communication equipment of other ranges.

6. This rule is violated when we are speaking about propagation of longitudinal electromagnetic waves, which exist in media (the ionosphere plasma, water, a soil, etc.). They have unique properties that have unfortunately not yet been fully studied, although longitudinal electromagnetic waves in the plasma (the *Langmuir waves* studied by *I. Langmuir* and *L. Tonks*) have been known since the 1920s.

7. The interest in the proper analysis of longitudinal waves in water (particularly, in sea water) has sharply increased in connection with a whole series of remarkable properties and the manifold applications in practice now and in the near future. Here, we can mention problems of electromagnetic and mechanical phenomena (see, for instance, Nefyodov, and Faraday [27, 42]), in addition to new types of waves, so-called, *knotted waves*, which have striking possibilities [31, 33]).

8. Increased attenuation in the absorption peaks allows information transmission on these frequencies at a low interference level from various services and enabling of secretive communication along the Earth surface at short distances. In addition, frequencies corresponding to absorption peaks in the atmosphere can be used in inter-satellite communication links of great length. In this case, the atmosphere fulfills the role of a suppression filter with respect to the Earth's interference. The millimeter-waves penetrate better through fog, smoke, rain, dust, etc., than the waves of optical and infrared ranges. They pass through the plasma with low attenuation; therefore, they are used for communication with missiles that pass through the ionized atmosphere. On the nontransparent frequency segment, the millimeter radio waves are fully absorbed and communication is impossible. Nevertheless, it is quite possible at these frequencies between two space radio electronic equipment that this communication channel will be simply shielded from observation from Earth.

9. In near-surface communication lines of the millimeter range, we observe nonstationary oscillations (fluctuations) of amplitude, phase, and directions of wave arrival, which are caused by atmosphere refraction and its irregularities, by the Earth's influence, and also by wave re-reflection from the Earth's artificial satellite surfaces, airplanes, and other objects on which the millimeter-range equipment is mounted. In the millimeter range, we can observe the effect of multi-pass propagation and the noticeable Doppler frequency shift.

Of course, definite attention is paid in this lecture course to classic transmission lines (a coaxial line, the rectangular, circle, elliptic, and dielectric waveguides, etc.). Obviously, the separate *basing elements* and connecting *transmission lines* between them can be implemented on the "traditional" plane (planar) IC.

Ultra-Fast Systems for Information Processing

Dynamics of time development, at all times, including *radio engineering*, *radio electronics*, radio physics, *electronics*, *microelectronics*, and some associated areas of science and technology, can be represented in the form of ascending step function. Its origin relates to the period of the Second World War with the development of metal–ceramic electronic valves and the application of printed boards for the guiding heads of anti-aircraft shells. The start of these activities was defined by notable discoveries in radio electronics based on the advantages in radio physics and solid-state physics; transistors and strip transmission lines, then acoustic–electronic devices, devices on magneto-static waves, etc. The latter steps relate to novel advantages in thin- and thick-film hybrid and monolith (semiconductor) technologies and their combinations, in addition to new approaches to the systems of information processing up to *nonlinear waves* and *solitary waves* (solitons). At present, these heterogeneous, so it would seem, directions are unified by the general concept of ultra-fast systems of information processing (UFSIP) directly in the super-high frequency (SHF) range and/or the millimeter range and in the optical range.

At the modern stage of development, ultra-fast systems are of paramount global and strategic importance. We remember the old truth: he/she who has all or almost all the information controls the world. At the beginning of the twenty-first century – *the century of information* – the struggle to hold data banks, systems of data processing, storing, transmission, protection (including "active") has become extremely severe. Each developed country tries to enter the new millennium with a serious back-log, having the large volume of the "*world-wide informational space.*"

Recently, there has appeared to be an interest not in the problem of radio signal transmission, but in the SHF energy transmission at long range. The science is returning to the famous experiments of *N. Tesla* [34], especially because the generalized electrodynamics allows for the first time a clear presentation of an analytical–numerical explanation of the sense of the Tesla experiment [34] (see Chap. 4).

Informational Space

The large circle of questions that touches upon, in essence, all human activity areas and the whole knowledge volume accumulated by humankind, is included in the synthetic concept of an *informational space*. At present, this knowledge "storeroom" is usually called the *databanks*. Each "civilized" society (not excluding, unfortunately, the criminal one) of any country of the world dreams about authorized access to these databanks and about the entrance (based on this access) into the worldwide informational space on equal terms. By the way, as was often the case in history, far from everybody can occupy a place under this "informational sun." The rich developed countries count themselves to be lucky men and not without foundation.

The process of redistribution (and simply, the capture, sometimes, even armed, the direct aggression) of the informational space began, of course, not today and, probably, we can consider the accurately prepared and excellently implemented operation on the discredit and further destruction of the highly promising large Soviet computer of the "BESM" series as the largest success of western countries in this connection in the informational space. After that, the USSR began at first slowly and later faster to be behind in the development level of prospective ultra-fast systems, connecting at first with an absolutely hopeless deal with "large" computers.

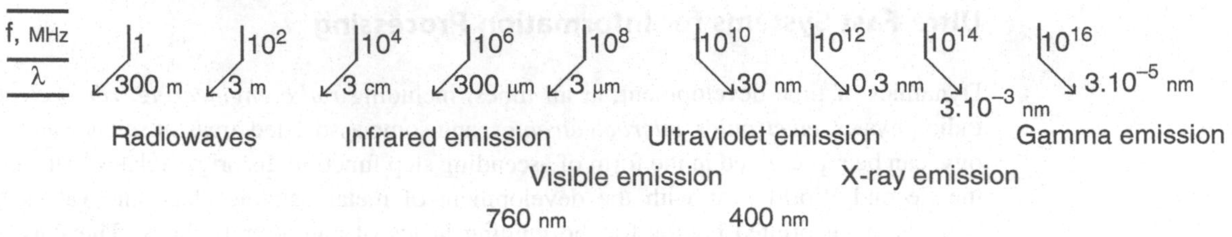

Fig. 1 The general wave scale, with which electrodynamics deals, and in particular, the techniques of antennas and radio wave propagation

The 3T System

As the main problem of the near future in the development of ultra-fast systems for information processing, we can consider the achievement of the known "$3T$" border (3-Tera) (processing speed 1 *Teraflop*, information transmission speed 1 *Terabyte/s*, the volume of operation memory 1 *Terabyte*). It seems to us that one of the possible ways out of the existing situation in Russia is related to the following proposal, which in the overwhelming majority is of Russian origin.

First of all, this is orientation to the sharp (by several orders) increase in the *clock speed* f_{cl} in ultra-fast systems, i.e., a transition to SHF and the millimeter range. In this way, the "jump" can be realized in qualitative improvement of native ultra-fast systems, because the other world feels itself well enough. The problem is that modern computing facilities and microelectronics for the frequency range of tens and hundreds of megahertz gives the possibility of having a definite level of operation speed, which meets the requirements of western countries and permits precisely this arsenal of means and technologies of large integrated circuits (LIC) and very large integrated circuits (VLIC) to be dealt with.

However, the growth of clock frequency f_{cl} from tens and hundreds of megahertz to units, tens, and in the near future to hundreds of gigahertz is the obvious way to increase the operation speed of the ultra-fast system, and we cannot pass it. Hence, we cannot manage without a cardinal variation of ideas and principles of implementations and technologies. Of course, we need to search for the most profitable, optimal systems of information processing. The general view of the frequency scale is presented in Fig. 1.

As we see, the radio range itself occupies a rather modest place in this figure. However, in modern radio engineering and, evidently, in UFSIP, both the lower and the higher frequencies are used (see Nefyodov [25, 28]).

Information Capacity and Resolution

"Pure" radio engineering systems intended for communication, radio navigation, radar technologies, radio astronomy, etc., are, of course, in the same position. First, growth of the operating (carrying) frequency allows an increase in the *information capacity C* (*C* is proportional to $\Delta\omega$, where $\Delta\omega$ is the frequency bandwidth occupied by the signal) of the communication channel (at the bottom, the information transmission [reception] speed). This, in turn, gives an opportunity to use the combined types of radio signal *modulation* (*amplitude phase*, *pulse code*, etc.), which essentially increases the secretiveness and noise immunity of the channel.

Another "radio engineering" circumstance, which dictates to society the necessity for a sharp increase in the operating frequency f (decrease in the wavelength λ), is the need to increase the system *resolution* (the directivity pattern width) θ, which, first of all, is defined by the dimensions of antenna system D and input devices for *spatial–temporal radio signal*

Fig. 2 Estimation of the resolution capability of the radio engineering systems according to the antenna pattern. (**a**, **b**) Classic patterns. (**c**) The knotted waves antenna pattern

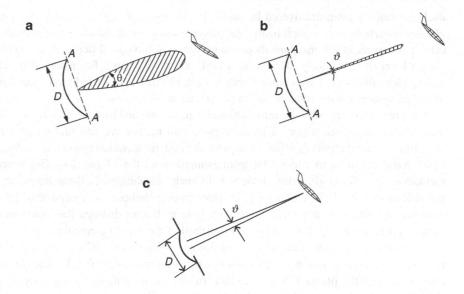

processing. At naturally restricted antenna dimensions D, the benefit in resolution θ is achieved at the expense of a decrease in λ, because, as we know, $\theta \approx \lambda/D$.

Antenna structures with so-called knotted waves [31, 33] have many more possibilities. Nevertheless, unfortunately, we are far from their industrial implementation.

The system in Fig. 2b, evidently, has better resolution because it has a smaller opening angle of the pattern compared with the system in Fig. 2a.

The choice of the *best* (optimal) *radio signal waveform* plays an important role in this circle of problems. In recent years, the problem of the application of *solitary waves* (solitons) for physical field interaction with the live substances for radio engineering tasks, in *bio-energy-informatics*, medicine, ecology, and other adjacent areas. A new term has even appeared: *electrodynamics of live systems* [6].

Integrated Circuits and Volumetric Integrated Circuits at Microwave and Millimeter-Wave Ranges

At present, there is an hour of triumph for microelectronics (in the direct sense), because assimilation and application of the cosmos and micro-cosmos are possible only with significant progress in science and technology in the area of signal transmission for ultra-long and ultra-small ranges and in the processing of the huge information streams, including on a real-time scale. The natural requirements for achievement of the limited possible physical parameters (in essence, the *fundamental limits*) are formulated precisely here to ultra-fast systems in microwave and in millimeter-wave ranges. Among the fundamental limits, we may use the concept of *potential noise-immunity*; the ultra-high bandwidth of operating frequencies at *digital signal processing*, high carrying (clock) frequency, multi-functionality, and minimal mass and dimensions. All these parameters correlate with each other, but they are controlled by the restriction condition of maximal efficiency at low power consumption. In addition, the developed *element base* of modern microelectronics, which is constructed based on hybrid and monolithic technologies, has fundamental restrictions. They relate not to the physical properties of each basing element, but with the electromagnetic field nature at implementation of the large systems, in particular, with requirement of the finite volume presence for energy transmission (canalization) of the high-frequency electromagnetic field.

The natural question arises: could we use the example of nature in the form of *biological structure*, for example, of the brain, where the processing performs at the *biological field level*? Although, the term *biological field*, in our opinion, is conditional because the biological field itself has not yet been discovered in nature, but the *electromagnetic field* exists together with *electromagnetic waves*, which justify the processes in the live media and organisms. Now, we cannot yet speak in full measure about the existence of biological fields, as we do not yet have such a level of technology. But the idea itself about volumetric formation of structures from basing elements, which are coupled with a weak electromagnetic field, has found its place in ultra-fast systems in microwave and in millimeter-wave ranges.

The processing speed of the informational signal is defined by the *clock frequency* f_{cl}, which moves with the development of technology toward microwave and millimeter-wave ranges. According to estimations of different experts, the system operation speed required today is 10^{12}–10^{15} bit/s, which is, of course, far from estimations of the $3T$ problem. Engineering implementation of radio engineering devices with such possibilities in these frequency ranges is possible only with 3D-IC technology, in which basing elements are inseparably linked, i.e., by electrical, constructive, and technological integration. ICs are distinguished into monolithic and hybrid types, according to the unification method of the basing elements.

Integrated circuits on its origin evolved from one-dimensional IC (at the end of the 1940s) to two-dimensional or planar IC technologies (at the beginning of 1980s). To date, we have clearly understood that the planar ICs achieve their fundamental limits technologically, in general, in reliability and in interconnection problems, and in *weight dimension parameters*. But the main problem is restriction in the operation bandwidth $\Delta\omega$, which, in principle, determines the system's operation speed. Solution of this problem in the last few decades was achieved by a multi-layer, multi-floor hybrid and other constructive–technological solutions in the form of LIC and VLIC. The transition of ideas and principles from low-frequency microelectronics into the SHF range introduced an insignificant widening of the operation frequency bandwidth and some improvement in mass-dimension indices, which plunged designers into despair and drove researchers to search for principally new solutions.

Discovery and detailed mathematical and physical modeling of a large set of new types of microwave transmission lines: *slot, coplanar, ridge-dielectric, dielectric, semiconductor*, lines with *acoustic* and *magneto-static waves*, the development of *monolithic technology* (from the low-frequency side) allowed a nontraditional insight into the designing of microwave devices in radio engineering and electronics. The historical development of microwave ICs and the nature of *information biological systems* formation itself suggests the best optimal variant – utilization of the third coordinate in the topology of microwave ICs, i.e., the transition to 3D-IC of microwave and millimeter-wave ranges with an indispensable account of features of information representation and processing [13–15].

Repeatedly, the question arose, and even now it sometimes arises: is this transition logical and reasonable? Electromagnetic fields are essentially *three-dimensional*. Their artificial reduction to two-dimensional structures is dictated by the necessity to simplify *mathematical models* to obtain solutions of the appropriate *boundary problems* on computers. In the planar ICs above the two-dimensional IC (predominantly in the microwave region), the shields are mounted for suppression of possible radiation into the external space and for avoidance of the spurious influence of IC. Achievement of selective properties is performed by a combination of the planar IC with metallic shields located in the vertical and/or horizontal planes. Thus, even in the planar ICs, latently, the third coordinate fulfills its useful functions. This is precisely why the transition to 3D-IC of microwave and millimeter-wave ranges is *logical*, which has been confirmed, in particular, by investigations into biological structures, where the concept of the planar distribution of functional units is completely absent. Hence, 3D-IC allows signal processing to be performed not only in the "horizontal" plane, but also in the "vertical" plane. This provides a decrease in the weight dimension indices by one, two (and even three) orders and leads to the essential growth of operating frequency bandwidth and to an increase in signal processing speed [13–15].

There is in our opinion one more very important comment. We should note that the 3D-IC idea *fits for any existing* (and particularly for future, for example, two-and-a-half-dimensional) *technology*. Evidently, the *hybrid integrated technology* for 3D-IC is not natural owing to the point-to-point wiring of active, nonlinear, and other devices. In this variant of construction, we need to create mounting panels or to assign them to higher "limits" at special 3D-IC floors. Ubiquitous assimilation and utilization, for instance, of monolithic and semiconductor ICs will open up huge perspectives before ultra-fast systems on 3D-IC for a decrease in weight–dimension parameters, an increase in operation speed, and increased reliability. This can be seen especially clearly in examples of LIC and 3D-IC. Thus, the development history of radio and computer technologies justifies the inevitable transition to ultra-fast systems on 3D-IC in the microwave and millimeter-wave ranges. From there, it is clear that 3D-IC modules for ultra-fast systems will be developed, in a certain sense, in "parallel" for LIC and VLIC. This is a natural method of radio engineering development.

The authors are sure that the lecture course in electrodynamics is the main fundamental course in the educational system for radio and electronics specialists and students.

As mentioned before, brief information about the most important researchers, scientists, and engineers with their contributions to electrodynamics and adjacent areas of science and technology is given, for example, in Appendix A. We would like to draw the reader's attention to the fact that the overwhelming majority of them worked practically simultaneously in several areas of mathematics, physics, astronomy, electrodynamics, etc. We remember the words of G.K. Lichtenberg: "...*Whoever does not understand anything except chemistry, he does not understand it enough*."

Contents

General Laws of Classic Electrodynamics and Elements of Macroscopic Generalized Theory

For your own acquaintance with our lecture course, you are starting by considering the main fundamental definitions and concepts (Sect. 1.1) of classical macroscopic electrodynamics. Under macroscopic electrodynamics, we understand such physical problems where we must analyze obstructions in the path of electromagnetic waves, when physical (not electrical) dimensions of these obstacles significantly exceed the wavelength of an electromagnetic field. In the opposite case: dimensions of the obstacle under consideration are much smaller than a wavelength. The understandable ambiguity of such definitions of macroscopic and microscopic electrodynamics is clear to any student. Therefore, more accurate estimations and definitions are discussed later in the book (see Chap. 4), when we describe the influence of electromagnetic fields upon the micro-particles (electrons, protons, atoms, ions, molecules etc.).

At first, we discuss the one following principal issue for electromagnetism. From the fundamental lecture course on physics, we perfectly remember the formal conformity (equivalence) of expressions for the gravitational force field and for the electrical Coulomb law concerning the force interaction of two separate electrical charges. This fundamental conformity consists in the quadratic dependence of these principally different forces on the distance between gravitation objects and between electrical charges. We notice the similar quadratic dependence in the Ampere law concerning the interaction force of two parallel conductors along which the electric current (DC or AC) flows.

Classical electrodynamics, beginning with Maxwell, Hertz, and many other scientists, had passed through huge development, and during these years a great number of examples of so-called paradoxes of electrodynamics had been accumulated. In these examples, the results of recurring experiments cannot be described by the conventional theory. First of all, this is the Biefeld–Brown effect consisting in the appearance of some additional force direct toward the positively charged plate of the conventional charged electrical capacitor.

Another example is connected to the non-equal attractive force of two conductors with the current flow, which destroys the well-known Newton law about the equivalence of action and counteraction forces, Similarly, the railotron effect, the Faraday unipolar engine and the unipolar generator do not have sufficiently clear scientific substantiation. The well-known Aharonov–Bohm effect can be attributed to paradoxes. All these (and many others) interesting and unusual issues are briefly described in the present Chapter with the aim of showing the reader that not all is absolutely clear and understandable in any branch of physics, and obliteration of any paradoxes is inadmissible as the precise investigation of any paradoxes opens new doors, not only in the theory, but also in the practice of our life.

The Stokes–Helmholtz theorem (Sect. 1.3.1), which describes the sense of the generalized electrodynamics' laws, is the basis of the content of Sect. 1.3. It is precisely the basis of the strict mathematical fundamentals of generalized macroscopic electrodynamics. We introduce a new concept of the scalar magnetic field (this is rarely discussed in the students' textbooks) and formulate the fundamental system of differential equations in partial derivatives of generalized electrodynamics (Sect. 1.3.2). It is shown that the classical system of Maxwell equations, which is discussed in many excellent textbooks on electromagnetism, is a special case in generalized electromagnetic equations (Sect. 1.3.3). From this generalized system, as a consequence, we obtain the generalized law of electromagnetic induction (Sect. 1.3.4). Accordingly, we must modify the law on the total current (Sect. 1.3.5) and the law on charge conservation (Sect. 1.3.6).

Various "simplifications," say, the transition to the so-called homogeneous (autonomous) equation system, which does not contain the right part with external constraining forces, are consequences of usually used approximations (assumptions) in an analysis of general electrodynamics equations. These assumptions of the initial equations, the goal of which is understandable – to study and to understand more exactly the physics of phenomena and processes – often lead to errors (not only quantitative,

© Springer International Publishing AG, part of Springer Nature 2019
E. I. Nefyodov, S. M. Smolskiy, *Electromagnetic Fields and Waves*, Textbooks in Telecommunication Engineering,
https://doi.org/10.1007/978-3-319-90847-2_1

numerical, but qualitative, fundamental and principal). To avoid such errors, we introduce the so-called "gold rule of electrodynamics," which recommends not tearing away the electric and magnetic fields from their sources during an analysis (Sect. 1.3.7). Section 1.3.8 is devoted to a new concept of the scalar magnetic field, and this Section is finished by derivation of the generalized continuity equation.

As the material of Chap. 1 is of principally general theoretical character, the Chapter does not contain material on computer modeling of processes and phenomena.

1.1   General Definitions

An *electromagnetic field* is the unified indivisible in the space and time process, i.e., it exists in any point of in space $(\vec{r})$, in any medium, and at each time moment (t). In modern physics, together with the electromagnetic field, the other fields are considered: gravitational, nuclear forces, etc. Another form of reality surrounding us, probably, is the more habitual concept of substance existing in the form of bodies, media, etc. Thus, it is assumed in advance that the existence of two kinds of matter is possible: a field and a substance. In a philosophical sense, it is reasonable to formulate a question: what is the field – the matter substance or the process happening in the material medium? This question can be still be only stated.

The concept of substance is familiar to us from secondary school courses in physics, and the specified and widened concept is familiar from the university course of general physics. We know that each body creates around itself a gravitational field, whose intensity at any point is the gravitational force, with which this field affects another body at this point. The magnitude of this gravitation force F is defined as follows:

$$F_m = \gamma\left(m_1 m_2 / r^2\right) \tag{1.1}$$

where m_1, m_2 are masses of interacting bodies located at the distant r, and γ is the gravitation constant.

The structure of Eq. (1.1) reminds us of the well-known *Coulomb law* about force interaction of two point charges q_1 and q_2 located at distance r:

$$\vec{F}_q = \vec{r}_0\left(q_1 q_2 / 4\pi\varepsilon_0 r^2\right). \tag{1.2}$$

Here ε_0 is the electric constant, which in the SI system has a dimension "farad per meter" and equal to $\varepsilon_0 = (10^{-9}/36\pi)$, F/m. The unit vector $\vec{r}_0$ is directed from the charge q_1 toward the charge q_2 (Fig. 1.1).

The Coulomb law was stated in 1785. However, in science history, we know the names of persons who came to the same results. For the first time, Georg Wilhelm Richmann suggested investigating experimentally the interaction law of electrically charged bodies in the course of his research (together with M.V. Lomonosov) into electric phenomena. In 1759, the professor of physics Franz Ulrich Maria Theodor Aepinus from St. Petersburg Academy of Sciences assumed for the first time that charges should interact inversely to the distance squared. In 1760, Daniel Bernoulli discovered the square law with the help of the developed electrometer. There was other research (Benjamin Franklin; John Robison; Henry Cavendish), who to some extent or other forestalled the Coulomb law [13, 14, 29].

Moreover, in accordance with the *Ampere law*, which was discovered experimentally, two parallel conductors with DC currents J_1, J_2, being in the distance r, interact between them with a force equal to[1]

Fig. 1.1 To the interaction of two point electric charges

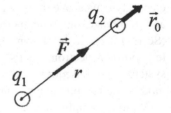

---

[1] This relationship still has some general character. Later, it is specified in the case of the arbitrary location of currents J_1, J_2 in the space (see Chap. 3).

$$\vec{F} \approx J_1 J_2 / r^2. \tag{1.3}$$

In some sense, Eqs. (1.1, 1.2, and 1.3) are "similar", which allows us to speak about the deep unity of the gravitational field and electromagnetic field in the total universe system. Sometimes, we can speak of the *inverse-square law*. For instance, in optics, the light intensity I with distance r from the point source is $I = A/r^2$, where A is some quantity proportional, in particular, to the cosine of the angle between the incident beam direction and the normally illuminated surface. The gravitation field, the field of magnetic poles, etc., can be attributed to this class of fields.

There is more detail than the formula (1.3) of the interaction of currents, which are located arbitrarily in relation to each other (the Ampere law; see, for example, Tomilin, and Nikolaev [12, 43]). In practice, this formula is almost never used; nevertheless, we would like to draw the reader's attention to the fact that, in spite of historical truth, the transverse magnetic force is attributed to the great A.M. Ampere. At the same time, we know about Ampere's famous experiments, from which it follows that there is a longitudinal force directed along the conductor current. In a number of experiments on conductor motion in the electromagnetic field, beginning with the aforementioned results of Ampere, it was shown that the presence of the longitudinal force allows explanation of the paradox of interaction of parallel currents [12, 43–45]. Obviously, this force must be called the *Ampere–Nikolaev's* force.

The electromagnetic field, in addition to a substance, is characterized by energy, a mass, and an impulse. However, a mass and an impulse are typical only for moving (propagating) the electromagnetic field. In contrast to a substance, which consists of separate atoms, molecules, electrons, protons etc., the electromagnetic field has no rest mass.

The year 1820 is considered to mark the birth of electrodynamics, when *Hans Christian Ørsted* discovered that an electric current creates a magnetic field around itself. In the same year, Ampere determined the interaction force between the electric current and a magnet. From these discoveries, the rapid study of electric and magnetic phenomena began as a manifestation of the unified electromagnetic mechanism, although, as we have just seen, observations and investigations of electric and magnetic demonstrations were started much earlier.

The widely known effect of deviation of the propagation of light waves away from the rectilinear trajectories under the influence, for instance, of the Sun's mass justifies the mutual influence of the electromagnetic field and the gravitational field. On the other hand, the electromagnetic field impulse is manifested under pressure, which is effected on the material bodies. Light pressure was first proven by the experiment in 1900 by Russian physicist *P.N. Lebedev*.

In our lecture course, we consider the processes of motion (propagation) of electromagnetic fields (waves) in various media and guiding systems, including resonance systems.

To date, we have been able to speak conditionally enough about two constituent parts of the unified electromagnetic field: electric and magnetic. The electric "part" – the *electric field* – is characterized by electric vectors $\vec{E} = \vec{E}(\vec{r}, t)$ and $\vec{D} = \vec{D}(\vec{r}, t)$, which are, relatively, the *strength of the electric field* and its *induction*, the dielectric strain. In the SI system, the electric field strength is expressed in volts per meter ([V/m]; the induction in Coulomb per square meter ([Q/m^2]). Similarly, its magnetic "part" – the *magnetic field* – is characterized by magnetic vectors $\vec{H} = \vec{H}(\vec{r}, t)$ and $\vec{B} = \vec{B}(\vec{r}, t)$, which are, relatively, the *magnetic field strength* and its *magnetic induction*. The magnetic field strength H is measured in *ampere* per meter and has a dimension (A/m), and its magnetic induction is measured in *tesla* (Tl).

We repeat: the aforementioned separation of the unified electromagnetic process into electric and magnetic "components" is conditional enough. Still, in 1820, one of the founders of the doctrine on electricity, A.M. Ampere wrote that "there is no magnetic field, there is only the electric field." Indeed, at present, we know publications in which the magnetic field is absent, but the "building" of electrodynamics, and, hence, of the whole of physics, remains stable and the main thing is that it allows description of many electric phenomena for which a clear physical analysis could not be permitted [46, 101]. In our lecture course, we shall pause at new ideas that add "correctness" and to give the more precise definitions to the classic macroscopic electrodynamics.

1.2 Some Paradoxes of Classic Electrodynamics

The main equations of classic electrodynamics – Maxwell equations (see later in this Chapter) – loyally served and continue to be the main "weapon" of researchers and engineers/designers of electrodynamics elements and units in many areas of radio electronics. Moreover, these equations cannot explain the whole series of physical phenomena and experiments (they are called paradoxes). Among them, for example, is the Biefeld–Brown effect[2] consisting in the fact that some force arises in the direction of the positive plate of the capacitor. We have no answer to the question why two conductors with currents are attracted to each other with different forces, violating *Newton's law*. Some researchers see the way out by introducing an *ether,* which is a concept that is long forgotten and supposedly disproved by the special relativity theory, in the introduction of the *physical vacuum* concept, which has some of the properties of the usual material medium, etc. A detailed description of these new concepts would divert us away from the main direction of our lecture course, although new ideas are always interesting, especially if we have the experimental results and approaches and devices implemented in practice. Nevertheless, the classic Maxwell theory will be duly served for many years and decades in various areas of modern and future radio electronics, and in other fields of physics.

From a historical perspective, and from today's position, special attention must be paid to a book by the great mathematician and mechanical expert *Edmund Taylor Whittaker, "History of ether and electricity theory"* [41], the first edition of which appeared in 1910, with the second edition being published in 1951. In the Russian language, this excellent book only saw light for the first time in 2001, i.e., more than 40 years, and even then it was not issued by any of the central Russian publishing houses.

Below, we draw the reader's attention to some well-known experiments, the sense of which is impossible to explain within the limits of classical theory. Many such examples are contained in the works of *G.V. Nikolaev* [43, 44], *A.K. Tomilin* [12], and *V.V. Erokhin* [10]. We describe some of these experiences only briefly.

Obviously, the most well-known is the aforementioned *Biefeld–Brown effect* (experiments carried out between 1925 and 1965), consisting in the translation motion of the plain high-voltage capacitor toward the positive plate at a connection with a high enough voltage to plates equal to 30–50 kV (Fig. 1.2a).

After long-term investigation, Brown created the disk-shaped film capacitors, which, being charged to about 50 kV, are capable of being lifted into the air and performing the circular motions with the angular speed of 50 rotations/s. In a series of literature sources, the Biefeld–Brown effect is treated as the electric phenomenon of the *ion wind* arising, which transmits its impulse to the surrounding neutral particles, promoting the formation of the pressure observed. This phenomenon is also known as *electro-hydro-dynamics* (evidently, analogous to *magneto-hydro-dynamics.* At the same time, it is rather obvious that we are far from having a full understanding of this phenomenon. Maybe, in the future, generalized electrodynamics will be able to provide a noticeable "push" with regard to efforts in this direction.

Fig. 1.2 Schemes of some of the experiments, which cannot be explained from the point of view of classic electrodynamics laws. (**a**) The Biefeld–Brown effect. (**b**) The railotron effect. (**c**) The Faraday unipolar motor. (**d**) The unipolar generator

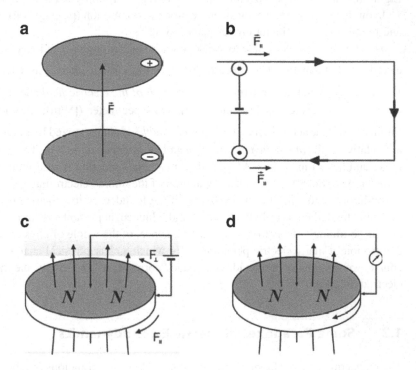

---

[2] Up to this time, we have accumulated many publications devoted to paradoxes of classic macroscopic electrodynamics.

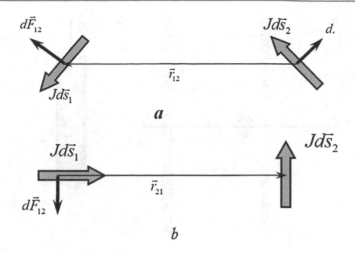

Fig. 1.3 The interaction scheme between two elements of nonparallel conductors with currents $\vec{jds_1}$ and $\vec{jds_2}$. (**a**) Nonparallel currents. (**b**) Perpendicular currents

On *violation of the Newton law*, from the point of view of classic electrodynamics, we cannot to get an answer to the question why two segments of the conductor with current, having violated the third Newton law, are attracted (pushed away) with different forces: $d\vec{F}_{12}^{A} \neq -d\vec{F}_{21}^{A}$ (Fig. 1.3) [12, 43, 44].

An interaction scheme between two elements of nonparallel conductors with currents is calculated according to the known Ampere formula. In the first case (Fig. 1.3a) the interaction forces between current elements $d\vec{F}_{12} = (\mu J_1 J_2/4\pi)\left(d\vec{s}_1 \times (d\vec{s}_2 \times \vec{r}_{12})/r_{12}^3\right)$ are directed not to the parallel lines of action; hence, $d\vec{F}_{12}^{A} \neq -d\vec{F}_{21}^{A}$, i.e., the *action–counteraction law* is violated.

An extreme case is when currents are perpendicular to each other (Fig. 1.3b). Here, we see the paradoxical result; namely, $d\vec{F}_{21} = (\mu J_1 J_2/4\pi)\left(d\vec{s}_2 \times (d\vec{s}_1 \times \vec{r}_{12})/r_{21}^3\right)$, i.e., when $\vec{F}_{12}^{A} \neq 0$ the interaction force of the second current upon the first one $\vec{F}_{21}^{A} = 0$ (owing to $d\vec{s}_1 \times \vec{r}_{21} \equiv 0$).

Later, various variations and additions were introduced into the Ampere formula, for instance, from the research of *W.E. Weber* on the interaction of two moving charges, but the results were not satisfactory [43, 45].

On *the railotron effect (the railotron "gun," the railotron engine, accelerator)*, the simplest scheme for its demonstration is shown in Fig. 1.2b. When the energy is connected to this scheme, two longitudinal forces $\vec{F}_{\parallel}$ arise, which act along the conductors (rails). The force $\vec{F}_{\perp}$ acts on the bridge (the bullet, the shell, the electron clot, etc.) between conductors, which pushes out the bridge into the space. Investigations into the railotron effect had begun during the First World War and are still continuing today ("General Atomics" company, San-Diego, CA, USA).[3] It is known that acceleration of small bodies (down to 100 g) is possible at very significant velocities to the order of 6–10 km/s. It is interesting that instead of the transverse conductor, we may use the plasma clot, the "piston," which has a fantastic speed of 50 km/s, which can have an application.

However, within the framework of classic electrodynamics, some explanation of physics of this gun operation has not yet been found. Unfortunately, in connection with the dissolution of the USSR, the sharing of research in the USSR on the railotron effect was interrupted. We see, that in the west, these investigations are continuing.

The *unipolar Faraday motor* (Fig. 1.2c) and the *unipolar generator* (Fig. 1.2d) are concerned with the same class of paradoxical effects. In the first case, ambiguities are connected with determination of the driving force of nature in the unipolar motor, in which the rotating magnet (the rotor) is used. Investigation shows that in this type of unipolar motor, the magnet–rotor rotates by the longitudinal forces $\vec{F}_{\parallel}$ only. The transverse force $\vec{F}_{\perp}$ applied to the side conductor (Fig. 1.2c) is the reaction. In the case of the unipolar generator (Fig. 1.2d), the paradox can be linked with the place of the electromotive force arising with the rotating magnet–rotor and because of the absence of a reaction in the case of fixed magnet usage. Experiments show

---

[3] *Kluev E.* Electromagnetic gun (in Russian)// Supernew reality, 2007, № 16, pp. 40–44.

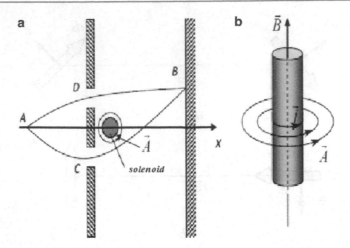

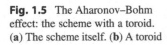

Fig. 1.4 The Aharonov–Bohm effect: the scheme with a solenoid. (**a**) the scheme itself. (**b**) The solenoid (upper view)

Fig. 1.5 The Aharonov–Bohm
effect: the scheme with a toroid.
(**a**) The scheme itself. (**b**) A toroid

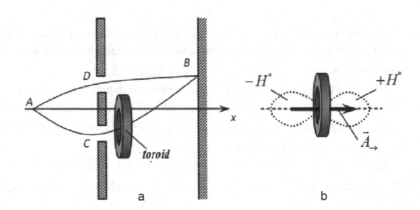

that electromotive force is induced in the rotating magnet–rotor only. An attempt at analysis of the railotron effect analysis is given in Purcell and Morin [45].

The famous *Aharonov–Bohm effect* is in the same situation. This sense (very briefly) consists of the following. Schemes of classic experiments on this effect are shown in Figs. 1.4a and 1.5a [43]. There is a shield with two small holes C and D located a small distance from each other. Behind the small shield, between holes, there is a solenoid with a very small diameter, in order not to introduce a noise into the interference picture due to the electron's interaction with the solenoid itself. When the DC current passes through the solenoid, the interference picture displaces (the B point on the continuous shield). It is clear that the effect of the $\vec{A}$ vector field manifests in various ways on the electron motion characteristics, which fly through the C and D holes. The trajectories of the particles do not coincide with $\vec{A}$ vector lines, even though they are curved.

Another example of a scheme of the Aharonov–Bohm effect is shown in Fig. 1.5a. Here, instead of a solenoid (compare Fig. 1.4a), a toroid is used.

In none of the cases presented, the classic electrodynamics theory (namely: the transverse Lorentz force, which acts on the point q charge, moving with $\vec{v}$ velocity, defined as $\vec{F} = q\left(\vec{E} + [\vec{v}\vec{B}]\right)$, does not give an answer to the question of electron motion and their deviation from the "true" trajectory. Later (Sect. 1.5), we (without great detail) discuss the influence of the new concept of the transverse and longitudinal forces (the Ampere forces and the Ampere–Nikolaev forces) on the motion processes of charged particles in the medium and in the conductors.

There are very many such examples of paradoxes that cannot be explained from the point of view of the classic macroscopic theory of electromagnetism, but those mentioned clearly justify our insufficient knowledge in this seemingly routine area of electrodynamics and physics as a whole.

1.3 Main Laws of Generalized Macroscopic Electrodynamics

1.3.1 The Stokes–Helmholtz Theorem

At present, in most textbooks on classic electrodynamics, the deductive approach to description is accepted. This method assumes a description of the general system of laws, which will be explained in the future and used for the basis of physics. Precisely such an approach is used by us in our lecture course, in contrast, for instance, to such excellent courses on classic electrodynamics, in which the theory is described sequentially, beginning from the static phenomena up to the light range, as in Born and Wolf, Nikolsky and Nikolskaya, and Tamm [46, 53, 54]. We are reminded of the excellent lecture course in physics by *O.D. Khvolson* (1852–1934).

Although we should certainly note that many results of classic electrodynamics are obtained by means of the static data (or experiments at very low frequencies) in electric and/or magnetic static. We repeatedly meet such examples in our course.

In these sections, we follow the general approach of the lecture course, studying and again using the deductive method: we immediately write the equation system of generalized macroscopic electrodynamics. Strictly speaking, we should begin from the results of *N.P. Khvorostenko* [5], who was founded the microscopic theory of generalized electrodynamics, taking into consideration the quantum phenomena. However, this would lead us far away from the main idea of the course. Therefore, we pause at the theory developed by *A.K. Tomilin* [12]. It is noteworthy that at the limit, for macroscopic cases, equations from Tomilin [12] coincide with Khvorostenko's results. We would note that Tomilin's results essentially develop Nikolaev's results [43, 44]; similar results were also obtained by *K. J van Vlaenderen* and *Dale A. Woodside*.

Starting the description of a rather new look at the most interesting areas of science, we should note that this is not a final variant of the generalized theory. Many things must be done in the future. Nevertheless, this is a clear and serious step forward because we get in our hands an "instrument" capable of giving answers to many "traditionally" complicated questions and problems, which are difficult for the classic theory of electromagnetism.

The further description is substantiated by the *main theorem of vector analysis* (sometimes called the expansion theorem, the Helmholtz splitting theorem, the Stokes theorem, or the Stokes–Helmholtz theorem) [54, 56]:

The Stokes–Helmholtz Theorem Any continuous vector field $\vec{F}(\vec{r})$ given in the whole space and disappearing into infinity together with its divergence and rotor, may be uniquely represented (to a vector constant) as a sum of potential (scalar) $\vec{F}_{grad}$ $(\vec{r}) \equiv \vec{F}_g(\vec{r})$ and solenoid (vector) $\vec{F}_{rot}(\vec{r}) \equiv \vec{F}_r(\vec{r})$ fields, i.e.,

$$\vec{F}(\vec{r}) = \vec{F}_g(\vec{r}) + \vec{F}_r(\vec{r}) \tag{1.4}$$

$$\text{where rot}\,\vec{F}_g(\vec{r}) = 0, \quad \text{div}\,\vec{F}_r(\vec{r}) = 0. \tag{1.5}$$

1.3.2 Generalized System Equation of Electrodynamics

According to the Stokes–Helmholtz theorem, the equation system of generalized macroscopic electrodynamics (in the differential form, i.e., concerned with the single point of the $\vec{r}$ space at a given moment in time t), i.e., vectors of the unified electromagnetic field $\vec{E}, \vec{D}, \vec{H}, \vec{B}$, and its sources $\vec{j}, \rho$ are linked between them by the following system of equations [12]:

$$\text{rot}\,\vec{H} + \underline{\text{grad}\,H^*} = \vec{j} + \varepsilon(\partial\vec{E}/\partial t), \tag{1.6}$$

$$\text{rot}\,\vec{E}_r = -\mu(\partial\vec{H}/\partial t), \tag{1.7}$$

$$\text{div}\,\vec{D} = \rho + \underline{\varepsilon\,\partial H^*/\partial t}, \tag{1.8}$$

$$\text{div } \vec{H} = 0, \tag{1.9}$$

$$\vec{B} = \mu \vec{H} \ , \tag{1.10}$$

$$B^* = \mu H^*. \tag{1.11}$$

In this system, the new terms linked with so-called *scalar (potential) magnetic field H** are emphasized. We touch on the sense of the scalar magnetic field later.

Before analysis of the presented equation system (1.6, 1.7, 1.8, 1.9, 1.10, and 1.11) and its physical consequences that may be obtained, we must make sure that it satisfies the main principle of physics; namely, the *Bohr conformity principle*. The essence of this principle consists in the following: *any new theory pretending to give deeper description of physical reality and a wider application area than the old theory should include the latter as a limited specific case.* In our specific case, we are to the limit of small frequencies.

1.3.3 The Maxwell Equation System

First, we must compare the equation system (1.6, 1.7, 1.8, 1.9, 1.10, and 1.11) with the classic system of Maxwell equations. Eliminating from the (1.6, 1.7, 1.8, 1.9, 1.10, and 1.11) system the new terms grad H^* and $\varepsilon(\partial B^*/\partial t)$ related to the scalar magnetic field $H^*(\vec{r})$, we come directly to the known system of Maxwell equations:

$$\text{rot } \vec{H} = \vec{j} + \frac{\partial \vec{D}}{\partial t}, \tag{1.12}$$

$$\text{rot } \vec{E} = -\frac{\partial \vec{B}}{\partial t}, \tag{1.13}$$

$$\text{div } \vec{D} = \rho, \tag{1.14}$$

$$\text{div } \vec{B} = 0. \tag{1.15}$$

The system of *electromagnetic field equations* (1.12, 1.13, 1.14, and 1.15) was obtained by Maxwell as a result of the generalization of accumulated knowledge about electrical and magnetic phenomena up to that time. It was published in 1873 (see, for instance, Stratton, and Faraday [29, 42]). Later, this system was presented by *Oliver Heaviside* and *Gustav Ludwig Hertz* in its present form (1.12, 1.13, 1.14, and 1.15). However, there is a point of view that this modernization was useless for classic electrodynamics.

We would like to note that Maxwell, when writing initial variants of his equation system, of course, relied on the results of the Faraday and Ampere electrical experiments [2]. However, to a great extent, he used some hydrodynamic, gaseous, and mechanical representations and analogies. Some of them are known under the name *Maxwell gear wheels*. Some variants of moving pictures of these gear wheels can be found on the Internet. Hydrodynamic analogies were successfully used, for example, by *P.A. Zhilin* (1942–2005) [55].

In addition, we should evidently take into consideration that the fundamental experiments of Faraday and Ampere were performed at relatively low "electrical engineering" frequencies. Now, the theory must describe the processes whose speed is higher by 10–12 orders, and thus, description of fast processes should correspond to the reality.

1.3.4   The Generalized Law of Electromagnetic Induction

Equation (1.7) is the differential form of the *Faraday's generalized law of electromagnetic induction* [42, 85, 88]: the electromotive force $\mathcal{E}$ in the closed loop, which is pierced by time-varying magnetic flow Φ, is equal to the variation speed of the flow

$$\mathcal{E} = -d\Phi/dt. \tag{1.16}$$

The sign "–" in the last equation corresponds to the Lentz rule: the induction current arising in the closed loop L is directed so that the magnetic induction flow created by it $\Phi = \int\limits_{S} \vec{B} \cdot d\vec{s}$ through the S area, restricted by the L loop, tends to contradict those variations of the Φ flow, which causes this current. In the picture on the plane it looks like Fig. 1.6a, b, and the three-dimensional view is shown in Fig. 1.6c: the induction current I_{ind} flows toward the positive direction of the loop bypass.

The Lentz rule has a deep physical sense because it expresses the energy conservation law.

Faraday's law of electromagnetic induction in the form (1.16): $\mathcal{E} = -d\Phi/dt$ fast entered into the arsenal of modern physics and electrodynamics, and it is beyond any doubt, at first glance. Here, $\mathcal{E}$ is the electromotive force $\mathcal{E} = IR$, and the I current defines the charge $\Delta q = \Delta\Phi/R$, which passes into the circuit during Δt time. However, a closer look at (1.16) reveals that here there are problems with the golden rule of electrodynamics; namely, in the law (1.16) there are no field sources – currents and charges – and according to Eq. (1.16), we may conclude that the time-alternate magnetic flow creates the *electric field* in the conductor. But this is principally a newer phenomenon than Faraday's description of the process: the alternate magnetic field piercing the closed conduction loop causes the electric current in the loop. The phenomenon of electromagnetic induction was formulated by Faraday as: the Δq charge passing through the closed circuit is proportional to the variation of the $\Delta\Phi$ magnetic flow and inversely proportional to the circuit resistance R: $\Delta q = \Delta\Phi/R$.

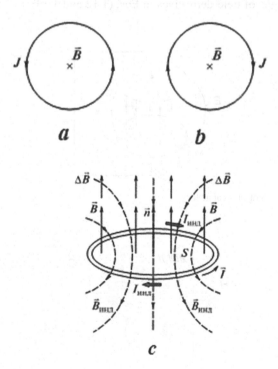

Fig. 1.6 Explanation of the Lentz rule: the magnetic field (**a**) increases in time or (**b**) decreases. (**c**) The fixed conduction loop is located in the uniform magnetic field when $\Delta\Phi/\Delta t > 0$

1.3.5   The Total Current Law

Equation (1.16) is the differential form of the *net current law* (the *Ampere–Maxwell law*): $\oint_L \vec{H}\,\vec{dl} = \int_S \vec{j}\,\vec{ds}$. The net current

law relates to the magnetic field circulation over the L loop and the net current

$$\vec{j} = \sigma\vec{E} + \vec{j}_{out} + \partial\vec{D}/\partial t, \tag{1.17}$$

which is enclosed by this loop. The so-called *outside current* $\vec{j}_{out}$, which is the field source and is considered as given (known), is included in the concept of the net current. However, the $\vec{j}_{out}$ current itself is not the result of considered real electromagnetic fields. In turn, the outside current $\vec{j}_{out}$ is the consequence of the *electromotive forces* $\mathcal{E}_{out}(\vec{r}, t)$ action, which are of non-electromagnetic origin. These are processes of a chemical, bio-electromagnetic, cosmic, diffusion, etc., character. Thus, for the outside current, the differential form of the *Ohm law* is true: $\vec{j}_{out} = \sigma\vec{E}_{out}$ (see Eq. (1.25)). The term $\vec{j}_{out} = \partial\vec{D}/\partial t$, the so-called *offset current*, representing the *Maxwell hypothesis* on the existence of this current, which adds the *conduction current* $\vec{j}$ (Fig. 1.7) to the free space (or in the medium at $\sigma = 0$), is included in the net current $\vec{j}$. Thus, the net current lines are closed in the space either to themselves, or to the sources (electric charges).

We should not fail to note that in spite of the external similarity in names, *conduction currents* and *offset currents*, these currents are also equivalent in the sense of similarity of the magnetic field $\vec{H}(r, t)$ excitation. They differ in the fact that conduction currents correspond to the electric charge motion, whereas the offset current in the vacuum corresponds to a *variation* of electric field $\vec{E}(r, t)$ strength.

We also note that the lines of force of the magnetic field $\vec{H}(r, t)$ excited by the offset currents $\vec{j}_{off} = \partial\vec{D}/\partial t$ form the right triplet of vectors with directions of these currents (Fig. 1.8a), but vectors $\vec{E}(r, t)$ and $\partial\vec{B}/\partial t$ form the left triplet (Fig. 1.8b). This is clearly seen from different signs of field derivatives in Eqs. (1.12 and 1.13).

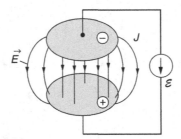

Fig. 1.7 Explanation of the offset current effect in a capacitor

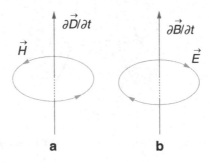

Fig. 1.8 Excitation of electromagnetic fields by (**a**) electric and (**b**) "magnetic" offset currents

Sometimes, instead of Eq. (1.17), the following is called the *net current*

$$\vec{j}_{\text{full}} = \vec{j} + \vec{j}_{\text{off}} = \vec{j} + \partial\vec{D}/\partial t. \qquad (1.18)$$

Using the first Eq. (1.12), the last equation can be presented as rot $\vec{H} = \vec{j}_{\text{full}}$, and having calculated the divergence from it:

$$\text{div}\,\vec{j}_{\text{full}} = \text{div}\,\vec{j} + \partial\rho/\partial t = 0. \qquad (1.19)$$

1.3.6 The Charge Conversion Law

The equation obtained is the classic *continuity equation* (see Sect. 1.3.8) representing the differential form of the *charge conservation law*. As a matter of fact, the electric charge conservation law states that the algebraic charge sum of the electrically closed system remains unchanged: $\sum q_n = \text{const}$ and it is exactly fulfilled.

From the latter equation, we can draw the following conclusions concerning the equality of divergence of some vector. In this case, the lines of such a vector are *continuous*. Thus, the lines of the net current density are continuous whereas the lines of conduction current density $\vec{j} = \sigma \cdot \vec{E}$ and the offset current density may have a beginning and an end. Thus, for instance, the lines of conduction current density have started from those points of the volume under consideration where the density of electric charges decreases, and have ended where the charge density increases. The lines of conduction current density may transfer into the lines of offset current density and vice versa. Details of the proof, explanation, and examples can be found, for instance, in Tomilin, Tamm, and Sommerfeld [12, 54, 132].

From the Maxwell equation system (1.12, 1.13, 1.14, and 1.15), it directly follows that the outside currents $\vec{j}_{\text{out}}$ and the outside charges ρ_{out} are the sources that excite the electromagnetic field. The statement that any spatial variation of vectors $\vec{E}$, $\vec{H}$ (the *rot* operation) corresponds to variations of field $\vec{H}$, $\vec{E}$ in time (the $\partial/\partial t$ operation) is another significant consequence of the Maxwell equations. In other words, electric and magnetic fields are tightly coupled. The independent existence of one field without another is *presumably* possible in the static case only ($\partial/\partial t \equiv 0$).

1.3.7 The Gold Rule of Electrodynamics

In the history of electrodynamics, as, in essence, in other sciences, various paradoxes and contradictions concerning simplifications of the main system of Maxwell equations (1.12, 1.13, 1.14, and 1.15), are known. Consideration of homogeneous equations (1.12, 1.13, 1.14, and 1.15) is one of them: the field sources $\vec{j}_{\text{out}}(\vec{r}, t) = 0$, $\rho_{\text{out}}(\vec{r}, t) = 0$ are absent from these equations. Such a simplification is sometimes useful and would seemingly lead to clear physical results, but this conclusion may be both nonphysical and incorrect.

From this, the *golden rule of electrodynamics* follows: on consideration of practical (and model) problems and systems, we cannot (without sufficient substantiation) separate $\vec{j}_{\text{out}}(\vec{r}, t)\,\rho_{\text{out}}(\vec{r}, t)$ from the electromagnetic field $\vec{E}$, $\vec{H}$ itself.

Thus, classic macroscopic electrodynamics at present is the specific case of the generalized macroscopic electrodynamics represented by Eqs. (1.6, 1.7, 1.8, and 1.9).

Introduction into electrodynamics of the concept of the scalar magnetic field is linked to Austrian professor *Stefan Marinov* (1931–1997) and Russian researcher (Tomsk-town) *George Nikolaev*. They proposed the introduction of some scalar function $H^*(\vec{r})$ that is connected to the vector potential $\vec{A}$ as:

$$H^* = -(1/\mu_0) \cdot \text{div}\,\vec{A} . \qquad (1.20)$$

In textbooks of electrodynamics, we usually encounter the statement that the vector potential $\vec{A}$ has no any physical sense, it is used as an auxiliary function, and the condition of its normalization is

$$\mathrm{div}\ \vec{A} = 0. \tag{1.21}$$

Equation (1.21) is the *Coulomb calibration condition*, which eliminates ambiguity of the introduction of vector potential $\vec{A}$ into the theory. Similar to the introduction of the *Lorentz calibration* condition, condition (1.22) eliminates such an ambiguity for the nonstationary electromagnetic field. According to Eq. (1.21), in magnetostatics, the lines of the $\vec{A}$ vector should be closed, i.e., the field of this vector is *vortical*.

In the classic macroscopic electrodynamics, the *Lorentz calibration condition* is usually used for the net field:

$$\mathrm{div}\ \vec{A} + \varepsilon_a \mu_a \partial \phi / \partial t = 0. \tag{1.22}$$

Thus, we note that utilization of the Coulomb and Lorentz conditions essentially restricts the completeness of the solution of the Maxwell equation. Moreover, in the educational literature (see, for example, Tamm [54]), it is usually stated that the vector potential $\vec{A}$ has no physical sense and it is introduced as some auxiliary function, whereas the normalization condition allows the ambiguity of its introduction to be eliminated.

In accordance with the Stokes–Helmholtz theorem (1.4), the electric field can be both vortical and potential, i.e., it may have the vortical ($\vec{E}_{\mathrm{rot}} \equiv \vec{E}_r$) and the potential ($\vec{E}_{\mathrm{grad}} \equiv \vec{E}_g$) components, whereas the net field $\vec{E}(\vec{r},t)$ is its sum:

$$\vec{E} = \vec{E}_r + \vec{E}_g. \tag{1.23}$$

Accordingly, the electric field induction is:

$$\vec{D} = \varepsilon \vec{E}_r + \varepsilon \vec{E}_g. \tag{1.24}$$

The main equation system of the generalized electrodynamics (1.6) justifies that the conduction current $\vec{j}(\vec{r},t)$ creates both the vector (solenoid) magnetic field $\vec{H}(\vec{r},t)$, and the scalar (potential) magnetic field $H^*(\vec{r},t)$. In general, both components of the unified magnetic field $\vec{H} = \vec{H}(\vec{r},t)$ are nonstationary and heterogeneous. Owing to variation of $\vec{B}(\vec{r},t)$ induction (namely, $\partial \vec{H}/\partial t$), the vortical electric field $\vec{E}_r(\vec{r},t)$ (1.7) is formed. Variation of the SMF induction $B^*(\vec{r},t)$ (grad H^*), together with electric charges $\rho(\vec{r},t)$, gives rise to sources and drains of the potential electric field (Eq. 1.8).

The electric field $\vec{E}(\vec{r},t)$ generally includes the potential and vertical components (Eq. 1.23). Therefore, the derivative $\partial \vec{D}/\partial t$ "creates" the vortical (solenoid) and potential (scalar) components of the magnetic field (Eq. 1.6). As a whole, the offset currents, like the conduction currents, give rise to both components of the magnetic field: scalar and vector. Equation (1.9) justifies the vortical character of the magnetic field $\vec{H} = \vec{H}(\vec{r},t)$.

1.3.8 The Scalar Magnetic Field

Usually, in classic electrodynamics and physics, considering fields of infinitely lengthy conductors is accepted, which contradicts the Stokes–Helmholtz theorem (Eqs. 1.4 and 1.5). In other words, such a scheme (a model) does not satisfy the requirements of the unicity theorem of the solution of Maxwell equations. Therefore, in reality, it has a sense of analyzing the field picture from the conductor with a finite length. Such an example is presented in Fig. 1.9 [12]. It is important that the lines of the vector potential $\vec{A}$ are directed along the z-axis – $\vec{A}_g$. The vortical component $\vec{A}_r$ of this field is caused by its heterogeneity along x and y.

Equation (1.8) is the law of the nonvortical electromagnetic field in the differential form: *the space point in which the nonstationary scalar magnetic field was created is the source (or the drain) of the electric field* [12].

Thus, the potential electric field can be created both by electric charges $\rho(\vec{r},t)$, and with the help of nonstationary scalar magnetic fields $H^*(\vec{r},t)$.

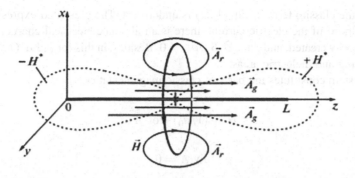

Fig. 1.9 Vector potential $\vec{A}$ lines of the vector magnetic field $\vec{H}$ and the scalar magnetic field H^*, which is created by the rectilinear conductor segment with L length

Equations (1.10 and 1.11), as appropriate relations in classics, are additional constitutive equations that connect the characteristics of the vector and scalar magnetic fields (separately).

The general equation system of generalized electrodynamics should be added by Ohm's law (in differential form for fixed media:

$$\vec{j}(\vec{r},t) = \sigma \cdot \vec{E}(\vec{r},t),$$ (1.25)

where $\sigma(\vec{r})$ is the electric conductivity of the medium, and under $\vec{E}(\vec{r},t)$ we understand the full strengths of the electric field (Eq. 1.23), containing the vortical $\vec{E}_r(\vec{r},t)$ and potential $\vec{E}_g(\vec{r},t)$ strengths, in contrast to the content sense of the $\vec{E}(\vec{r},t)$ concept in classics. Thus,

we have a generalization of Ohm's law in Eq. (1.25).

The *continuity equation* is important, and should evidently be changed compared with the classic:

$$\partial \rho / \partial t + \operatorname{div} \vec{j} = 0$$

To determine the generalized continuity equation, we obtain from Eq. (1.8) the expression for the effective electrical charge [12]:

$$Q_{\text{eff}} = \int_V \rho_{\text{eff}} \cdot dV = \int_V (\rho + \varepsilon \, \partial B^* / \partial t) \cdot dV.$$ (1.26)

1.3.9 The Generalized Continuity Equation

The electric current $\vec{j}(\vec{r},t)$ flowing through the S surface, which restricts the V volume, is connected with variation of the effective electric charge Q_{eff} by the obvious equation $\int_X \vec{j}(\vec{r},t) \cdot d\vec{s} = -\int_V (\partial Q_{\text{eff}} / \partial t) \cdot dV$. Having applied to its left side part of the *Ostrogradsky–Gauss* theorem for the flow of the $\vec{A}(\vec{r})$ vector through the arbitrary closed S surface, equal to the integral from the divergence of this vector over the V volume, which is restricted by the S surface $(\oint_S \vec{A} \cdot d\vec{s} = \int_V \operatorname{div} \vec{A} \cdot dV)$,

we obtain the required equation – the *generalized continuity equation*:

$$\frac{\partial \rho}{\partial t} + \varepsilon \frac{\partial^2 B^*}{\partial t^2} + \operatorname{div}\left(\vec{j} - \frac{\partial \vec{D}}{\partial t}\right) = 0.$$ (1.27)

The new (compared with the classic) term in Eq. (1.27) is underlined. The presented expression (1.27) states that at the point that is the source (the drain) of the electric current, there is an alternate electrical charge, and in it, the nonstationary scalar magnetic field is necessarily created, and thus, $\partial^2 B^*/\partial t^2 \neq 0$. Besides, in this form, Eq. (1.27) is applicable for media in which there are both conducting and dielectric areas.

The following equation system constitutes the basis of classic magnetostatics:

$$\vec{H} = (1/\mu_0) \cdot \text{rot } \vec{A}, \tag{1.28}$$

$$\text{div } \vec{A} = 0. \tag{1.29}$$

Equation (1.29) (see also Eq. (1.21)) is the Coulomb calibration condition, which excludes the ambiguity of the vector potential $\vec{A}$, similar to the introduction of the Lorentz calibration condition (1.22) eliminating such an ambiguity for a nonstationary electromagnetic field.

1.4   The Energy of the Electromagnetic Field, the Umov–Poynting Vector, the Equations for Energy Balance – the Umov Theorem

As we already mentioned, the action of the electromagnetic field is manifested by its influence on the moving charge q, and also by the effect on any body located in the area occupied by the field. This action is demonstrated through the electromagnetic energy W distributed over some V area restricted by the S surface (Fig. 1.10; the S surface may be both real and virtual):

$$W = (1/2) \int_V \left(\varepsilon_{\text{abs}} \vec{E}^2 + \mu_{\text{abs}} \vec{H}^2 \right) dV. \tag{1.30}$$

where "abs" means the absolute value.

Thus, the electromagnetic energy W is distributed in the space with the volume density

$$w = (1/2)\varepsilon_{\text{abs}}E^2 + (1/2)\mu_{\text{abs}}H^2. \tag{1.31}$$

From Eq. (1.31), it directly follows that the electromagnetic field energy W represents a sum of electric W^{el} and magnetic W^{mag} fields:

$$W^{\text{el}} = (1/2) \int_V \varepsilon_{\text{abs}} \vec{E}^2 dV = (1/2) \int_V \vec{E} \, \vec{D} \, dV, \tag{1.32}$$

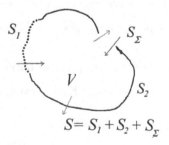

$$S = S_1 + S_2 + S_\Sigma$$

Fig. 1.10 Diagram of the arbitrary V volume surrounding by some "combined" surface $S = S_1 + S_2 + S_\Sigma$. Surfaces S_1 and S_2 differ in their properties, and the S_Σ surface represents a hole through which the energy may depart from the V volume or enter into it

$$W^{\text{mag}} = (1/2)\int_V \mu_{\text{abs}}\vec{H}^2 dV = (1/2)\int_V \vec{H}\,\vec{B}\,dV. \tag{1.33}$$

The electromagnetic field energy W in the V volume may change in time at the expense of at least two processes. It can be transferred into other forms of energy of "non-electromagnetic" nature, for example, into the heat energy on body heating with σ conductivity, chemical energy, biophysical energy, etc. Variation of energy quantity in the V area may also be provided at the expense of its departure (arrival) from this volume through some S_Σ holes or through the S surface itself by virtue of its partial or complete transparence (for instance, at the expense of the mentioned fancy) or owing to some features of the field itself (see Sect. 1.3). The energy may not only disappear from the V volume, but enter it through the same S surface or through part of it (S_Σ).

The first of the aforementioned processes of electromagnetic field energy W variation in the V volume (transferred to other energy types) is described as the power delivered (obtained) by the field in the time unit as follows

$$P = \int_V \vec{j}\cdot\vec{E}\,dV. \tag{1.34}$$

In Eq. (1.34), the integrand is the *volume power density*

$$p = \vec{j}\cdot\vec{E}. \tag{1.35}$$

The energy departing from the V volume (or arriving in it from outside) is defined as follows:

$$P_\Sigma = \oint_S \vec{U}\,d\vec{S}, \tag{1.36}$$

where via $\vec{U}$ we designated the *Umov–Poynting vector* representing the *electromagnetic energy flow density*.

The Umov–Poynting vector $\vec{U}$ in classics is defined as the cross-product of vectors of electric $\vec{E}$ and magnetic $\vec{H}$ field strengths in each point of some S surface:

$$\vec{U} = \left[\vec{E},\vec{H}\right] \tag{1.37}$$

Quantities W, P, and P_Σ are connected by the "integral" relation – by *equation of the power balance* – the *Umov theorem*:

$$dW/dt + P + P_\Sigma = 0. \tag{1.38}$$

In essence, this is the *energy conservation law* for the electromagnetic field.

As we see from the presented Eqs. (1.26, 1.27, 1.28, 1.29, 1.30, 1.31, 1.32, 1.33, 1.34, 1.35, 1.36, 1.37, and 1.38), the field energy characteristics represent the quadratic functions of field strengths $\vec{E}$, $\vec{H}$. Hence, for these characteristics, the *principle of superposition of partial solutions* (the *superposition principle*) is violated, i.e., we must substitute in Eqs. (1.26, 1.27, 1.28, 1.29, 1.30, 1.31, 1.32, 1.33, 1.34, 1.35, 1.36, 1.37, and 1.38) (and in similar equations) the *net* field only. At the same time, we note that we are within the limits of *linear electrodynamics*, i.e., for the field quantities, the superposition principle is always true, and it is widely used in theory and in practice.

In addition to the *free charges* included, for example, into Eq. (1.2), it is accepted to distinguish the *bound charges*. *Free charges* we understand to be those capable of moving relatively large distances under the influence of the electric field. Under *bound charges* we understand the electrical charges of particles included in the structure of the atoms and molecules of dielectrics and also ion charges in the crystal dielectrics with an ion lattice.

We should note that in the vacuum and also in liquid and gaseous media, the ordered motion of charged particles, not linked with the media conductivity, is possible, which is described by Ohm's law $\vec{j} = \sigma\cdot\vec{E}$ (Eq. 1.25). The motion of charged particles in the vacuum or in the medium, which does not have conductivity (i.e., $\sigma = 0$), is called the *convection current*, for whose characteristic we introduce the concept of convection current density $\vec{j}_{\text{conv}}$, and then the third Maxwell equation (1.14) will be:

$$\text{div } \vec{D} = \rho + \rho_{\text{conv}}. \tag{1.39}$$

As in classic macroscopic electrodynamics, its generalized variant should be added by a series of additional conditions sufficient for satisfaction of the unicity theorem of the generalized Maxwell equations (1.6, 1.7, 1.8, 1.9, 1.10, and 1.11). It is evident that, as a whole, the total energy of the electromagnetic field in some V volume, which is restricted by the S surface, is wasted on the transition into other energy types, into radiation (reception), etc. In appropriate classic equations (1.12, 1.13, 1.14, and 1.15), we need the presence of the scalar magnetic field and associated fields from (1.6, 1.7, 1.8, 1.9, 1.10, and 1.11).

Then, instead of the "classic" Umov–Poynting vector (1.33), we must use the relation

$$\vec{U} = \left[\vec{E}, \vec{H} \right] + \underline{\vec{E} \cdot H^*}, \tag{1.40}$$

in which the first term is "traditional" (1.33) and determines the energy transfer in a direction perpendicular to the plane of $\vec{E}$ and $\vec{H}$ vector location. The second term in (1.40) determines the energy transfer along the direction of the $\vec{E}$ vector.

Now, for the volume density of the energy flow, we have (compare Eq. (1.27)):

$$w = (1/2) \cdot \left(\vec{E} \cdot \vec{D} + \vec{H} \cdot \vec{B} + \underline{H^* \cdot B^*} \right). \tag{1.41}$$

As a result, we obtain the same (in form and in physical content) general equation of the power balance – the Umov theorem (1.41), as in the classic theory, but in it we take into account the energy transfer in the direction of the electric field $\vec{E}$ vector.

The disadvantage of the traditional expression of the energy flow density: $w = (1/2) \cdot \left(\vec{E} \cdot \vec{D} + \vec{H} \cdot \vec{B} \right)$ obtained from Eq. (1.41) by omitting the underlined term, is the utilization of particularly vortical field characteristics. Thus, it turns out that electric and magnetic components of the total electromagnetic field vary in-phase whereas from the usual physical representations, the arguments of $\vec{E}$ and $\vec{H}$ ·functions should be shifted by $\pi/2$, but this does not follow from the Maxwell equations. The generalized electrodynamics theory eliminates this paradox.

1.5   The Generalized Law of Electromagnetic Interaction

In classic electrodynamics, the above-mentioned transverse Lorentz force is widely known, which acts on the point charge q moving with the $\vec{v}$ velocity, and defined as

$$\vec{F} = q \left(\vec{E} + \left[\vec{v} \vec{B} \right] \right). \tag{1.42}$$

Such a representation does not give an answer to the question about the motion of the electrons and their deviation from the "true" trajectory, which has been discussed many times in the scientific and educational literature. Nevertheless, relying on representations defined by the equation $\text{rot } \vec{H} + \text{grad} H^* = \vec{j}$ (see later Eq. (2.49)), we should take for the density of the Lorentz transverse force as follows

$$\vec{f}_{\text{Lor}} = \text{rot } \vec{H}_{\text{own}} \times \vec{B}, \tag{1.43}$$

where $\vec{H}_{\text{own}}$ is the strength of the magnetic field vector of the conductor.

Then, for the longitudinal electromagnetic force (Ampere–Nikolaev), if interaction with separate charged particles with an external scalar magnetic field (the Lorentz force analog) occurs, we can write:

$$\vec{F}_{\text{own}} = qB^* \cdot \vec{v}. \tag{1.44}$$

From Eqs. (1.43 and 1.44), it follows that

$$\vec{f} = \text{rot}\,\vec{H}_{\text{own}} \times \vec{B} + B^* \text{grad}\,H^*_{\text{own}}. \tag{1.45}$$

The equation obtained is the generalized law of electromagnetic interaction [12]. The distinctive feature of such a formulation of the generalized law of electromagnetic interaction is an account of transverse and longitudinal components of the full electromagnetic force and its vortical character.

1.6 Classification of Media

1.6.1 Natural Media

From *constitutive equations* (1.10, 1.11, and 1.12) (*equations of state*), we see that media, in which electrodynamics processes occur, are characterized by their parameters ε, μ, and σ. Very generally, media can be divided into *linear* and *nonlinear*. In the linear medium parameters ε, μ, and σ, by definition, do not depend on the magnitude of electric or magnetic fields: $\varepsilon \neq \varepsilon(|\vec{E}|), \mu \neq \mu(|\vec{B}|), \sigma \neq \sigma(\vec{E},\vec{H})$. In the nonlinear medium, at least one of its parameters (or all together) depends on electric or magnetic fields: $\varepsilon = \varepsilon(|\vec{E}|), \mu = \mu(|\vec{B}|), \sigma = \sigma(\vec{E},\vec{H})$. In reality, all media in nature are nonlinear; however, at weak enough (not strong) fields, in many practical cases, we may neglect (in the first approximation) by the dependence of ε, μ, and σ parameters on the field intensity. Obviously, regarding its physical content, the nonlinear media and processes in them are much richer and more manifold than in the linear media. However, the analysis is much more complicated. In our lecture course, we consider only linear media. However, for their assimilation, a large amount of *continuous* labor is necessary (we emphasize this once more).

Concerning t time, the media can be *stationary*, when the parameters do not depend on time t, and *nonstationary*, when this dependence exists at least for one media parameter among ε, μ, and σ.

The class of linear media is wide enough and manifold and includes the *homogeneous*, *heterogeneous*, *isotropic*, and *anisotropic* media. The e, μ, and σ parameters of *homogeneous* media do not depend on coordinates $(\vec{r})$, and their properties are the same at all points and in all directions or in the part of the space under consideration. If at least one of its parameters depends on coordinates, such a medium concerns the class of *heterogeneous* media.

In practice, the media, whose properties are similar in different directions, are of special interest. Such media can be concerned with the class of *isotropic* media, and they are typical in that vectors $\vec{D}$ and $\vec{E}$, $\vec{B}$, and $\vec{H}$, and also $\vec{M} = \chi_m \vec{H}$ and $\vec{H}$, $\vec{P} = \chi_e \vec{E}$, and $\vec{E}$ are parallel in these media. Here, χ_e is the dielectric susceptibility of the medium, and χ_m is the magnetic susceptibility. Besides, the parameters ε, μ, and σ are scalar quantities. In anisotropic media, at least the one of these parameters is a tensor $\|\varepsilon\|$ or $\|\mu\|$, which, for example, takes the form:

$$\|\varepsilon\| = \begin{Vmatrix} \varepsilon_{xx} & \varepsilon_{xy} & \varepsilon_{xz} \\ \varepsilon_{yx} & \varepsilon_{yy} & \varepsilon_{yz} \\ \varepsilon_{zx} & \varepsilon_{zy} & \varepsilon_{zz} \end{Vmatrix}, \|\mu\| = \begin{Vmatrix} \mu_{xx} & \mu_{xy} & \mu_{xz} \\ \mu_{yx} & \mu_{yy} & \mu_{yz} \\ \mu_{zx} & \mu_{zy} & \mu_{zz} \end{Vmatrix}. \tag{1.46}$$

The anisotropic medium may be homogeneous: in this case, all tensor components do not depend on coordinates.

The equation system of electrodynamics remains unchanged, but with natural replacement of scalar media parameters ε, μ, and σ with the tensor quantities $\|\varepsilon\|$, $\|\mu\|$, $\|\sigma\|$. We note, by the way, that fundamental results on the properties of anisotropic bodies were obtained by the Russian scientist *V.K. Arkadiev* (1884–1953).

The following equations of state should be added to the general equation system of electrodynamics:

$$\vec{D} = \|\varepsilon\| \cdot \vec{E}, \quad \vec{B} = \|\mu\| \cdot \vec{H}, \quad \vec{j} = \|\sigma\| \cdot \vec{E} \tag{1.47}$$

Crystalline media, for example, have anisotropic properties. They are typical in their structure ordering, whose polarization sometimes depends on the external field direction. Another example of an anisotropic medium is, for instance, an ionosphere – an ionized medium located in the permanent magnetic field of the Earth. It has interesting physical properties and essentially affects the radio wave propagation of some wave ranges. Here, many scientific results are attributed to Russian scientists, including *B.A. Vvedensky, V.A. Fock, A.N. Shchukin,* and *M.V. Shuleikin.*

Later (see Chap. 4), we provide more details of the description of waves in the gyrotropic medium, i.e., the anisotropic medium, for which the tensor $\|\varepsilon\|$ components in в (1.46) have the form: $\varepsilon_{xx} = \varepsilon_{yy}$, $\varepsilon_{yx} = -\varepsilon_{xy}$, $\varepsilon_{xy} = -\varepsilon_{xy}$, $\varepsilon_{xz} = \varepsilon_{yz} = \varepsilon_{zx} = \varepsilon_{zy} = 0$. For instance, magnetized ferrites widely used in microwave and millimeter-wave ranges have gyromagnetic properties [21, 47, 57, 58].

Summary of Some Media Properties

Engineering electrodynamics deals with the manifold media and materials. Here, it is important that in our lecture course on macroscopic electrodynamics, the media parameters are *given*; their determination is concerned with the big branches in microscopic electrodynamics, physics and other disciplines that are beyond the limits of our course. Some information on the parameters of practically used substances is listed in Table 1.1 in Nefyodov [28], p.46.

1.6.2 Artificial Media, Chiral Media

A large branch of modern electrodynamics is devoted to artificial media. Far from always, at development of the specific device, we can use the set of materials available to the engineer/designer. The simplest example is the development of matched devices (covering), say, for dielectric lens antennas, optical instruments, etc. Far from always, we can select the matched layer with the necessary values of ε, μ, and thus, it is necessary to replace it by some other construction whose integral characteristic satisfies the necessary requirements.

The last decade has been characterized by essential attention being paid to artificial media with *spatial dispersion*. They are called *chiral media* and have a set of unique properties, that serve as the basis for the development of the wide class of transmission lines and basing elements in microwave and mm-wave ranges. We will become acquainted with some of them, but for now we note their main features.

It is usually assumed that there is an ambiguous dependence between the $\vec{D}(\vec{r},t)$ induction and the electric field $\vec{E}(\vec{r},t)$ strength, which is defined by relations $\vec{D} = \varepsilon \vec{E}$ and $\vec{B} = \mu \vec{H}$ (see Eqs. (2.29), (2.30)) for the isotropic homogeneous medium, or Eq. (1.42) for the anisotropic medium. Nevertheless, the statement that there is ambiguous correspondence between $\vec{D}(\vec{r},t)$ and, $\vec{E}(\vec{r},t)$ requires at least two specifications. First of all, the electric field induction $\vec{D}(\vec{r},t) = \varepsilon \vec{E}(\vec{r},t)$ depends not only on $\vec{E}(\vec{r},t)$, but on the time-derivative of this vector. For arbitrary time dependence $\vec{E}(\vec{r},t)$, the equation $\vec{D}(\vec{r},t) = \varepsilon \vec{E}(\vec{r},t)$ is wrong. We can use the formula (1.6) for only sine time dependence and neglecting dispersion $\varepsilon = \varepsilon(\omega)$. The second specification is concerned with the *spatial dispersion* phenomenon, i.e., the fact that $\vec{D}(\vec{r},t)$ depends not only on $\vec{E}(\vec{r},t)$, but on its spatial derivatives. Therefore, equation $\vec{D}(\vec{r},t) = \varepsilon \vec{E}(\vec{r},t)$, strictly speaking, is wrong. This equation only remains true when fields in the space change, as in the plane wave.

In general, for sine oscillations, in those points where there are no outside sources, the constitutive equations for the chiral media are:

$$\vec{D} = \varepsilon \vec{E} - i\aleph \vec{H}, \quad \vec{B} = \mu \vec{H} + i\aleph \vec{E}. \tag{1.48}$$

Here ε, μ, $\aleph$ are constitutive parameters that do not depend on the field structure; hence, they must be given for problems of macroscopic electrodynamics. When the chirality is absent ($\aleph \equiv 0$), expressions (1.48) transfer to known $\vec{D} = \varepsilon \vec{E}$, $\vec{B} = \mu \vec{H}$ (see Eqs. (2.29 and 2.30). For practical tasks, the most important case, at least, at present, is when ε, μ, $\aleph$ parameters are scalar quantities, i.e., isotropic chiral media.

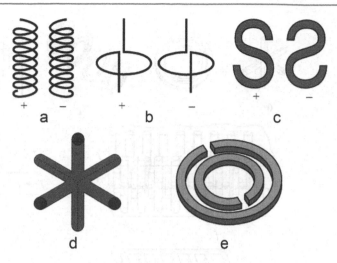

Fig. 1.11 (**a**, **b**) Three-dimensional and (**c**) two-dimensional chiral elements ("particles"). (**d**, **e**) The isotropic chiral particle. (Taken from Nefyodov and Kliuev [73], p. 40)

The physical analysis of chiral media properties and devices based on it is still awaiting a detailed investigation. However, running a few steps ahead, we note that the remarkable property of chiral media (three- and two-dimensional) is the presence of *resonant phenomena*, although they are small compared with the wavelength.[4]

For example, the electrodynamics structure created by means of uniform right-screw and left-screw conduction spirals in the homogeneous dielectric layer (Fig. 1.11a) is an example of chiral media.

The chirality effect is demonstrated if sizes of conducting spirals are essentially less than the wavelength, whereas the distance between them is commensurable with the wavelength of electromagnetic emission. Other examples of chiral elements are spheres with spiral conductance, cylinders with conductance along screw lines, etc. Media composed of the single or double spiral particles have "natural" anisotropy (Fig. 1.11a). The medium consisting of isotropic chiral particles (Fig. 1.11d) is isotropic.

It is interesting to note that various elements of the alive nature have chiral properties. By the way, the chirality of the alive nature attracted the attention of researchers much earlier than their application today. Apparently, the famous French biologist *Louis Pasteur* (1822–1895) paid attention to them, when in 1860 he explained the chirality nature of some isomers. An essential contribution to the construction of chiral structure models was made by outstanding Russian scientist *N.A. Umov*. Different bio-molecules have various forms of chirality. Thus, the DNA molecule has right-screw orientation, amino-acids have the left orientation, the ferment pepsinogen has right orientation, etc. [6, 50, 51].

1.6.3 Media with Negative Permeability (Metamaterials)

In the last few years, the problem of electrodynamics and physics of media with negative permeability and negative refraction has arisen. The prefix "meta" itself means in Greek "out," which allows utilization of the term "metamaterials" for structures whose effective electromagnetic properties go beyond the limits of the properties of the components that form this material. From the very origin of the media class, which was the end of the eighteenth century, the creation of media with $\varepsilon < 0$, $\mu < 0$ was performed in an artificial way by means of various "usual" elements, for example, metal rods, open wire windings, etc. (Fig. 1.12). Then, the finite wire segment can play the role of capacitor, and open winding the role of inductance. Together, therefore, they can form a resonant circuit. Experiments with metamaterials showed their narrowband properties, and then the "complication" of structures began (Fig. 1.12b–h).

First experiments with these twisted structures were apparently carried out by *J.C. Bose*. He investigated the polarization properties of structures developed by him and found out their unique properties. After that, a series of publications by *K.F.*

---

[4] This problem is rather relevant for understanding many phenomena, in particular, in millimeter-range therapy, when the cellular biological structures manifest their resonant properties in the millimeter-range (for $\lambda \approx 4\ldots9$ mm), whereas the cell sizes are of the order 100–200 μm (see, for instance, [1.33–1.36]).

Fig. 1.12 Elements of structures with negative ε and μ: the triplet element for creation of (**a**) isotropic ε negative (ENG) structures, (**b**) the double ring resonator and (**c**) the media element with it, the (**d**) metasolenoid, and (**e**) the system of thin $kr < <1$ metal conductors, (**f**) the interlayer joint and (**g**) its equivalent circuit, (**h**) the three-dimensional electromagnetic bandgap crystal (ENG means ε negative)

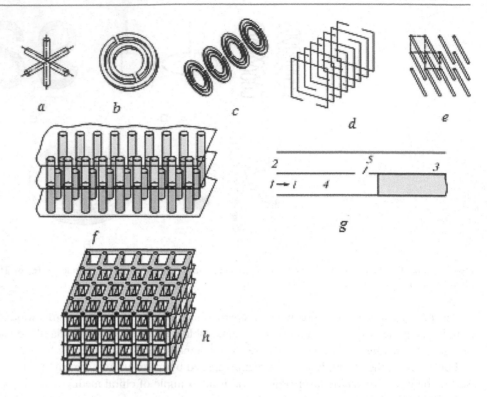

Lindman and *W.E. Kock* followed. In Russia, *L.I. Mandelstam* discussed the possibility of negative group velocity of the electromagnetic field for the first time. For media with negative permeability, the new properties in the Doppler effect we predicted, in the *Vavilov–Cherenkov* radiation, the specific optical devices were offered, etc. The various composite materials with the negative group velocity in super-high frequency (SHF) and in the optical range were created.

The brief classification of physical media depending on the sign of the dielectric (ε) and magnetic (μ) permeability is given in Table 1.1. The wide class of materials and metamaterials, which is being broadly investigated at present and, more importantly, is being used in practice. So, for instance, the application of metamaterial allows a reduction of the antenna sizes by 50 times. Other applications are known [123].

To date, a huge number of different "constructions" that manifest the properties of metamaterials, is suggested and studied (in various detail). The simple examples of these media elements are shown in Figs. 1.12 and 1.13.

These figures show part of the medium (Fig. 1.12c) from double ring resonators (Fig. 1.12b). The metasolenoid (Fig. 1.12d) and corresponding media can be concerned with this class. The scheme of interlayer transition using structures of the type shown in Fig. 1.12e is presented in Fig. 1.12f and its equivalent circuit in Fig. 1.12g. An example of the metamaterial of ε negative type consisting of the equidistant set of thin metal conductors located, for instance, in the dielectric or between two metal surfaces, is shown in Fig. 1.12h. In the last few years, the three-dimensional electromagnetic bandgap crystals have found wide application (Fig. 1.12h). Eliminating the separate layer from these crystals, we may obtain analogs of waveguides, their irregularities (break, bends, etc.), basing elements (inter-layer transitions, cavities, couplers, etc.). Serious research into these structures was performed by *S.E. Bankov*, and by the scientific school of *A.D. Shatrov* [50].

1.6.4 Concept of Below-Cutoff Media

Consideration of natural and engineering media (Table 1.1) will be not complete if we do not mention, at least, briefly, the so-called below-cutoff media and waveguide structures. In the last few years, they have provoked a significant interest and have a wide field of activity for research and practical application. If a situation with processes described in the below-cutoff transmission lines is more or less clear, then the properties of the below-cutoff media are known much less. We should immediately note that description of media by nondisperse parameters ε, μ is, in general, the far-reaching idealization of physical properties of the real media (see Sect. 4.2). Even with the "usual" loss of presence in the medium, let alone the mode of wave (signal) amplification, the structure properties are far from "ideal," which is usual for classic electrodynamics. Usually, it is considered that losses in media are small enough, and we use the *method of small perturbations* that is habitual

Fig. 1.13 Examples of transmission lines, slow-wave structures ($p/\lambda \ll 1$): the lattice from (**a**) circular and (**b**) rectangular conductors, (**c**) the two-dimensional lattice from rectangular waveguides, (**d**) the multi-wire line, (**e–n**) lattices of various profiles; the mutual arrow indicates the direction of signal transmission (along z), (**o**) the lattice from isotropic bands, (**n**) the two-dimensional periodic lattice from ideal cylindrical conductors, (**p**) the two-dimensional double-element lattice, (**q, r**) the two-element lattice from anisotropic conduction bands. (Taken from Nefyodov and Kliuev [73], p.47)

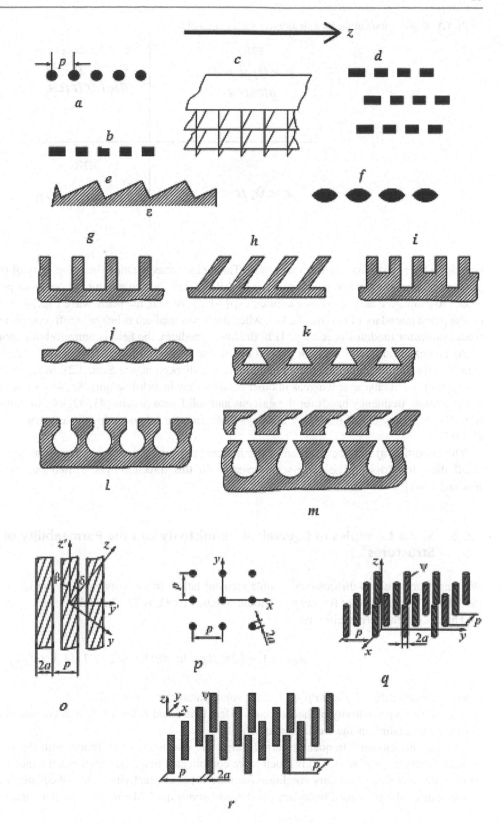

for physicists and engineers. However, as accurate analysis shows, this method happens to be not effective enough and noncomplete for the mentioned media with noticeable losses and with amplification.

From the second and fourth quadrants of Table 1.1, the condition of the below-cutoff medium are: a) $\varepsilon' > 0$, $\mu' < 0$ and b) $\varepsilon' < 0$, $\mu' > 0$. In the simplest case, the plane wave in the infinite homogeneous space propagates with the wave-number $k\sqrt{\varepsilon\mu}$,

Table 1.1 Classification of media with different values of ε and u

ENG $\varepsilon < 0, \mu > 0$ *plasma*	DPS (double positive) $\varepsilon > 0, \mu > 0$ *dielectrics*
DNG $\varepsilon < 0, \mu < 0$	MNG $\varepsilon > 0, \mu < 0$

ε

i.e., along the z-axis it has the form $\exp.\{ikz\}$. Taking into consideration the complexity of permittivity and permeability ε and μ, we may present the wave-number as $k = k' + ik.''$ When the wave-number k becomes purely imaginary ($k' = 0$), the wave ceases to propagate and becomes damping: $\exp\{-k''z\}$. At that, the plane wave $\exp.\{ikz\}$, which normally incidents ($\varphi = 0$) on the plane boundary of two media, say, when the lower medium is below cutoff, completely reflects from the boundary: the reflection factor modulus is equal to 1. In the lower medium, the field is nonzero but exponentially damps: $\exp\{-k''z\}$.

At examination of the medium properties for the below-cutoff case, we should take into account that these properties significantly depend on the frequency (i.e., there is a dispersion, see Sect. 4.2). In some frequency bands, the medium may have a traditional character whereas in other bands it may be below-cutoff. So, for example, conditions b) above are fulfilled in the definite frequency bands for the gaseous and solid-state plasma [41, 42, 44]. In particular, for the collisionless plasma ($\varepsilon' = 0$), the below-cutoff frequency area lays in the frequency band, which is less than the plasma frequency: $\omega < \omega_p$, where $\varepsilon' < 0$.

The condition a) may be satisfied in the frequency bands of ferromagnetic resonance for the gyromagnetic media with small dielectric losses. The large-scale research in this direction was carried out by *A.G. Gluschenko* [95] and *A.A. Rukhadze* [61].

1.6.5 Some Examples of Equivalent Permittivity and the Permeability of Widespread Wire Structures[5]

We start from the two-dimensional double-element lattice from cylinders with the ideal surface conductivity (Fig. 1.13п) [60, 109]. The condition of frequently periodic structure (see Eqs. (2.18 and 2.19): $kp \ll 2\pi$ is applied to the lattice parameters.

The equivalent permeability is:

$$\varepsilon_{\text{eff}} = 1 + \left(2\pi/(kp)^2 \ln(\pi q)\right), \quad \mu_{\text{eff}} = 1/\left(1 + \pi q^2/2\right), \tag{1.49}$$

where the period filling $q = 2a/p \ll 1$ is the small parameter of our task.

The effective permittivity of the lattice ε_{eff} is less than 1 and at low enough frequencies becomes negative (the supercritical plasma); the medium is the weak diamagnetic.

The two-dimensional frequently periodic structure consisting of cylinders with right-screw and left-screw (anisotropic) surface conductivity (Fig. 1.13r) is much more complicated in analysis, but much richer in its physical content. The general form of the two-sided boundary condition for anisotropic conductivity is described later (see Sect. 2.1.8). For the case under consideration, the two-sided boundary conditions (from Eq. (2.24)) on the cylinder surface (for $r = a$) have a form:

---

[5] *Maltsev V.P., Shatrov A.D.* (in Russian) Radiotekhnika i Electronika, 2009.-Vol.54.-№ 7. Pp. 832-837; 2011.-Vol.56.№ 6.-Pp.689-693; 2012. Vol.57.-№ 2. P.187.

$$E_z^+ = E_z^-, \quad E_\varphi^+ = E_\varphi^-$$

$$E_z \cos\psi + E_\varphi \sin\psi = 0,$$

$$\left(H_z^+ - H_z^-\right)\cos\psi + \left(H_\varphi^+ - H_\varphi^-\right)\sin\psi = 0, \tag{1.50}$$

where signs "+" and "−" correspond accordingly to $r > a$ and $r < a$, ψ is the twisting angle.

For right-screw lines we have $\psi > 0$, for left-screw $-\psi < 0$.

We know that at diffraction of the plain wave on the thin ($ka << 2\pi$) anisotropically conducted cylinder with the small twisting angle ($|\psi| << 1$) on the frequency satisfying the condition

$$ka = |tg\psi| \tag{1.51}$$

the resonance arises. It is manifested in sharp increase of the first azimuth harmonic amplitude for the surface current. Obviously, in the lattice of such cylinders and especially in the medium from these elements, this resonance is more pronounced, intensive.

In this case, the equivalent permittivity of the medium is determined by (1.45), while the.

permeability is:

$$\mu_{\text{eff}} = \left(1 - tg^2\psi/(ka)^2\right)/\left(1 - tg^2\psi/(ka)^2 + \pi q^2/2\right). \tag{1.52}$$

The dielectric properties of the structure are the same as for the lattice from conducting cylinders; namely, $\varepsilon_{\text{eff}} < 1$ and has a weak frequency dependence.

Some data for specific medium parameters for frequently periodic, anisotropically conducting cylinders (Fig. 1.13r) are shown in Figs. 1.11 and 1.12.

Data on μ_{eff} "sharply" depend on the spiral twisting angle ψ, and on the initial portions upon the frequency (ka) (see Program P.1.1).

1.6.6 Note on Negative Permeability

Simultaneous sign variation for permittivity ε and permeability μ (from positive to negative) leads to interesting physical consequences. For instance, the energy flow, which is carried by the wave, is determined by the classic Umov–Poynting vector $\vec{U}$ from (1.37) (for a more complete definition of the Umov–Poynting vector see in Eq. (1.36)) and always forms the right triplet with field $\vec{E}$ and $\vec{H}$ vector (the so-called "right" medium). If $\varepsilon > 0$, $\mu > 0$ in the medium, vectors $\vec{U}$ and $\vec{k}$ have the same direction (+ in Fig. 1.14a). Another situation when $\varepsilon < 0$, $\mu < 0$ (the "left" medium). In this case, vectors $\vec{U}$ and $\vec{k}$ are directed in different directions (Fig. 1.14b). Because the $\vec{k}$ vector coincides in direction with the phase velocity, it is clear that "left" media (substances) are the media with the so-called negative phase velocity. In other words, the phase velocity is opposite to the energy flow in the left media.

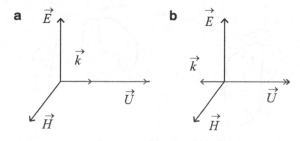

Fig. 1.14 Determination of (**a**) right and (**b**) left media

Here, the paradox arises connected with the fact that if $\varepsilon < 0$, $\mu < 0$ simultaneously, the wave (process) energy, according to Eq. (1.37) $w = (1/2)\varepsilon_a E^2 + (1/2)\mu_a H^2 < 0$ is negative, which evidently contradicts the usual physical representations. Elimination of this contradiction is possible if the left medium has a *frequency dispersion*, i.e., when we have nonzero derivations $\partial\varepsilon/\partial\omega$ and $\partial\mu/\partial\omega$ (see, for instance, Gluschenko and Zakharchenko [95]).

1.7   The Lorentz Lemma, the Reciprocity Theorem

In many problems of electrodynamics and, particularly, at analysis and calculation of the emitting systems (antennas), the *Lorentz lemma* finds a wide application. It is formulated in two forms,: integral and differential, and connects two fields, $\vec{E}_1, \vec{H}_1$ and $\vec{E}_2, \vec{H}_2$, between them in the mutual V region of the space. These fields are caused by independent outside sources $\vec{j}_{out}^{(1)}$ and $\vec{j}_{out}^{(2)}$, which are concentrated in some arbitrary noncrossing areas V_1 and V_2 (Fig. 1.15).

Writing Maxwell equations in the differential form (1.12) and (1.13), performing the simplest operations for these two fields, we obtain the following equation

$$\text{div}\left[\vec{E}_1, \vec{H}_2\right] - div\left[\vec{E}_2, \vec{H}_1\right] = \vec{E}_2 \, \vec{j}_{out}^{(1)} - \vec{E}_1 \, \vec{j}_{out}^{(2)}, \tag{1.53}$$

which is the *Lorentz lemma* in the differential form.

The integral form of the *Lorentz lemma* directly follows from Eq. (1.49), which must be integrated over the whole V area surrounded by the S surface. Having applied the *Ostrogradsky–Gauss theorem* (about transformation of the volume integral from the arbitrary vector into the integral over the S surface), we get

$$\oint_S \left\{ \left[\vec{E}_1, \vec{H}_2\right] - \left[\bar{E}_2, \vec{H}_1\right] \right\} d\vec{S} = \int_V \left(\vec{j}_{out}^{(1)} \vec{E}_2 - \vec{j}_{out}^{(2)} \vec{E}_1 \right) dV. \tag{1.54}$$

The latter Eq. (1.54) is the *integral form of the Lorentz lemma*. From Eq. (1.54), transferring the S surface into infinity and requiring the fulfillment of the *emission condition* (namely, fields $\vec{E}_1, \vec{H}_1$ and $\vec{E}_2, \vec{H}_2$ decrease in infinity faster than $1/r$; we come to the expression for the *reciprocity theorem*:

$$\int_{V_1} \vec{j}_{out}^{(1)} \vec{E}_2 dV = \int_{V_2} \vec{j}_{out}^{(2)} \vec{E}_1 dV. \tag{1.55}$$

The *reciprocity theorem* finds a wide application in an antenna theory and often significantly simplifies the solution of many tasks. We discussed the simplest form of the reciprocity theorem, assuming that a medium in the V volume is linear and isotropic. We can show that this theorem remains true for the linear anisotropic medium. Thus, tensors $\|\varepsilon\|$, $\|\mu\|$ should be necessarily symmetric.

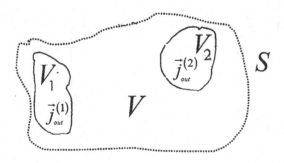

Fig. 1.15 Discussion of the Lorentz lemma

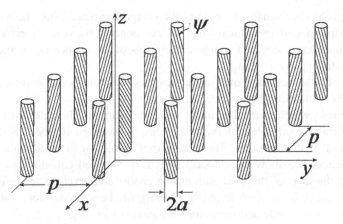

Fig. 1.16 Medium structure with narrow anisotropic cylinders

1.8 Examples of MathCAD Packet Utilization

In the present chapter, we use (see Appendix B – http://extras.springer.com/2019/978-3-319-90847-2) the only one example of MathCAD packet application for electromagnetic problems – Program P.1.1 – calculation of effective permeability for the medium composed of narrow anisotropic cylinders (system structure is shown in Fig. 1.16).

In this example, the effective permeability is given by Eq. (1.52):

$$\mu_{\text{eff}}(\psi, \text{ka}, q) = \frac{\left[1 - \dfrac{\text{tg}^2(\psi)}{(ka)^2} \right]}{\left[1 - \dfrac{\text{tg}^2(\psi)}{(\text{ka})^2} \right] + \dfrac{\pi q^2}{2}}$$

where ψ is the spiral twisting angle. This formula is programmed into the MathCAD medium (see P.1.1 in Appendix B – http://extras.springer.com/2019/978-3-319-90847-2). Calculation of the effective permeability in the MathCAD packet gives the required plots straight away as a function of ka with the $q = 2a/p \ll 1$ parameter. As we see from the calculation results, the effective permeability function has nonremovable discontinuity at a definite value of ka; thus, a position of the break point depends on the q parameter. On the left of the break point, the plot is at first almost constant. But approaching the break, μ_{eff} sharply increases to plus infinity. On the right of the break point, the plot begins from minus infinity and, reducing in modulo, tends toward the previous constant value.

Physical interpretation of the plots obtained consists in new wide possibilities, when varying the frequency, and we can realize different properties in the one device with very wide limits. In particular, it concerns nano-electronic problems, when it is very difficult (or impossible) to obtain a wide set of nano-material properties from the real device. Significant results on permeability were obtained by the scientific school of Russian professor *A.D. Shatrov*.

The MathCAD packet allows easy construction of clear three-dimensional pictures (see Fig. 3 in P.1.1). In this figure axes correspond to ka, μ, and μ_{eff}. As we see from the three-dimensional picture, at first, the total surface is practically constant ($\mu_{\text{eff}} = \mu_0$); then the plot sharply increases, before changing a sign, and again tends toward μ_0.

We recommend that the reader selects the problem parameters independently and calculates both two-dimensional and three-dimensional pictures.

1.9 Summary

Chapter 1 is devoted to the main definitions, concepts, and relationships of macroscopic electrodynamics. The idea about a unity of the electromagnetic field with other fields (the gravitational field, the field of nuclear forces, etc.) is especially important and fruitful. A little later, based on generalized electrodynamics, we show the connection between macroscopic electrodynamics (Chap. 4) with the electromagnetic field of the charged particle. Thus, the magnetic field arises at charged particle motion with respect to the vacuum medium and has two components: the vortex (vector) and the potential (scalar). In addition, we established a connection between a charge and a mass of the elementary particle. Moreover, we connected together the phenomena of inertia and gravitation.

We draw the reader's attention to several paradoxes of classical electrodynamics that require a detailed examination and are not independent of the methods of classical theory (Sect. 1.3.1).

The Stokes–Helmholtz theorem, which states that, under definite conditions, the field must be represented in the form of the summation of potential (scalar) and solenoid (vector) fields, is assumed as the basis of generalized macroscopic electrodynamics by the creators of this theorem. The generalized system of electrodynamics equations (Sect. 1.3.2) is obtained. For picture completeness, we show the classical system of Maxwell equations as well. The generalized law of electromagnetic induction and the laws of the total current and charge conservation are formulated (Sects. 1.3.4 through 1.3.6). Here, the golden rule of electrodynamics is formulated, which will be necessary for us at the end or our lecture course, when we discuss the problems of waveguide and resonator excitation (Chap. 12).

In this chapter, we paid great attention to an explanation of a new concept in electrodynamics: the scalar magnetic field (Sect. 1.3.8).

In accordance with the generalized theory, the concept of electromagnetic field energy, its characteristic Umov–Poynting vectors, and equations of energy balance (the Umov theorem) are formulated in a new way (Sect. 1.4).

A large part of this chapter is devoted to the description of many media in which electromagnetic waves (oscillations) propagate on boundaries where the electromagnetic processes (reflection, transmission, scattering, etc.) occur.

Checking Questions

1. What does the completeness of Maxwell equation system (1.6, 1.7, 1.8, and 1.9) mean?
2. Why should the Maxwell equation system (1.6, 1.7, 1.8, and 1.9) be supplemented by constitutive Eqs. (1.10 and 1.11)?
3. What is the difference between the natural medium and the chiral one?
4. How does the presence of the imaginary part of medium permittivity influence electromagnetic wave propagation?
5. In which cases must we use the two-side boundary conditions?
6. What are emission conditions and how may they be used?
7. Using Program P.1.1, calculate the two-dimensional function of the effective permeability of the medium with anisotropic cylinders for $ka = 0.05$ and $ka = 4$. What is the difference between these cases and previous parameter values? Explain the physical case of differences and similarities.
8. Using Program P.1.1, calculate the three-dimensional function of the effective permeability of the medium with anisotropic cylinders for $ka = 0.05$ and $ka = 4$. What is the difference between these cases and previous parameter values? Explain the physical case of differences and similarities.

Boundary Conditions, Integral and Complex Forms of Electrodynamics Equations, Classification of Electromagnetic Phenomena

2

It is known that to precisely solve differential equations with partial derivatives (such equations were discussed in Chap. 1 and should be considered as the base equations for various problems of electromagnetism), we need to formulate the so-called boundary conditions: the values of required solutions of differential equations at definite boundaries. Most often, the boundaries between media of different types are such boundaries for electrodynamics problems. For instance, at electromagnetic wave transfer from the air (where this wave propagates into free space) into, say, the specific dielectric medium, we meet with the necessity of solving the differential equations with partial derivations for the necessity of defining solutions in a two-media (air–dielectric) boundary (interface). The definition of such boundary conditions provides solution unambiguity and, therefore, they are important.

In Chap. 2, we first explain the necessity of introducing boundary conditions with examples of the various common and special boundary surfaces for typical electrodynamics' tasks (Sect. 2.1.1). Here, we undertake a general consideration of boundary conditions. Then (Sect. 2.1.2), the general types of boundary conditions are analyzed. To that end, we need to introduce the general medium parameters: a permeability, a permittivity, a conductivity, and the chiral parameter reflecting the chiral properties, which were discussed in Chap. 1. Most often, general boundary conditions lead to the significant complication of electromagnetic problems, which make it difficult to understand the physics processes. That is why the approximated boundary conditions can frequently be used, of which Schukin's boundary conditions (Sect. 2.1.3) are a classical example. Schukin's boundary conditions essentially simplify the solution of many boundary problems in electrodynamics. The anisotropic impedance boundary conditions are described further (Sect. 2.1.4), in addition to (Sect. 2.1.5) the boundary conditions on frequently periodic structures (Weinstein–Sivov conditions). After that, we offer students the opportunity to examine more complicated cases of different types of two-sided boundary conditions: at first, the general type of two-sided boundary conditions (Sect. 2.1.6), and then two-sided boundary conditions on frequently periodic structures (Sect. 2.1.7), and finally (Sect. 2.1.8) for anisotropic conductivity. Section 2.1 ends with a brief qualitative analysis of the impedance boundary conditions of the resonance type, which allow phenomena in the open resonance structures to be studied. These results are important to us for further engineering discussions.

A necessary physical condition of the general electromagnetic problem is the obligatory presence of the propagation medium for electromagnetic waves. Such media are described with the help of permeability and permittivity, which are the complex quantities in the general case. To that end, we additionally introduce the so-called constitutive equations (Sect. 2.2) with the sources of the electric field sources and the concept of the "golden rule of electrodynamics," which recommends that students connect the solution of electromagnetic tasks to sources of the electromagnetic field.

In many practically interesting cases, the solution of electrodynamics problems can be simplified at use of the integral or complex forms of electrodynamics equations (Sects. 2.3 and 2.4). Here, we discuss the principle of the permutation duality and the unicity theorem of electrodynamics equations (Sect. 2.5).

Section 2.6 is devoted to the classification of electromagnetic phenomena with extraction of a region of electrostatic problems, a region of geometric optics, and a definition of activity limits of the well-known Fresnel's and Fraunhofer's zones. Special attention is paid to promising artificial media and to chiral media.

A concept of crucial electromagnetic structures and crucial problems of various types is introduced in Sect. 2.7.

Section 2.8 is devoted to the description of numerical problems, which can be solved using the software packet MathCAD. Chapter 2 concludes with a brief summary and set of problems for students, which should remind them of the material studied.

© Springer International Publishing AG, part of Springer Nature 2019
E. I. Nefyodov, S. M. Smolskiy, *Electromagnetic Fields and Waves*, Textbooks in Telecommunication Engineering,
https://doi.org/10.1007/978-3-319-90847-2_2

2.1   Boundary Conditions

2.1.1   Note on Boundary Conditions of Generalized Electrodynamics

As we clearly see from Chap. 1, practical electrodynamics deals with manifold media. Thus, naturally, all media may change its parameters depending on coordinates and time in various manners. They may be continuous and discontinuous (spasmodic). Here we are limited to obtaining boundary conditions for tangent and perpendicular (normal) directions of two media boundaries, 1 and 2 (Fig. 2.1) for strengths and inductions of electric and magnetic fields in the case of classic and generalized electrodynamics. We start from a very general case, when there are homogeneous media with different properties on both sides of the S boundary. Thus, we can write:

$$E_\varsigma^{(2)} = E_\varsigma^{(1)}, \quad \varepsilon_2 D_\varsigma^{(2)} = \varepsilon_1 D_\varsigma^{(1)}. \tag{2.1}$$

In contrast to the classic case (which is considered later), here, on the boundary surface, two currents are possible, which flow in directions $\vec{\varsigma}$ and $\vec{\eta}$ (Fig. 2.1a). Thus, the $\vec{\eta}$ vector is directed perpendicular to the S area. The current in the direction $\vec{\eta}$ is

$$i_\eta = (1/l) \int_S \left(\vec{j} + (\partial \, \vec{D}/\partial t) \right) d\,\vec{s}. \tag{2.2}$$

The current i_η creates the vector magnetic field $\vec{H}$ on both sides of the boundary. From this, the boundary conditions for the tangent components of the magnetic field strength vector follow:

$$H_\varsigma^{(2)} - H_\varsigma^{(1)} = i_\eta. \tag{2.3}$$

Similarly, from (1.7), the boundary condition for scalar magnetic field strengths follows:

$$\left(H^{*(2)} - H^{*(1)} \right) l = J_n, \tag{2.4}$$

in which $j_n = l i_n$, and i_n is the surface current density, which flows on the normal to two media boundary.

Accordingly, for induction of the scalar magnetic field, the boundary condition is

$$\left(B^{*(2)}/\mu_2 \right) - \left(B^{*(1)}/\mu_1 \right) = i_n. \tag{2.5}$$

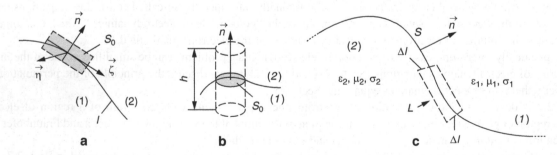

Fig. 2.1 Boundaries S of two media *1* and *2* with different values of permittivity: determination of conditions for (**a**) tangent and (**b**) normal components. (**c**) The area along the boundary L is shown by the dotted line, on which integration is fulfilled at derivation of the boundary conditions (2.9)

From Ohm's law (1.25), the boundary conditions for the tangent component of the current density directly follow:

$$j_\varsigma^{(2)}/j_\varsigma^{(1)} = \sigma_2/\sigma_1. \tag{2.6}$$

The boundary condition for the normal component of current density is obtained from the generalized continuity Eq. (1.27) by its integration over the small cylinder volume in Fig. 2.1b:

$$j_n^{(2)} - j_n^{(1)} = -\partial\delta_{eff}/\partial t, \tag{2.7}$$

where δ_{eff} represents the surface density of the effective charge, which is formed from the density of the usual electric charge δ and the charge density, induced by the nonstationary scalar magnetic fields in the first and second media:

$$\delta_{eff} = \delta + (\varepsilon_1/S_0)\int_{V_1}\left(\partial B^{*(2)}/\partial t\right)\partial V + (\varepsilon_2/S_0)\int_{V_2}\left(\partial B^{*(1)}/\partial t\right)\partial V. \tag{2.8}$$

It is important that the normal component of the current density on the media boundary becomes discontinued, not only because of the presence of the varying density of electric charges, but in the presence of the nonstationary scalar magnetic field: $\partial^2 B^*/\partial t^2$ [2, 12].

Further, we consider the classic electromagnetic field to help the reader to examine the huge sea of literature devoted to classic electrodynamics.

2.1.2 Boundary Conditions of the General Type

We have not yet applied any conditions to the medium parameters $\varepsilon, \mu, \sigma, \mathcal{N}$, i.e., they can be changed in the considered space or in the given volume V in the arbitrary manner: continuously, spasmodic, etc. Apparently, in engineering practice, we most often encounter cases in which these parameters change spasmodically at transition through the same S surface, which is the boundary of two or several media in the V volume. Evidently, that electromagnetic field $(\vec{E}, \vec{H})$ should also feel some change (for instance, the jumping of some its components because the field itself is continuous). These changes in the field on the S boundary can be taken into consideration by boundary conditions, the most general of which are conditions of continuity of tangent components to S at each point of its electric and magnetic fields. Having designated by indices (1), (2) the field components on both sides of the S boundary (Fig. 2.1c), we write the boundary conditions as:

$$E_t^{(1)} = E_t^{(2)}; \quad H_t^{(1)} = H_t^{(2)}. \tag{2.9}$$

Boundary conditions (2.9) follow from the integral equation form (see Sect. 2.3; Eqs. (2.32 and 2.33)) at realization of the limiting transition, when the L integration contour (in Fig. 2.1c), which is marked by the dotted line at the limit (the L rectangular sides, which are perpendicular to the S boundary, tend to zero: $\Delta l \rightarrow 0$), coincide with the S surface.

In the similar manner, we can obtain the boundary conditions for the field components, which are normal to S. They are of a "discontinuous" character, whose value relates to the difference in ε, μ parameters on both sides of the S boundary. We do not write these boundary conditions, because in practice, in the overwhelming majority of cases, it is enough to use conditions in the form (2.9).

Continuity conditions (2.9) look as if the surface electric currents (with $\vec{i}^e$ density) and magnetic currents (with $\vec{i}^m$ density) are absent on the S surface. At the presence of surface currents, we should use the following instead of Eq. (2.9):

$$\left[\vec{n}, \vec{E}^{(2)} - \vec{n}, \vec{E}^{(1)}\right] = -\vec{i}^m, \left[\vec{n}, \vec{H}^{(2)} - \vec{n}, \vec{H}^{(1)}\right] = \vec{i}^e \tag{2.10}$$

Here, the normal to S directing toward the medium with the index (1) (Fig. 2.1c) is designated as $\vec{n}$.

If the one from media, for instance, (2), is a good conductor, then the "usual" for us transverse electromagnetic field does not penetrate it; thus, fields are absent from the medium (2), i.e., $\vec{E}^{(2)} = \vec{H}^{(1)} \equiv 0$. Hence, on the surface of a very good conductor ($\varepsilon \to j\infty$, and in the overwhelming majority of practical cases, μ is the finite quantity), we obtain the so-called ideal boundary condition for the electric field component (1.32) tangent to S

$$E_t = 0 \quad \text{on} \quad S. \tag{2.11}$$

The boundary condition (2.11) on the ideally conducting surface can be written in the vector form as $[\vec{n}, \vec{E}] = 0$. This type of boundary condition is called Dirichlet condition. Another Neumann condition is:

$$\partial H_t / \partial_n = 0 \text{ on } S. \tag{2.12}$$

Here, n designates the normal to the S surface at each point.

We further consider the wide class of boundary conditions based on the classic representation. They essentially simplify the solution of diffraction and radio wave propagation tasks in various situations and allow reliable results to be obtained and the specific elements and devices to be described and developed. Obviously, in some cases, they must be added taking into consideration the generalized macroscopic theory.

2.1.3 Schukin's Impedance Boundary Conditions

The solution of many electrodynamics tasks is seriously simplified if instead of strict boundary conditions of continuity of tangents to the two-media boundary S (2.9) we use so-called *equivalent boundary conditions*. They are introduced, for example, when the medium properties in both sides of the S boundary (Fig. 2.1) strongly differ. Apparently, Schukin's boundary conditions on the S surface, when one of which, say, the air, and another has large but finite conductivity σ [59], are the "oldest", best-known, and most often used. They take the following form:

$$E_x = wH_y, \quad E_y = -wH_x, \quad x, y \in S. \tag{2.13}$$

Here, $w = \sqrt{\mu/\varepsilon}$ and for $|\varepsilon| \gg 1$ and finite μ, the quantity w is small and, hence, tangents to S – the components of the electric field – are near zero, and conditions (2.13) at the limit for $w \to 0$ transfer to (2.11). Otherwise, when $\mu \gg 1$ and ε is small or restricted, conditions (2.12) are satisfied on the S boundary.

The physical content of conditions (2.12) is explained in Fig. 2.2: the incidence of two plane waves i_1 and i_2 on the plane boundary of two media 1 and 2 under various angles φ_1 and φ_2. At large conductivity σ of the medium 2, the refraction angle ξ of the wave t passed into the second medium is practically the same for both waves i_1 and i_2.

As a result of the incidence of waves $i_{1,2}$, the plane waves are formed: reflected r_1 and r_2, and refracted t_1 and t_2. At high enough conductivity σ_2 of the medium 2, the refraction angle ξ is small and weakly dependent on the incident angle φ. Hence, in the medium 2, the plane wave t propagates, and its which front is almost parallel to the boundary ($z = 0$). Relations (2.13) are the connection between the field components of this passed wave (see later in Chap. 4).

2.1.4 Anisotropic Impedance Boundary Conditions

The anisotropic impedance boundary conditions [57]:

$$E_x = -Z_{11}H_x + Z_{12}H_y, \quad E_y = -Z_{21}H_x + Z_{22}H_y. \tag{2.14}$$

are more general than conditions (2.13). In (2.14) values Z_{ij}, $i, j = 1,2$ are elements of the input impedance matrix. Some data about Z_{ij} are reported, for example, in Kurushin and Nefyodov [57].

If the diagonal elements of the $\|Z\|$ matrix are such that $Z_{11} = Z_{22} = 0$, and nondiagonal elements are the same, $Z_{12} = Z_{21} = 0$, then from (2.14) Schukin's boundary conditions follow in the form of Eq. (2.13). In general, we can define a connection between the tangent components of fields $\vec{E}_t$ and $\vec{H}_t$ on the S surface as

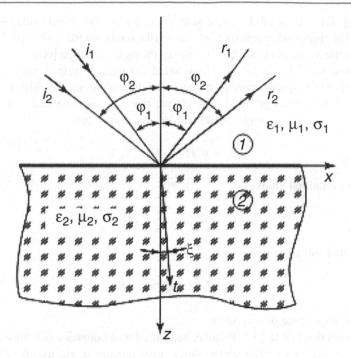

Fig. 2.2 Derivation of Schukin's boundary conditions

$$\vec{E}_t = \Im \vec{H}_t, \tag{2.15}$$

where we introduced the same integral-differential operator $\Im$, whose form should be discovered from the general equations of electrodynamics (1.6, 1.7, 1.8, 1.9, 1.10, and 1.11).

If we know the complete orthonormal system of eigenfunctions $\vec{E}_t = \Im \vec{H}_t$, of the $\Im$ operator from (2.15) on the S surface, then fields $\vec{E}_t$ and $\vec{H}_t$ can be represented in the form of expansion of the function system:

$$\vec{E}_t = \sum_k a_k \vec{E}_k, \vec{H}_t = \sum_k b_k \vec{H}_k. \tag{2.16}$$

The operator $\Im$ from (2.15) takes the form of the matrix operator:

$$\Im = \|Z\| = \begin{Vmatrix} Z_{11} & Z_{12} & \cdots & Z_{1n} & \cdots \\ Z_{21} & Z_{22} & \cdots & Z_{2n} & \cdots \\ \cdots & \cdots & \cdots & \cdots & \cdots \\ Z_{n1} & Z_{n1} & \cdots & Z_{nn} & \cdots \end{Vmatrix}. \tag{2.17}$$

Elements of the Z_{ij} matrix are defined by the physical–geometrical data of the object, in which (importantly) we include its characteristics in connection with the external space. Thus, matrices Z_{ij} are external volume parameters and are saved at its connection with any external "circuit." A significant contribution to studying these physical models and elements of the matrices of anisotropic impedance was made by *E.P. Kurushin* (1935–1987; see, for instance, Kurushin and Nefyodov [57]).

2.1.5 Boundary Conditions on Frequently Periodic Structures

The structures, whose period L is much shorter than the wavelength λ or even when $kL \ll 1$, are called the frequently periodic structures. The class of frequently periodic structures is extremely wide and manifold and it includes wire, lattice shields, slowing systems of electronic and accelerator devices, and various emitting (receiving) systems of the spiral antenna type. The typical property of the frequently periodic structures is the ability to support the surface wave existence. Various examples of such systems are shown in Fig. 1.13. They can be single-sided (Fig. 1.13g–l) or double-sided character (Fig. 1.13a, b). The incident field is reflected from the single-sided structure, whereas in the double-sided system part of the energy (sometimes, all the energy – the Maliuzhinets effect, see Sect. 8.7.2) – penetrates through the lattice [60].

Let us consider the simplest case. Let it be the single-sided corrugated surface with rectangular corrugations (see Fig. 1.13g), and the z-axis of the rectangular coordinate system be directed along the ridge of corrugations (perpendicular to them; the corrugations with longitudinal grooves). The groove size we designate as g, ant its depth as h. Then, for the corrugation with the transverse grooves, the single-sided boundary conditions are

$$E_z = \eta^m H_x, \quad E_x = 0, \tag{2.18}$$

and for corrugation with the longitudinal grooves

$$E_z = 0, \quad E_x = \eta^m H_z. \tag{2.19}$$

Here, the parameter of corrugated surface is defined as:

$$\eta^m = iZ_0(2g/L)\tan(k_0 h). \tag{2.20}$$

where $Z_0 = \sqrt{\mu/\varepsilon}$ is the wave impedance of a vacuum.

The equivalent boundary conditions (2.18), (2.19) are called impedance boundary conditions of *Weinstein–Sivov* and are widely used in practice, for example, in microwave electronics at calculations of electron flow interaction with the wave, in accelerators of charged particles, in waveguides with frequently periodic structures (see Sect. 8.6), etc.

2.1.6 General Double-Sided Boundary Conditions

Modern transmission lines and basing elements of microwave and millimeter-wave modules for ultra-fast systems of information processing often have several layers of magnetic–dielectric. In particular, three-dimension integrated circuit (3D-IC) of microwave and millimeter-wave ranges use the tens of layers. The straightforward analysis of such multi-layer structures based on the "standard" boundary conditions of continuity of the tangent field components on $S_{j,\,j+1}$ boundaries of adjacent layers $(i, i+1)$ in accordance with (2.9) usually leads to bulk expressions, which are not suitable for calculations and, moreover, physical analysis. The way out is the double-sided boundary conditions connecting the tangent components of $\vec{E}_{t1}, \vec{H}_{t1}$ fields with $\vec{E}_{t2}, \vec{H}_{t2}$ components on two rather close surfaces S_1 and S_2.

Presenting the expressions for $\vec{E}_{t1}, \vec{H}_{t1}$ and $\vec{E}_{t2}, \vec{H}_{t2}$ fields in the form of expansion over the full system of functions in the form of (2.16), we can obtain the following equation for double-sided boundary conditions:

$$\left\| \begin{matrix} \vec{E}_{k1} \\ \vec{H}_{k1} \end{matrix} \right\| = \|T_k\| \cdot \left\| \begin{matrix} \vec{E}_{k2} \\ \vec{H}_{k2} \end{matrix} \right\|. \tag{2.21}$$

The matrix $\|T_k\|$ is called the transmission matrix, and in general (such as the matrix $\|Z\|$ from Eq. (2.17)) it represents the same integral–differential operator. Elements of the transmission matrix $\|T_k\|$ have spatial dispersion.

Double-sided boundary conditions (2.21) find very wide application in the analysis of dielectric waveguide properties, light-guides, and basing elements created based upon them. Let us examine them in detail, taking into account the great importance of anisotropic electrodynamics structures at implementation of new technologies. If we are oriented to the practical purposes of various systems, we may distinguish several groups of tasks, for solution of which it is expedient to use relations (2.13 and 2.14).

These are tasks relating to the propagation of electromagnetic waves along the boundary of the vacuum – the anisotropic half-space over the gyrotropic or bi-gyrotropic medium. Here, we can mention the generalization of the classic *Sommerfeld* problem regarding emission of the thin wire with current (electric or magnetic) over the impedance surface onto the case of the anisotropic impedance. The problems of magneto-optics can be attributed to this class. Thus, the impedance approach to these problems is, to our opinion, more universal, because it allows deeper penetration into the physical sense of phenomena and leads the simplest way to connection of the effects observed with the medium parameters, for example, for tasks of electromagnetic field interaction with the plasma [61].

The second group of tasks relates to the propagation of electromagnetic waves in the plane transmission line with anisotropic walls. In particular, the waveguide of the Earth – an ionosphere – is described by this model. Many tasks of antenna-waveguide and ferrite engineering, microwave and millimeter-wave microelectronics are connected to this group of tasks.

Finally, the third group of tasks, for which the impedance approach (2.13) is expedient, relates to diffraction on the transparent and half-transparent objects, which have anisotropic conductivity, covering in layers of the material with tensor permittivity and permeability. The key task for them we may consider a problem of electromagnetic wave diffraction on the half-plane with the anisotropic impedance, which also generalizes the known *Sommerfeld* problem [57]. Of course, a selection of specific tasks in the groups mentioned is to a great extent subjective and is defined, in particular, by authors' interests and by applications that are interesting from practical and cognitive points of view. The tasks of electromagnetic wave propagation in chiral media (see Katsenelenbaum et al., and Neganov and Osipov [50, 51]), in addition to tasks of wave diffraction on the manifold chiral structures, can be attributed to this group.

2.1.7 Double-Sided Boundary Conditions on Frequently Periodic Structures

A number of practical electrodynamics structures include the wire lattices from circular, plain, rectangular, and other conductors. First of all, it is possible for spiral waveguides, for antennas, etc. There are several types of approximate boundary conditions for analysis of these structures in the presence of the small parameter of $\chi = L/\lambda \ll 1$ Then, if to designate the direction along conductors of the plain lattice as $\vec{s}$, and the perpendicular to conductors as $\vec{t}$ (in the same plane), then for the lattice from parallel metal bands (Fig. 1.13b), we may write the following [60]:

$$E_s = i\xi_1\left[H_t^{(2)} - H_t^{(1)} - \frac{1}{ik\partial s}\left(E_n^{(2)} - E_n^{(1)}\right)\right], \xi_1 = \chi \ln \sin\frac{\pi b}{2p},$$

$$H_s^{(2)} - H_s^{(1)} = -4i\xi_2\left(E_t - \frac{1}{ik\partial s}H_n\right), \xi_2 = \chi \ln \cos\frac{\pi b}{2p}. \tag{2.22}$$

In (2.22), b is the band width, $\vec{n}$ is a normal to the lattice surface directed from the side 1 toward side 2 of this surface; thus, the triplet of directions $\vec{t}$, $\vec{s}$, $\vec{n}$ is the right triplet.

If the lattice is formed from the parallel wires with the b-radius (see Fig. 1.13a), then at $b \ll L$ we have the conditions:

$$E_s = i\xi_0\left[H_t^{(2)} - H_t^{(1)} - \frac{1}{ik\partial s}\left(E_n^{(2)} - E_n^{(1)}\right)\right], H_s^{(2)} = H_s^{(1)}, \tag{2.23}$$

and $\xi_0 = \chi \ln(2\pi b/L)$. The boundary conditions (2.22 and 2.23) were suggested independently by USSR scientists *L.A. Weinstein* and *A.V. Sivov* [50, 60]. Together with the aforementioned boundary conditions, the averaged boundary condition of *Kontorovich* [60] may be used.

2.1.8 Double-Sided Boundary Conditions for Anisotropic Conductivity

In microwave electronics and in some other applications of engineering electrodynamics, at initial estimations, one can often use the model of the anisotropically conducting surface. Let the plane $z = 0$ be such a plane and let the lines of conductivity be parallel and directed along the x-axis. Then

$$E_x^+ = E_x^- = 0, \quad E_y^+ = E_y^-, H_x^+ = H_x^-. \tag{2.24}$$

The physical sense of (2.24) conditions consists in the fact that in the plane $z = 0$, there are no (we assume) magnetic currents, and the surface electric current has only the x-component.

The plane linearly polarized wave $E_x = \exp\{ikz\}$, which normally falls on the surface $z = 0$ with condition (2.24), fully reflects (the transmission factor is zero: $T = 0$). The plane wave with orthogonal polarization $E_y = \exp\{ikz\}$ passes through the boundary $z = 0$, not noticing it, because its reflection factor is equal to zero ($R = 0$).

In practical systems, the frequent lattice (the lattice period L is essentially shorter than the wavelength λ) from the band conductors with the correctly chosen filling factor q may serve as a surface with boundary conditions (2.22). Coefficients T and R for the frequent band lattice (see Fig. 1.13b) are expressed as:

$$T = \frac{-2i(p/\lambda)\ln\sin(\pi q/2)}{1 - 2i(p/\lambda)\ln\sin(\pi q/2)}, R = \frac{2i(p/\lambda)\ln\cos(\pi q/2)}{1 - 2i(p/\lambda)\ln\cos(\pi q/2)} \tag{2.25}$$

From this, it follows that the optimal value of the filling factor, which simultaneously provides the smallness of T and R, is $q = 1/2$. Thus,

$$T = R \approx (L/\lambda)\ln 2. \tag{2.26}$$

For the lattice from circular conductors (Fig. 1.13a), the optimal value $q \approx 0.3$, which is near to a half-value of the inverse "gold proportion" $\overline{\Phi}/2 = 1/2\Phi = 0.309$, where $\Phi = 2.618$ [62].

Dependences of coefficients $R(p/\lambda)$ and $T(p/\lambda)$ from the relative period value $P = (p/\lambda)$ are calculated according the Program P.2.1 (Appendix B – http://extras.springer.com/2019/978-3-319-90847-2). We used the following designation in this Program: $P = L/\lambda$, $q = b/\lambda$, $R1(q,P)$ is the reflection coefficient, and $T(q,p)$ is the transmission coefficient for the case of electric polarization of the incident field ($H_\xi = 0$). For surfaces with a complicated form, the boundary conditions of anisotropic conductivity also have the form (2.23), where variables x, y should be considered the local coordinates.

2.1.9 Equivalent Impedance Boundary Conditions of the Resonance Type

This type of boundary conditions is widely used at analysis and calculations in the open resonators (or their systems) and in the open waveguides [23, 126]. An example of band open resonator formation (Fig. 2.3b; left) from the half-infinite planar waveguide is shown in Fig. 2.3b, right.

If the frequency of the wave (i), which arrives at the open end of the waveguide, is near to the critical value, it will almost fully reflect from the open end of this waveguide (r), i.e., the modulus of the wave reflection factor $|R| \approx 1$ (the *Weinstein* effect). In other words, if the almost integer number of the half-wavelengths remains within the distance $2l$, i.e., the *resonance conditions*, are satisfied, we have

$$2kl = \pi q + 2\pi p, \quad p << 1, \quad q = 1, 2, \ldots \tag{2.27}$$

Also in Fig. 2.3, the other cases of final structure formation are presented (Fig. 2.3, left) from the half-infinite or key structures (Fig. 2.3, right).

In the case of the fulfillment of Eq. (2.27), we may imagine that at the waveguide end the same boundary is mounted (vertically) with such an impedance that the reflection factor coincides with those in a real situation. Let us designate as $2l$ the distance between parallel planes that form the plane waveguide (Fig. 2.3b). Let the transverse coordinates be x, y. Then, on the default plane boundary, the field components are connected as follows:

$$f_{E,H}(x) = (\beta' + i\beta_{E,H})\sqrt{l/2k}\, df_{E,H}/dx. \tag{2.28}$$

The latter equations are the equivalent boundary conditions of the resonance type. The numerical β coefficients, including in Eq. (2.28), can be found, for instance, in Nefyodov and Fialkovsky, and Paul [58, 126].

Fig. 2.3 The finite (in cross-section) transmission lines and corresponding key (half-infinite) structures: (**a**) is a band ⇔ half-plane; (**b**) is the band open resonator ⇔ open end of the plane waveguide; (**c**) is an asymmetric stripline (microstrip line) ⇔ the open end of the plane waveguide with magneto-dielectric filling; (**d**) is a symmetric double slot line ⇔ open end of the plane waveguide with magneto-dielectric filling (in essence a mirror reflection of the structure (**c**))

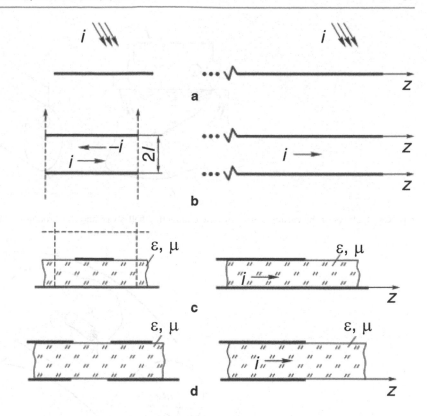

2.2   Constitutive Equations, Sources of the Electrical Field, the Golden Rule of Electrodynamics

A series of the same conditions should be added to the main system of electrodynamics Eqs. (1.6, 1.7, 1.8, and 1.9). The wide class of such conditions; namely, the boundary conditions, are examined in the Sect. 2.1. The other conditions are the constitutive equations, i.e., the equations connected to the field $\vec{E}$, $\vec{H}$ strengths and their inductions $\vec{D}$, $\vec{B}$ through the media parameters ε, μ of permittivity and permeability. In general, ε, μ are complex quantities: $\varepsilon = \varepsilon' \pm j\varepsilon''$, $\mu = \mu' + j\mu''$ The $\vec{D}$ vector is sometimes referred to as the vector of electric offset. The ε parameter (the relative medium permittivity) is connected to $\varepsilon_{abs} = \varepsilon\varepsilon_0$, which is the absolute permittivity measured in Farad/meter; ε_0 is the vacuum permittivity. In engineering practice, the relative quantities ε, μ, which define the medium refraction index $n^2 = \varepsilon\mu$, are often used.

The last relation (for n^2) is called the Maxwell law and it directly follows from Eqs. (1.6 and 1.7). The first measurements of $n = 3$ (for $\mu = 1$) were performed in 1856 by *V. Weber* and *R. Kohlrausch* [46, 99]. In Russia, similar investigations were carried out by *N. Shiller* by means of the method developed by him. We discuss the sense of signs before imaginary parts of permeability and the refraction index.

The constitutive equations in the simplest case (ε, μ, σ are scalar quantities) have the following form:

$$\vec{D} = \varepsilon\vec{E} \tag{2.29}$$

$$\vec{B} = \mu\vec{H} \ . \tag{2.30}$$

Usually, Ohm's law $\vec{j}\ (\vec{r}, t) = \sigma\ \vec{E}\ (\vec{r}, t)$ is added to conditions (2.29), (2.30), in which the quantity σ is the differential conductivity (measured in Siemens/meter) at the given point of the space (the body), and Eq. (1.25) itself is the differential form of Ohm's law (1826).

The aforementioned differential form of Ohm's law (Eq. 1.25) is necessary in cases when the current in the conductor section is distributed in the same volume non-uniformly (see, for example, Fig. 2.4a). It is clear that the current $\vec{j}$ from electrodes spreads over the conductor volume non-uniformly, and the net current in the total circuit should be determined by integrating this differential distribution over the surface S (Fig. 2.4).

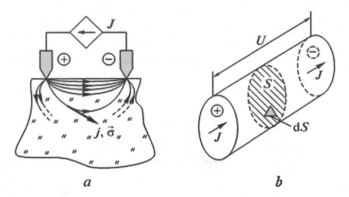

Fig. 2.4 Pictures of the current flow in (**a**) the conducting half-space and (**b**) through the conductor segment

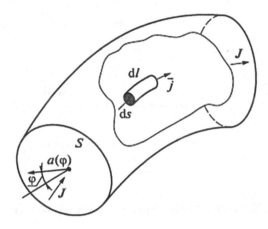

Fig. 2.5 Explanation of Ohm's law

Regarding the direct current J in the same conductor, at whose ends there is the potential difference U, then Ohm's law takes the more usual integral form

$$J = U/R, \tag{2.31}$$

in which R is the conductor resistance depending on the length l, the size of its cross-section S, and the material type: $R = l/\sigma S$.

Conditionally, the connection scheme between Ohm's law in the integral and differential forms (1.21, 1.22, 1.23, 1.24, and 1.25) is shown in Fig. 2.5.

Here, in the conductor with unchanged cross-section S, which average size is a, we select the elementary cylinder with the length l and the cross-section ds, such that $ds \ll S$, and its average cross size is much less than a. Then, obviously, the current dI, flowing through the cylinder, is $dI = dU/dR$, where U is the voltage between its butts. Thus, taking into account that directions $\vec{j}$ and $\vec{E}$ are parallel, and $dI = jds$ and $dU = Edl$, we obtain:

$$\vec{j} = \sigma \vec{E}.$$

The coefficient of proportionality μ between vectors $\vec{B}$ and $\vec{H}$ of the magnetic field is called the absolute magnetic permeability of the medium and it is measured in Henry/meter. In general, $\mu = \mu_0 \mu_r$, where the constant multiplier μ_0 is the magnetic constant, which in SI system is $\mu_0 = 4\pi \cdot 10^{-7}$ Hn/m. The quantity μ_r is the relative magnetic permeability; sometimes (when it does not lead to confusion) index "r" is omitted.

In classic electrodynamics, it states that the magnetic field is always vortical, but the electric field may be both vortical and potential. Hence, In general represents the "sum": superposition of vortical and potential components. The electrical field can be purely potential only in the static case ($\partial/\partial t = 0$). In generalized macroscopic electrodynamics, the magnetic field (besides vertical) acquires additionally the potential character (see Chap. 1).

(It is important that in the Maxwell equation system (1.6, 1.7, 1.8, 1.9, 1.10, and 1.11) itself, the medium parameters (ε, μ, σ) are not included, and as these equations are true for any media, we must add them to constitutive Eqs. (2.29 and 2.30) (they are sometimes called the *state equations* because they characterize the medium).

At the end of this section, we consider the *continuity Eq.* (1.19): $\partial\rho/\partial t + \text{div}\,\vec{j} = 0$ and apply it to the infinite uniform isotropic medium with low conductivity σ. From Ohm's law $\vec{j} = \sigma\,\vec{E}$, it directly follows that: $\text{div}\,\vec{j} = \text{div}(\sigma\,\vec{E}) = (\sigma/\varepsilon)\text{dil}\,\vec{D} = (\sigma/\varepsilon)\rho$. Then, the continuity equation is as follows: $\partial\rho/\partial t + (\sigma/\varepsilon)\rho = 0$, and its solution is $\rho = \rho_0 \exp\{-(\sigma/\varepsilon)t\}$, where $\rho_0(\vec{r})$ is the volume charge density in the initial time moment $t = 0$. Thus, in the lossy medium (at each point or $\rho \neq 0$), the volume charge density $\rho(t)$ decreases over time according to the exponential law. The time interval τ, during which the charge in the same small volume decreases by e times, is called the *relaxation time*: $\tau = \varepsilon/\sigma$. For metals, τ is to the order 10^{-18}, and for dielectrics it is to the order of several hours.

Concluding the more than unpretentious review of the basics of modern electrodynamics – Maxwell's equations – we would like to note some of the merits. First of all, we need to note that they are of an empiric character and were obtained based on the experiments of Faraday, Ampere, and other researchers in the distant past. However, the thing is not in the time, because the aforementioned founders performed their experiments with currents where the charge speed was by 12–15 orders less than the speed of light. Now, the actions of equations and their consequences are sometimes extended (without sufficient proof and substantiation) into the relativistic region.

On the other hand, Faraday's experiments covered a series of the simplest cases only, which results in the fact that Maxwell's equations leads, for example, to infringement of the Newton's third law in the case of nonclosed currents, to incorrect numerical results in the case of the non-uniform flow of magnetic induction through the contour area, and to many other contradictions [10, 63].

2.3 Integral Form of Classic Electrodynamics Equations

For the solution of some electrodynamics tasks, it is convenient to use Maxwell's equations not in the differential (Eqs. 1.12, 1.13, 1.14, and 1.15) form, but in the integral form:

$$\oint_\Gamma \vec{H}\,\overleftarrow{dl} + \int_S \vec{j}\,\vec{ds} + \int_S \partial\vec{D}/\partial t\,ds, \tag{2.32}$$

$$\oint_\Gamma \vec{E}\vec{dl} = -\frac{d}{dt}\oint_S \vec{B}\,\vec{ds}, \tag{2.33}$$

$$\oint_S \vec{D}\vec{ds} = \int_V \rho\,dV = Q, \tag{2.34}$$

$$\oint_S \vec{B}\vec{ds} = 0. \tag{2.35}$$

Equation (2.32) represents Maxwell's first equation in the integral form and expresses, evidently, the net current law (1.6) or the Ampere–Maxwell law. Figure 2.6a, b illustrates Eq. (2.32).

Equation (2.31) represents the integral formulation of the *Faraday's* generalized law of induction (Eq. (1.6)). When writing these equations, we assumed that the Γ contour is arbitrary and closed; it is fixed and does not vary in time, and S is some arbitrary surface that is stretched on the Γ contour. The Q charge is the total charge concentrated in the area (volume) V, included in the arbitrary closed surface S (Fig. 2.6c).

The *Maxwell's* third equation (Eq. (2.34)) is the *Ostrogradsky–Gauss* law (theorem). It is assumed that the Q charge inside the arbitrary closed surface S is distributed in an arbitrary manner (Fig. 2.6c). Special attention should be paid to the case when the Q charge represents the system of separate point (for instance) charges. In this case, the integral over the volume V in Eq. (2.34) should be replaced by the sum of all N charges located inside the S surface:

Fig. 2.6 Diagram explaining (**a**) the net current law (1.16), (**b**) the generalized Faraday law of electromagnetic induction, and (**c**) the Ostrogradsky–Gauss law

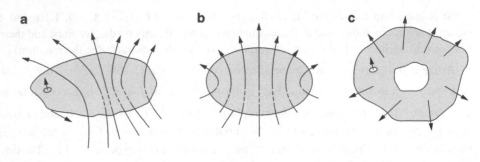

$$\oint_S \vec{D}\vec{ds} = \sum_{i=1}^{N} q_i. \tag{2.36}$$

It is accepted to assume so in the static case ($v = 0$) and, probably, at very low speeds of charges. At large motion speeds of charge, we should add the appropriate supplements to the theory, which sometimes turn out to be rather significant compared with the static case [10, 12].

From the third equation (Eq. (1.14)), it directly follows that the divergence of the $\vec{D}$ vector is nonzero only at those points of the space, where the free charges are situated. The lines of the $\vec{D}$ vector start on the positive charges and finish on the negative ones. As we mentioned earlier in Chap. 1, this occurs if we do not take into account the scalar magnetic field.

The Maxwell's fourth equation (Eq. (1.9)), Eq. (2.35) states that the flow of the magnetic induction vector $\vec{B}$ through any arbitrary closed surface S is by definition equal to zero. Thus, the lines of the magnetic induction vector $\vec{B}$ are continuous.

In other words, *there are no magnetic charges in nature*. Usually, in this place, we can "traditionally" say: "they have not yet been discovered." However, from Chap. 1 it follows that they cannot be discovered because we cannot meet them in nature. This is proved especially earnestly in publications by *Erokhin* [10] (see also Vaganov and Katsenelenbaum [65]). Nevertheless, sometimes, magnetic charges (following traditions) are taken formally into consideration and, in some cases, facilitate understanding of the essence of the task and its solution. As an example, the reader evidently knows well (from usual physics) the concept of the "mirror" electric charge – its "reflection" in an ideally conducting infinite plane. The system from two charges can be considered as the elementary dipole; we may enter the concept of the dipole moment $\vec{p}$, etc.

2.4   Electrodynamics Equations in the Complex Form, the Principle of Interchangeable Duality: Electrodynamics Equations and Monochromatic Processes

If the electromagnetic field $(\vec{E}, \vec{H})$ is the harmonic function of time, it is more convenient to use the so-called complex form of Maxwell's equations. In general, the function of time can be chosen in the form of $\exp(\pm i\omega t)$. Here, i is an imaginary unit, and ω is the oscillation radian frequency, which is given by the field source (for instance, by the oscillator, the arriving wave, the emission of the cell group, etc.). Thus, the sign of the time function ($\pm$) may be any, but it is important that it is not changed throughout one book (or paper). For our description, we choose the function $\exp(-i\omega t)$. As all field components obviously depend on time in a similar manner (we consider only linear media), the time function $\exp(-i\omega t)$ may be excluded from all intermediate expressions and "remembered" in the final results.

Replacing the time differentiation operation $\partial/\partial t$ in Eqs. (1.6 and 1.7) with the multiplication operation by $-i\omega$, we obtain the complex forms:

$$\text{rot }\vec{E} = i\omega\mu_a \vec{H}, \quad \text{rot }\vec{H} = \vec{j}_{\text{out}} - i\omega\varepsilon_a \vec{E};$$
$$\mu_a = \mu_0\mu, \quad \varepsilon_a = \varepsilon_0\varepsilon. \tag{2.37}$$

In a vacuum, say, the first equation from the system (2.37) has the form rot $\vec{E} = ik\vec{H}$, the wave number of the free space for wavelength λ, and the light speed c is:

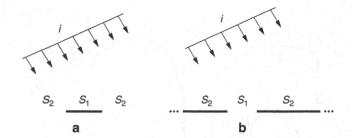

Fig. 2.7 Illustration of the duality principle: incidence of the wave I to the S_1 area. (**a**) A band, a disk. (**b**) A slot, a hole

$$k = \omega/c = 2\pi/\lambda \tag{2.38}$$

or in a more general view, the disperse equation for determination of the wave vector $\vec{k}$ has the form:

$$k^2 - (\omega/c)^2 n^2 = 0 \tag{2.39}$$

where $n^2 = \varepsilon\mu$ is the already known square of the refraction index of the medium (see Sect. 2.2).

At first glance, Eqs. (2.38 and 2.39) represent the same thing. Nevertheless, we see from Eq. (2.39) that a simultaneous change in signs of permittivity ε and permeability μ do not influence the susceptibility of the medium. The duality principle corresponds to it. Indeed, if in Eq. (2.37), in the simplest case of the absence of an outside current, $\vec{j}_{\text{out}} = 0$, we perform the formal change

$$\begin{aligned} \vec{E} \to \vec{H}, \ \vec{H} \to \vec{E} \\ \varepsilon_a \to -\mu_a, \quad \mu_a \to -\varepsilon_a. \end{aligned} \tag{2.40}$$

then the first equation of Eq. (2.40) transfers into the second, and the second transfers into the first. From this, an important condition follows, also called the *principle of interchangeable duality* of electrodynamics equations, or simply the duality principle.

The area of action of the duality principle can be essentially widened if we formally take into consideration the fictitious magnetic currents $\vec{j}^{(m)}$ and charges $\rho^{(m)}$.

From the properties of the duality principle, a useful consequence occurs. Namely, if there are two boundary tasks for geometrically similar areas, then all conditions that the vectors $\vec{E}$, $\vec{H}$ must fulfill for the first task transfer to conditions for the vectors $\vec{E}$, $\vec{H}$ for the second task at aforementioned replacements. Thus, having constructed the solution of one task (for example, the task of wave i diffraction on the S_1 band; Fig. 2.7a), the solution of the second task (diffraction of the same wave on the slot in the infinite plane S_1; Fig. 2.7b) can be obtained by means of corresponding replacements in accordance with Eq. (2.40).

2.5 The Solution Uniqueness Theorem for Electrodynamics Tasks

Manifold electrodynamics tasks can be conditionally divided into two classes: internal and external tasks. They are shown schematically in Fig. 2.8.

In the usual statement, the internal task (Fig. 2.8a) is stated as: to determine the field within some arbitrary surface $S = S_1 + S_2$ limiting the V volume, in the separate part of which the same area $V_{j,\,\rho} = V_1$ is located with field outside sources $\vec{j}_{\text{out}}, \rho_{\text{out}}$. The limiting surface S may be "combined", i.e., on one of its parts, for instance, on S_1, the tangent component of the electric field E_t is given, and on S_2, the tangent component of the magnetic field H_t is given. Evidently, the S surface may be formed from an arbitrary number of parts S_i with different values of E_t and H_t.

For the external task (Fig. 2.8b), it is typical for the $V_{j,\rho}$ area with sources to be located outside the V volume, and we can meet several such "scattering" objects (for example, V_i, $i = 1,2,$). In each of them, the tangent components of field E_{ti} or H_{ti} are given. Obviously, on S_1 surfaces, we may also define the boundary conditions of the impedance type by Eqs. (2.13, 2.14, and

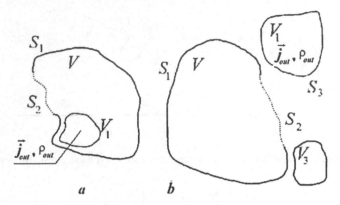

Fig. 2.8 Classification of electrodynamics tasks. (a) The internal task. (b) The external tasks

2.15). Thus, for instance, in the external task (Fig. 2.8b), the field inside V_i areas need not be found: they are given on its external surfaces by the corresponding boundary conditions.

The uniqueness theorem for both formulated internal and external tasks can be proved only with the assumption that in each point of the areas under consideration, the arbitrarily small electrical and magnetic losses occur: everywhere either $\varepsilon'' > 0$, or $\mu'' > 0$.

We do not describe the full proof of this theorem, but we draw up a plan of such a proof. To do this for internal tasks, we introduce the difference between two assumed solutions of Maxwell's equations: $\vec{E} = \vec{E}_1 - \vec{E}_2,\ \vec{H} = \vec{H}_1 = \vec{H}_2$.

It is clear that differential field $\vec{E}$, $\vec{H}$ satisfies Maxwell's equations (1.6), (1.7) and the ideal (zero) boundary conditions (2.11 and 2.12) on the appropriate S_i surfaces. Applying the theorem for power balance (1.34) to the difference field $\vec{E}$, $\vec{H}$, we can show that for $\varepsilon'' > 0$ and $\mu'' > 0$ the difference field is $\vec{E} = \vec{H} \equiv 0$, which proves the theorem.

To prove the uniqueness theorem for the external task, besides the assumption about the presence of small losses at each point, we must introduce additional conditions. First, all outside sources $\vec{j}_{out}, \rho_{out}$ are located on the finite distance, and, second, fields on the infinite distance from sources and "scatterers" dampen faster than $1/R$, where R is the distance from any point in the region occupied by sources and scatterers.

There are several forms of radiation condition. We consider the following:

$$\left|\vec{E}\right| < M/R^{1+\alpha}, \left|\vec{H}\right| < M/R^{1+\alpha} \text{ for } R \to \infty, \tag{2.41}$$

where M and α are positive numbers.

These conditions are sufficient for the uniqueness theorem proof for the external task. The physical sense of the uniqueness theorem for the external task consists in the fact that there should not be secondary waves, which arrive from infinity to the V area under consideration.

Thus, if we find the solution of electrodynamics equations for both the aforementioned main tasks that meets the general equations and boundary conditions, it will be unique.

Finally, we would like to note the so-called condition on the edge. In both internal and external tasks we need to deal with ideally conducting shields, strips, corner areas, dielectric wedges, etc. (see, for instance, Fig. 2.4). In the presence of such formations inside of the V area under consideration, the formulation of the boundary problem is ambiguous. In this case, the additional physical condition, which is called the condition on the edge, is introduced. Usually, it is formulated as:

$$\int_V \left(\vec{E}\,\vec{D} + \vec{H}\,\vec{B}\right) dV \Rightarrow 0 \tag{2.42}$$

as some volume V' tends to zero in the close vicinity of the edge. From this condition, it follows that in the vicinity of the edge, any component of the electromagnetic field cannot increase faster than $\rho^{-1+\alpha}$ ($0 < \alpha < 1$), where ρ is the distance to the edge. More precisely, components of the electrical field $\vec{E}$, which are parallel to the edge, are always limited or equal to zero in the vicinity of the edge. Perpendicular components have a singularity to the order of $\rho^{-1+\alpha}$. In the case of infinitely thin, ideally conducting strips located in the homogeneous isotropic medium $\alpha = 1/2$.

In engineering practice, the fulfillment of the condition on the edge does not usually require key significance. It is true, first of all, owing to unreality, there are no such sharp edges, as they are "rounded" to a certain extent. Another situation occurs for obtaining strict solutions. Here, fulfillment of this condition is strictly obligatory, because it leads to direct violation of the conditions of the uniqueness theorem. The condition on the edge is widely used in computer practice, as it allows, for example, faster solution convergence to be provided.

However, we know cases when nonfulfillment of the edge (the wedge) conditions led to great errors at investigation of tasks of wedge falling onto the elastic half-space (see, for instance Kostrov [56]).

2.6 Classification of Electromagnetic Phenomena, Zones of Geometric Optics, Fresnel and Fraunhofer Zones

2.6.1 Electrostatics

Systems of Maxwell equations in differential (Eqs. 1.6, 1.7, 1.8, and 1.9) and integral (Eqs. 2.32, 2.33, 2.34, and 2.35 forms cover practically all electromagnetic phenomena concerning engineering electrodynamics. In some specific cases, Maxwell's equations are simplified. The very simplest case happens when the field does not depend on time ($\partial/\partial t = 0$,) and, besides, the motion of charged particles is absent, although this is a far-reaching idealization, because there are always moving particles, for instance, in the conductor [12].

Accepting such an idealization, we ascertain that the system (Eqs. 1.12, 1.13, 1.14, and 1.15) can be divided into two independent systems:

$$\text{rot }\vec{E} = 0, \quad \text{div }\vec{D} = \rho, \quad \vec{D} = \varepsilon\,\vec{E}\,; \tag{2.43}$$

$$\text{rot }\vec{H} = 0, \ \text{div }\vec{B} = 0, \ \vec{B} = \mu_0\,\vec{H}. \tag{2.44}$$

This system has an interesting feature. The equations in (2.43) contain the electric field vectors $\vec{E}$, $\vec{D}$ only, whereas Eq. (2.44) contain the magnetic field vectors $\vec{H}$, $\vec{B}$. This means that in this case ($\partial/\partial t = 0, j = 0$), the electric and magnetic phenomena are *independent*.

Here we are tightly confronted with the well-known paradox of the classic theory. On the one side, the electromagnetic field is unified as it follows from general equations (1.12, 1.13, 1.14, and 1.15), but on the other hand, in the static, these equations are divided into two groups (2.43 and 2.44). The famous *R. Feynman* wrote on this theme: "As soon as in static charges and currents are constant, the *electricity and magnetism are different phenomena.*" [64]. Here, a large number of questions arise; the generalized classic theory gives answers to some of them (see Chaps. 3 and 5).

Satisfying the same conditions $\partial/\partial t = 0, j = 0$ for the full equation system, we obtain:

$$\begin{aligned} \text{rot }\vec{E} &= 0, \quad \text{div }\vec{D} = \rho, \\ \text{rot }H + \underline{\text{grad}H^*} &= 0 \quad \text{div }\vec{B} = 0. \end{aligned} \tag{2.45}$$

The presence of terms with the scalar magnetic fields (H^*) in both lines of (2.45) makes them *dependable*; thus, the Feynman note can be eliminated.

Keeping to the accepted representations, we discuss phenomena described by the system (2.43), which are called electrostatic in classic theory. Electrostatic fields are the fields created by charges, which are fixed, unchanged in time and in value. The equation system (2.43) is the general system of electrostatic equations.

2.6.2   Magnetostatics

We know that the fundamentals of classic magnetostatics are formed by the following equations:

$$\vec{H} = (1/\mu_0)\mathrm{rot}\,\vec{A}\,, \tag{2.46}$$

$$\mathrm{div}\,\vec{A} = 0 \tag{2.47}$$

where $\vec{A}$ is the vector potential; $\vec{H}$ is the magnetic field strength; μ_0 is the magnetic constant.

Equation (2.47), as we already mentioned, is the *Coulomb* normalization, which is introduced for unambiguity of the determination of vector potential $\vec{A}$. According to condition (2.47), in magnetostatics, the lines of the $\vec{A}$ vector $\vec{A}$ must be closed, i.e., the field of this vector is vortical.

Introduction of the scalar magnetic fields H^* concept, which relates to the vector potential $\vec{A}$ as

$$H^* = -(1/\mu_0)\mathrm{div}\,\vec{A}\,, \tag{2.48}$$

eliminates the necessity in artificial *Coulomb* normalization (2.47), meets the conditions of the *Stokes–Helmholtz* theorem (1.4, 1.5), and forms the basis of generalized magnetostatics [12].

Equations (2.46 and 2.47) characterize the fields created by the permanent magnets. They can also be used for analysis of magnetic field properties, created by direct currents in the region, when the conduction current density is equal to the theory and which does not couple with the current (does not envelop its lines). Phenomena described by the system (2.46), (2.47) are called magnetostatic, and Eqs. (2.46 and 2.47) are called magnetostatic equations.

Two circumstances, at least, are interesting here. First, the physical processes at very low frequencies, when the length of the magnetostatic wave is essentially less than that of the electromagnetic wavelength. The small size of the wavelength, or (the same) small size of body dimensions, gives the possibility of neglecting delayed terms and using the magnetostatic Eqs. (2.46 and 2.47) during the study and application of magnetostatic waves.

On the other hand, problems of determination of the permanent magnetic fields [51] are concerned with the class of magnetostatic tasks.

2.6.3   Stationary Electromagnetic Processes

In the presence of direct current $(\vec{j} = \mathrm{const})$, the electric and magnetic fields cannot already be considered as independent as in the classic case; they are connected by Ohm's law $\vec{j} = \sigma\,\vec{E}$ (1.25) and, accordingly, by the scalar magnetic field. In this case, the general equation system of electrodynamics takes the form:

$$\mathrm{rot}\,\vec{H} + \underline{\mathrm{grad}H^*} = \vec{j} \tag{2.49}$$

$$\mathrm{rot}\,\vec{E}_0 = 0, \tag{2.50}$$

$$\mathrm{div}\,\vec{D} = \rho, \tag{2.51}$$

$$\mathrm{div}\,\vec{H} = 0. \tag{2.52}$$

From Eqs. (2.49, 2.50, 2.51, and 2.52) we see that the connection between the electric and magnetic fields is defined by the electric current $\vec{j}$ and the scalar magnetic field $(\mathrm{drad}H^*)$.

The so-called quasi-stationary processes, i.e., processes that flow slowly enough $(\partial/\partial t \ll 1)$, are a separate class of processes. In this case, in the classic Maxwell's first equation, the presence of the conduction current can usually be neglected, but the offset current: $\mathrm{rot}\,\vec{H} = \vec{j}$. Nevertheless, in this case we automatically neglect the scalar magnetic field variation $(\mathrm{grad}\,H^{*})$. In cases when conduction currents are absent (for example, the capacitor in the AC circuit), the offset currents must be taken into consideration; thus, $\mathrm{rot}\,\vec{H} = -\partial\,\vec{D}/\partial t$.

Maxwell's equation for analysis of quasi-stationary processes may be written in the usual form:

$$\mathrm{rot}\,\vec{E} = -\partial\,\vec{B}/\partial t.$$

In general, it is necessary to use the full system of electrodynamics (Eqs. 1.6, 1.7, 1.8, and 1.9).

2.6.4 Potential and Vortical Fields

In Chap. 3 we described a great deal about these types of fields; therefore, we now make some preliminary notes. The field of some arbitrary vector $\vec{A}$ is called *potential* if it satisfies the following conditions in the considered V region:

$$\mathrm{rot}\,\vec{A} = 0, \quad \mathrm{div}\,\vec{A} \neq 0. \tag{2.53}$$

Thus, we assume that $\mathrm{rot}\,\vec{A} = 0$ in the whole region, and $\mathrm{div}\,\vec{A} \neq 0$ at least in the one point of the V region. The potential field is defined by the scalar function (the potential) φ, so that

$$\vec{A} = -\mathrm{grad}\,\varphi. \tag{2.54}$$

The lines of the $\vec{A}$ vector in the potential field have the beginning and/or the end, starting (and ending) at points, where $\mathrm{div}\,\vec{A} \neq 0$. When $\mathrm{div}\,\vec{A} = 0$, the lines of the $\vec{A}$ vector are continuous.

The field of some arbitrary vector $\vec{A}$ is called vortical (solenoid, tubular) if it satisfies the following conditions in the V region under consideration:

$$\mathrm{rot}\,\vec{A} \neq 0, \quad \mathrm{div}\,\vec{A} = 0. \tag{2.55}$$

Thus, we assume that $\mathrm{div}\,\vec{A} = 0$ everywhere in the V region, and $\mathrm{rot}\,\vec{A} \neq 0$ is true at least in one point of this region.

In the case of "purely" sinusoidal oscillations, the system (Eqs. 1.6, 1.7, 1.8, and 1.9) is simplified with the help of the artificial approach, which was called the *method of complex amplitudes*, and the system of Maxwell's equations (1.6, 1.7, 1.8, and 1.9) proceeds to relations (2.37).

2.6.5 Geometric Optics

At very high frequencies, it is acceptable to distinguish electromagnetic phenomena by comparing the wavelength λ or the wave-number $k = \omega/c = 2\pi/\lambda$ with the dimensions of the obstacle a. At $ka \rightarrow \infty$ (or for $ka \gg 1$). This is the zone of phenomena of geometric optics (diffraction (scattering) on the wide structures, large bodies). In essence, the diffraction on small narrow obstacles for $ka \ll 1$ can be attributed to quasi-static or even static phenomena. And finally, this is the resonance zone, when dimensions of the zone coincide (or almost coincide) with the integer number of half-wavelengths $k = \pi q$, $q = 1,2,$ ... (see Sect. 1.9 for more detail on the resonance conditions). Thus, all phenomena in electrodynamics certainly take into account the ratio between wavelength λ and obstacle dimensions a.

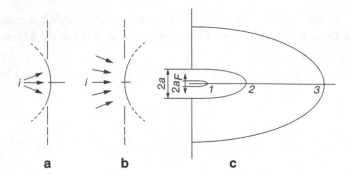

Fig. 2.9 Diffraction on the slot in the nontransparent shield: incidence on the slot of (**a**) divergent and (**b**) convergent waves. (**c**) Zones of geometric optics convenience (*1*), Fresnel approximations (*2*), and Fraunhofer approximations (*3*)

Table 2.1 Zones of diffraction

$a^2/z\lambda$	$a^2/R\lambda$		
	$\ll 1$	~ 1	$\gg 1$
$\gg 1$	Geometric optics	Geometric optics	Geometric optics
~ 1	Geometric optics	Fresnel	Fresnel
$\ll 1$	Geometric optics	Fresnel	Fraunhofer

2.6.6 Fresnel and Fraunhofer Zones

In the theory and practice of wave diffraction on obstacles, antennas, and other applications, it is acceptable to distinguish the near, intermediate (Fresnel), and far (Fraunhofer) zones. Thus, besides values of a and λ, it is necessary to take into account the distance from the shield to the observation point z. Thus, in the classification of phenomena behind the shield (and in front of it) such dimensionless parameters as $a^2/(R\lambda)$ and $a^2/(z\lambda)$ participate. Here, R is the radius of the "divergent" cylindrical (Fig. 2.9a) (or spherical) wave, which falls on the half-plane or the hole in the shield ($R > 0$); if the wave has a convergent front, then $R < 0$ (Fig. 2.9b). It is clear that for the plane wave ("spotlighting" beam) $1/R = 0$.

Clearer interpretations of the Fresnel and Fraunhofer zones are presented in Fig. 2.9, and also in Table 2.1 [65, 66].

In Table 2.1 and in Fig. 2.9 we see zones in which laws of geometric optics and the approximations of Fresnel and Fraunhofer act. a_F is the size of the first Fresnel zone:

$$a_F = \lambda^{1/2}(1/z - 1/|R|)^{-1/2}. \tag{2.56}$$

To be more precise, the division into zones (near, intermediate, and far) is sufficiently conditional (see also Neganov and Nefyodov, and Neganov et al. [18, 19]).

We would like to discuss here one more very useful circumstance for the further understanding of phenomena physics that is closely connected to the duality principle of Maxwell"s equations (see Sect. 1.6).

Field pictures near the wire (Fig. 2.10a), say, in the two-conductor line (Fig. 2.10a), when the second conductor is quite far displaced, are shown in Fig. 2.10b. We see the electric field picture near a slot in an infinitely large shield (in essence, this is also the two-conductor line formed by two half-infinite shields). In a more general case, for instance, in diffraction theory, the duality principle forms the connection between electromagnetic fields, which are diffracted on the S hole, cut in the infinitely thin, ideally conducting plane shield, and on the plane plate. that coincide in the form with the S hole. This is often called the generalized *Babinet's principle* or simply the duality principle. In particular, this principle allows the theory development of so-called plain diffraction radiators, including narrow slots in the plain shield, which are equivalent to the thin electric vibrator.

The principle of complementarity – *Babinet's* principle is effectively used, for example, to solve tasks of irregularity – basing elements in strip-slot structures [16]. It turns out to be a rather useful and effective means of solving some tasks, because we can use with its help the huge experience of diffraction task solution in waveguide irregularities [67, 77].

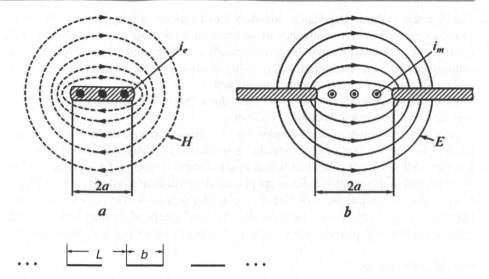

Fig. 2.10 (**a**) The magnetic field near the conducting strip and (**b**) the electric field near the slot in the infinite shield

Fig. 2.11 Geometry and sizes of the lattice

2.7 Classification of Electrodynamics Problems, the Concept of Key Structures and Tasks

We should note once more that equivalent impedance conditions of various types significantly simplify the analysis of complicated electrodynamics problems. The *Sommerfeld* problem of the plane wave falling onto the ideally conducting half-plane (Fig. 2.3a; right) is the classic key problem, in which the boundary conditions (2.11) or (2.12) are fulfilled. By the way, the half-plane was the specific case of the ideally conducting two-dimensional wedge that was examined by *Sommerfeld* as well.

The much more general tasks on the wedge with *Schukin*'s impedance boundary conditions (2.13) were studied in the 1940s by *G.D. Maliuzhinets* [6, 81]. He introduced the special functions that were named after him. The diffraction task on the half-plane with anisotropic impedance boundary conditions is in this group [57].

In many cases, the key structures can be considered by strict methods [69–71]. The transfer to the solution of the finite structure task, especially when it is long (Fig. 2.3, left; $kw \gg 1$) is not simple. Then, using the sequential diffraction method, for example, we can construct the asymptotic series over the inverse powers of a large kw parameter: $a(1/kw) + b(1/kw)^2 \ldots$, where $a, b, \ldots$ are constants [71, 126]. Sometimes such a series turns out to be suitable almost up to the static region [65, 126].

2.8 Examples of MathCAD Packet Utilization

In the present chapter, we use (see Appendix B – http://extras.springer.com/2019/978-3-319-90847-2) the only example of MathCAD packet application for electromagnetic problems – Program P2.1 – study of reflection and transmission of the plane electromagnetic wave through the frequently periodic lattice (see Sect. 2.1.8) (the system structure is shown in Fig. 2.11).

As initial data, we define the period of the lattice, the wavelength, and the parameter of metal.

We use the formulas (2.25) for calculation in the MathCAD packet. Calculation results show the following. With growth of the q-parameter, the reflection factor at first is rather small, but then begins to increase sharply. If the lattice period increases, the reflection factor begins to grow fast. The transmission factor behaves itself in the opposite way: at first it is large but begins to decrease with g growth.

Three-dimensional pictures are shown in Appendix B (http://extras.springer.com/2019/978-3-319-90847-2), P.2.1.

2.9 Summary

The wide class of fundamental problems of classical and generalized electromagnetic is considered, which includes the final formulations of boundary electrodynamics tasks in two forms: differential and integral. The main classes of boundary conditions are given, including both its' complete form and various approximate impedance boundary conditions, beginning with Schukin's classic impedance boundary conditions up to boundary conditions of the resonance type. These various boundary conditions are suitable in various practical conditions, and they differ in the fact that with its utilization, first, the solution of electrodynamics problems is essentially simplified, and second, understanding of the physical sense of the solution obtained can be achieved. Unfortunately, using purely numerical methods for solutions, the revealing of the physical features of the task can sometimes be highly complicated.

Obtaining of generalized Fresnel formulas for a two-media boundary (Sect. 5.6.1, Chap. 5) is a classic example of the application of anisotropic boundary conditions.

Auxiliary electrodynamics principles are considered; namely, the principle of interchangeable duality and the unicity theorem for the solution of electrodynamics problems. By the way, the first principle permits (knowing the solution, for instance, to the task with Dirichlet boundary conditions) to construct a solution to the task with Neumann conditions.

At the end of this chapter, the concept pf crucial structures and tasks is given. This allows us to know, say, a strict solution to the crucial (semi-infinite) structure, to construct a solution to the finite length structure, which is usually interesting for practice. On this basis, we construct the almost "strict" theory of the microstrip line (Sect. 7.3.7). In particular, the theory of open resonators with planar mirrors (Sect. 11.2.1) can be developed with application of resonance boundary conditions.

Checking Questions

1. What are differences between boundary conditions in classic and generalized electrodynamics?

2. Which physical phenomenon serves as the basis for the formulation of Schukin's impedance boundary conditions?
3. What is necessary to demand in the case to ensure the transfer of the impedance boundary conditions (2.14) to Schukin's impedance boundary conditions (2.13)?
4. Which physical considerations were laid based on the formulation of boundary conditions on the frequently periodic structures (Weinstein–Sivov boundary conditions)?
5. How are the resonance conditions (2.27) and the Weinstein effect connected?
6. What is the sense of the Ostrogradsky–Gauss theorem?
7. In which form can we represent the solution obtained by the method of subsequent diffractions?
8. Why are the concepts of the Fresnel and Fraunhofer zones introduced and how are they defined?
9. What is the sense of the introduction of Schukin's boundary conditions (2.13)?
10. Using MathCAD Program P.2.1, calculate the two-dimensional pictures of the refraction factor and the transmission factor for $L = 0.05$ and $L = 0.5$. Explain the physical sense.
11. Using the MathCAD Program P.2.1, calculate the three-dimensional pictures of the refraction factor and the transmission factor for $L = 0.05$ and $L = 0.5$. Explain the physical sense.

Wave Equations in Classic Electrodynamics, Electrodynamics Potentials, the Green's Function, the Main Classes of Solutions

3

In this chapter, we begin the consecutive description of the main accepted approaches to electromagnetic problems. Traditionally, instead of Maxwell's equations (Chap. 1), it is acceptable to investigate the so-called wave equations for electrodynamics problems (Sect. 3.1.1, electromagnetic fields and waves) and for acoustics problems (Sect. 3.1.2, sound and ultrasonic acoustic fields and waves). The principal convenience for us of such a transfer to wave equations consists, in particular, in the uniformity of the wave equations for electrodynamics and acoustics. On the other hand, instead of two equations of the first order, we obtain one equation of the second order, which simplifies the task.

The medium parameters, which are used for field determination, are included in the right parts (an external force) of wave equations together with outside sources of this field. The complexity of the problem consists in the fact that a functional view of the distribution of the field sources is inaccurate, and for outlet, we must introduce the vector $\vec{A}$ and scalar φ potentials for consideration (Sect. 3.2). Having known the potential values, the necessary transition to the required functions of vectors of electrical $\vec{E}$ and magnetic $\vec{H}$ field strengths is performed using simple differentiation.

In the classic electromagnetism theory, the potential is the relative concept, and the potential difference only has a physical sense. Therefore, the addition of any constant to the potential does not change the physics process, i.e., the potential itself is not the unambiguous concept. Ambiguity of potential introduction can usually be usually eliminated by the so-called Coulomb and Lorentz calibration relationships (Chap. 4). In the future, we will show that the introduction of the mentioned calibration relationships essentially narrows a range of physical consequences of the main electrodynamics equations (Sect. 4.2, Chap. 4).

Examples of electrostatic systems are further presented in Sect. 3.2; namely, the field pattern of two similar or opposite charges, the field of the charged square plate and, finally, the field of the plane charged capacitor. All these examples are supported by MathCAD programs for the modeling and a better understanding of the physical phenomena.

Understanding and analysis of the processes in electrodynamics structures are significantly simplified by the introduction of the so-called sourcewise Green's function (Sect. 3.3). The Green's function is a fundamental concept in electromagnetism, the solution's sense of the wave equation or the Helmholtz equation with the right-hand part in the form of a delta-function $\delta(\vec{r})$. An example of the utilization of Green's function is presented for solution of the wave equation with the given right-hand part. In the future, we will be acquainted with the Kisun'ko–Feld theory of waveguide excitation, which essentially simplifies the analysis of excitation tasks (Chap. 12) of many electrodynamics structures.

A small section (3.3) of our lecture course is devoted to the most widely functions used in practice, the so-called model Green's functions, when the emission sources are distributed over the infinite plane or over the conducting filament, and in the form of the point source. We assume initially that these sources are placed in the infinite uniform space, and then the Green's function has a form of the plane, cylindrical, or spherical wave. Russian scientists Kisun'ko and Feld constructed the famous theory of waveguide excitation using the eigenwave of the given waveguide as the Green's function.

The use of the Green's function in the form of the spherical wave allows substantiation and creation (Sect. 3.5) of the mathematical representation of the Huygens–Fresnel–Kirchhoff principle, which finds a widespread use in the theory and practice of electromagnetism.

In the whole series of practical cases, we can reduce the wave diffraction problem on obstacles to the parabolic equation (Sect. 3.6.1) instead of the wave equation. This allows us to obtain right away the clear physical interpretation of the picture of

© Springer International Publishing AG, part of Springer Nature 2019
E. I. Nefyodov, S. M. Smolskiy, *Electromagnetic Fields and Waves*, Textbooks in Telecommunication Engineering,
https://doi.org/10.1007/978-3-319-90847-2_3

such a complicated electromagnetic phenomenon as wave diffraction. We accurately examine the diffraction pictures of the plane wave on the conducting half-plane (Sect. 3.6.2) and on the slot in the infinite lane shield (Sect. 3.6.3).

The next Sect. (3.7) is devoted to consideration of the properties and characteristics of the simplest plane electromagnetic waves. Like the other simplest waves, namely, cylindrical and spherical, they are convenient models for investigation of more complicated electrodynamics processes and devices. The Helmholtz equation is the basis for the construction of the wave propagation picture; the plane waves (forward and backward) are its solution. A classical picture of plane wave propagation in the free space is presented here as well.

Chapter 3 ends with examples of the usage of the MathCAD packet program: the electric field picture of two charges (Sect. 3.8.1), the field of the charged square plane (Sect. 3.8.2), and finally, the field of the plane capacitor (Sect. 3.8.3).

3.1   Wave Equations in Classic Electrodynamics

3.1.1   Wave Equations, Electrodynamics

Wave equations in electrodynamics are the main "weapon" for both a researcher and an engineer. They are equally present in classic and generalized electrodynamics theories. There are several reasons for it, with which we will sequentially become acquainted with in our lecture course. We begin with the case of classic electrodynamics.

In general, *Maxwell's equations* (1.12, 1.13, 1.14 and 1.15) contain six components of $\vec{E}$, $\vec{H}$ fields, which are subjects for determination. It is evidently desirable to simplify the initial equation system, for instance, to reduce it to a smaller number of equations. One of simplifications, as we saw in the previous chapter, is the elimination of time dependence ($\exp\{-j\omega_0 t\}$) (for monochromatic fields) in addition to the introduction of ideal or impedance boundary conditions of various types, allowing us not to consider a field in the same V volume, replacing this field with the knowledge (accurate or approximate) by $\vec{E}$, $\vec{H}$ field (or, more precisely, their ratios) in the limited V volume of the S surface.

Sometimes, it is convenient to transfer to the one equation from the two first-order equations (Eqs. 1.12 and 1.13), but, of course, this equation will have a second order. Let the field sources in (1.12 and 1.14) be outside, i.e., $\vec{j}_{\text{out}}, \rho_{\text{out}}$, whereas the medium is homogeneous and isotropic. For Eq. (2.37) in the complex form, for non-uniform (in the general case) by means of affecting the rot operator on the left- and right-hand sides of the first equation in Eq. (2.37) and replacing rot $\vec{H}$ from the second equation in (2.37), we obtained the heterogeneous equation:

$$\Delta \vec{E} + k^2 \varepsilon\mu \, \vec{E} = i\omega\varepsilon\mu \, \vec{j} + (1/\varepsilon)\text{grad} \, \rho_{\text{out}} \tag{3.1}$$

In a similar manner, we obtain the equation for the magnetic field $\vec{H}$:

$$\Delta \vec{H} + k^2 \varepsilon\mu \, \vec{H} = -\text{rot} \, \vec{j}_{\text{out}}. \tag{3.2}$$

In Eqs. (3.1 and 3.2), which have the same type, the $\Delta \equiv \nabla^2$ operator designates the *Laplace operator*, which, for example, in the Cartesian coordinate system, is written as $\Delta = \partial^2/\partial x^2 + \partial^2/\partial y^2 + \partial^2/\partial z^2$.

Equations (3.1 and 3.2) are called *Helmholtz equations*. If we start from Maxwell's equations (1.6, 1.7, 1.8, 1.9, 1.10, and 1.11), then equations that are obtained as a result of the same approach for $\vec{E}$, $\vec{H}$ field components are called *wave equations*. In engineering practice, we often deal with monochromatic fields ($\exp\{-j\omega t\}$) and, therefore, with the Helmholtz equations (3.1 and 3.2). The transfer to nonstationary processes is performed with the help of the Fourier transform or the Laplace transform. This approach is well-known from electrical engineering.

Sometimes, by the way, the application of Fourier and Laplace transforms becomes inconvenient or does not correspond to the statement of the physical problem. In this case, we need to solve the nonstationary electrodynamics problem in its full volume. It was shown on the series of impressive examples, for instance, that a form of the parabolic antenna is not optimal in some cases, and the mirror form should be matched with radio signal features. Unfortunately, this interesting circle of questions is completely omitted from our lecture course.

To analyze the electrodynamics structures, the heterogeneous Eqs. (3.1 and 3.2) are slightly inconvenient as usually the field sources $\vec{j}_{\text{out}}, \rho_{\text{out}}$ are not precisely but approximately (or very approximately) known. Data for analysis of $\vec{j}_{\text{out}}, \rho_{\text{out}}$ can be taken from either experiment or from other calculations, or from more or less reliable considerations. Thus, an error is introduced into the right-hand parts of Eqs. (3.1 and 3.2) in the initial situation. Furthermore, being differentiated on coordinates $(\vec{r})$, they introduce one more error into the right-hand parts of Eqs. (3.1 and 3.2), and thus, into the solution obtained. The usual way out of this situation in classic theory is with the introduction of the scalar $\varphi(\vec{r}, t)$ and vector $\vec{A}(\vec{r}, t)$ potentials. We discuss these in Sect. 3.2.

3.1.2 Wave Equation, Acoustics

The wave equations in the form of Eqs. (3.1 and 3.2) or similar to them describe the very manifold wave processes in nature. We draw the reader's attention to only one phenomenon; namely, to *acoustic wave* propagation. We note the great contribution to the theory and engineering of acoustic waves and oscillations made by Russian scientists *N.N. Andreev, L. M. Brekhovskikh, S.I. Vavilov, G.D. Maliuzhinets,* etc.

The main characteristics of the acoustic field are its propagation velocity $\vec{v}$ and the pressure p. The complex amplitudes of $\vec{v}$ and p are linked to each other by the following differential equations:

$$\operatorname{div} \vec{v} = (i\omega/\rho c^2)p, \operatorname{grad} p = i\omega\rho \vec{v}, \tag{3.3}$$

in which c is the light speed in the medium with the density ρ. Assuming the ρ density as a constant that does not depend on coordinates, and eliminating from (3.3) the $\vec{v}$ velocity, we obtain the required wave equation

$$\Delta p + k^2 p = 0, \quad k = \omega/c. \tag{3.4}$$

This wave equation (3.4) is the *scalar equation*, in contrast to *vector equations* (3.1 and 3.2) for electromagnetic field components. Frequently, the solution of the scalar acoustic equations helps us to understand electromagnetic phenomena.

3.2 Electrodynamics Potentials

At non-exact knowledge of the right-hand parts of Eqs. (3.1 and 3.2), we have the way out from this situation, as we already mentioned, consisting in transfer to the scalar $\varphi(\vec{r})$ and the vector $\vec{A}(\vec{r})$ *electrodynamics potentials.* Formally, they are introduced as:

$$\vec{B} = \operatorname{rot} \vec{A}, \quad \vec{E} = i\omega \vec{A} - \operatorname{grad}\varphi. \tag{3.5}$$

Strictly speaking, such an introduction of φ, $\vec{A}$ potentials is not an unambiguous action because of the addition of any scalar to the scalar potential φ: $\varphi + \text{const}$ does not lead to any changing, as $\operatorname{grad}(\text{const}) \equiv 0$. Similarly, if we add the gradient of any scalar function θ to the vector potential $\vec{A}$, we again will have nothing new because rot grad $\theta \equiv 0$. The mentioned ambiguity is eliminated by the *potential calibration condition.* There are several such conditions.

So, in classic magnetostatics, the magnetic field strength $\vec{H}$ is defined through the vector $\vec{A}$ potential as:

$$\vec{H} = (1/\mu_0)\operatorname{rot} \vec{A}. \tag{3.6}$$

Ambiguity of the $\vec{A}$ function is eliminated by *Coulomb normalization*

$$\operatorname{div} \vec{A} = 0. \tag{3.7}$$

In classic macroscopic electrodynamics, for the full electromagnetic field, we can use usually the *Lorentz calibration condition*:

$$\text{div } \vec{A} + \varepsilon_{\text{abs}} \mu_{\text{abs}} \partial \varphi / \partial t = 0. \tag{3.8}$$

Later (in this chapter) we discuss in detail the sense of normalizations (3.7 and 3.8). Thus, we note that utilization of aforementioned Coulomb and Lorentz conditions essentially restricts the completeness of the solution of Maxwell's equations. Moreover, in the educational literature (see, for example, Tamm, and Landau and Lifshitz [54, 72]), authors usually affirm that the *vector potential* $\vec{A}$ *has no physical sense* and it is introduced exclusively as an auxiliary function, and the normalization condition allows elimination of its ambiguity. We discuss this later in this chapter.

Substituting field components from the determining relations (3.5) into Maxwell's equations (1.12 and 1.13), we obtain equations for scalar φ and vector $\vec{A}$ potentials:

$$\Delta \varphi + k^2 \varepsilon \mu \varphi = -\rho_{\text{out}} / \varepsilon, \Delta \vec{A} + k^2 \varepsilon \mu \vec{A} = -\mu \vec{j}_{\text{out}} \tag{3.9}$$

As we see, the outside field sources (not its derivatives, as in Eqs. (3.1 and 3.2) for $\vec{E}$, $\vec{H}$ fields) are directly included in the right-hand parts of the Helmholtz equations (3.9). for more complete wave equations, see later (Eqs. 4.9, 4.14, 4.15, 4.16 and 4.17).

The equation system (3.9) consists of not six (as in Eqs. (3.1 and 3.2)), but four unknown functions to be determined. Having determined the potentials φ, $\vec{A}$ from the system (3.9), it is easy (by means of simple differentiation) to transfer directly to the required true $\vec{E}$, $\vec{H}$:

$$\vec{H} = (1/\mu)\text{rot } \vec{A}, \quad \vec{E} = -\text{grad}\varphi + ik \vec{A}. \tag{3.10}$$

At the limit for very low frequencies, or for $\omega \to 0$ ($k = \omega/c \to 0$), the following electrostatic equations can be derived from (3.9):

$$\Delta \varphi = -\rho_{\text{out}} / \varepsilon_{\text{abs}} \tag{3.11}$$

Poisson's equation and

$$\Delta \varphi = 0 \tag{3.12}$$

Laplace's equation.

Similarly, Poisson's equation for magnetostatics is

$$\Delta \vec{A} = -\mu_0 \vec{j}, \tag{3.13}$$

Solution of Poisson's equation (3.13) is in general case is

$$\vec{A} (\vec{r}') = (\mu_0/4\pi) \int_V \frac{\vec{j} (\vec{r})}{r} dV, \tag{3.14}$$

where $r = \sqrt{(x' - x)^2 + (y' - y)^2 + (z' - z)^2}$ is the modulus of the radius vector, which defines the distance between the volume element dV, in which the $\vec{j}$ current flows, and the point of the $\vec{A}$ potential determination.

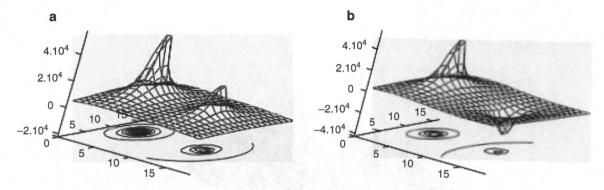

Fig. 3.1 Pictures of the static electric field of (**a**) two equal-sign and (**b**) opposite-sign charges. (Taken from Nefyodov and Kliuev [73], p. 87)

Applying the rot operator to (3.14), we have

$$\vec{H} = (1/4\pi) \int_V \frac{\vec{j} \times \vec{r}}{r^3} dV. \tag{3.15}$$

The latter equation represents the *Biot–Savart law*. Sometimes, this law includes three names, adding Laplace, who proved it in the general case.

Below, we present examples of some electrostatic task solutions.

The field of two charges First of all, this is an example of the solution of *Poisson*'s equation for two charges q_1, q_2, located at a distance L from each another (Fig. 3.1). Fields are calculated according to Program P.3.1 (see Appendix A) for their potentials and fields [54, 73]. Program P.3.1 allows the changing (at the will of the researcher) of values and signs of charges, the distance between them, and their positional relationships. Pictures of the electric field of two equal-sign (Fig. 3.1a) and two opposite-sign (Fig. 3.1b) charges are shown in Fig. 3.1, and pictures of equipotential lines are presented below. In the P.3.1 program, the ε_0 values are in Far/m, $q_{1,2}$ are in Q, and L is in m.

The charged plate The field of the charged square plate (a side of the square is D, m; see Fig. 3.2) is another example.

The electric field is defined by the expression (compare with Eq. (3.15)):

$$\vec{E}(\vec{r}) = \frac{\rho_s}{4\pi\varepsilon_0} \int \frac{\vec{r} - \vec{r}'}{\left|\vec{r} - \vec{r}'\right|} ds', \tag{3.16}$$

in which the $\vec{r}, \vec{r}'$ is the observation point ($\vec{r} = \vec{e}_y y + \vec{e}_z z$) and the point on the plate surface is $\vec{r}' = \vec{e}_x x + \vec{e}_y y$, values ε_0 and ρ_s are determined in Program P.3.2 (see Appendix B – http://extras.springer.com/2019/978-3-319-90847-2).

Computation of the integral over the plate surface in (3.16) for y- and z-components of the electric field is fulfilled according to Program P.3.2, allowing determination of the field in the place surrounding the plate [74]. Figure 3.2 shows some of the results. Using Program P.3.2, we may build the field picture in any sections and have the complete representation of the electric field of the uniformly charged square plate in the surrounding space.

The described Program P.3.2 allows representation of calculations and results for any set of sizes and charge values, which gives a sufficiently complete picture of the element of active physics under investigation.

The plane capacitor The next example is the solution of Poisson's equation for the electrostatic potential and for the electric field of the plane capacitor. For this, we use Program P.3.3 for the solution of Poisson's equation for the plane capacitor (see Appendix B – http://extras.springer.com/2019/978-3-319-90847-2) [73, 74], and the results of some numerical calculations are shown in Fig. 3.3b, c.

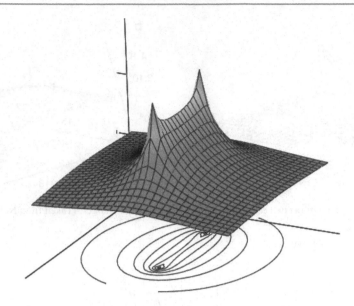

Fig. 3.2 The charged square plate $D \times D$ and its electrical field for $D = 0.5$ m, $\rho_s = \frac{K\pi}{M^2}$

Fig. 3.3 (**a**) The plane capacitor and (**b**) the electric field E in the plane $x = 0$ with oppositely charged plates for $a = 30$ cm, $d = 45$ cm and (**c**) pictures of charge density distribution on the upper and lower plates. (Taken from Nefyodov and Kliuev [73], p. 89)

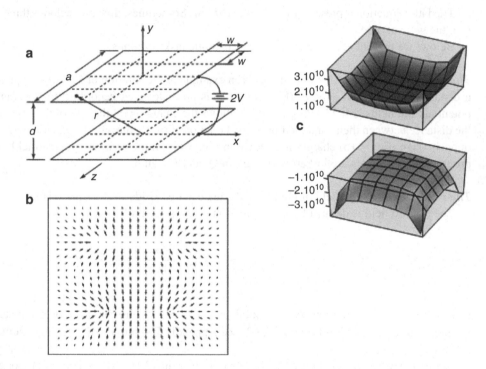

3.3   The Green's Function Concept

Let the Helmholtz equation be for any component of electric $\vec{E}$ or magnetic $\vec{H}$ fields, or electrodynamics potentials φ, $\vec{A}$. We designate this component, for instance, as $u(\vec{r})$. Then the Helmholtz equation for $u(\vec{r})$ has the usual view $\left(\varepsilon(\vec{r}) = \text{const}, \mu(\vec{r}) = \text{const} \right)$

$$\Delta u(\vec{r}) + k^2 \varepsilon(\vec{r}) \mu(\vec{r}) u(\vec{r}) = F(\vec{r}), \tag{3.17}$$

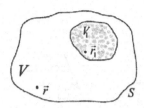

Fig. 3.4 Deriving the reciprocity theorem using the Green's function $G(\vec{r}, \vec{r}')$, and for representation of the δ field according to the *Huygens–Fresnel principle* (in this case the observation point $\vec{r}$ is located outside the S surface)

where the same known function of the field sources $F(\vec{r})$ is on the right-hand side.

The Green's function $G(\vec{r} - \vec{r}')$ of Eq. (3.17) is the solution of the Helmholtz equation (3.17) with the right-hand part in the form of δ-function:

$$\Delta G(\vec{r} - \vec{r}') + k^2 \varepsilon(\vec{r}) \mu(\vec{r}) G(\vec{r} - \vec{r}') = \delta(\vec{r} - \vec{r}'). \qquad (3.18)$$

In Eq. (3.18), the observation point and the source point (the point at which the δ source is located) are designated as $\vec{r}$, $\vec{r}'$. It should be remembered that the δ-function (sometimes, it is called the *Dirac* function), in addition to the widely used in electronics and electrodynamics *step function* or the *Heaviside unit function*, relate to the class of general functions and is a suitable mathematical image. It is known that the δ-function is equal to zero everywhere except for the $\vec{r}'$ point, in which it is equal to infinity (i.e., at the coincidence of the observation point and the point at which the source is located: $\vec{r} = \vec{r}'$). Of course, the Green's function $G(\vec{r} - \vec{r}')$ should meet the appropriate boundary conditions on the S boundary in the region of interest or the radiation conditions. Nevertheless, there is a definite arbitrary rule in the boundary condition choice for $G(\vec{r} - \vec{r}')$: they do not necessarily have to coincide with boundary conditions for the required function $u(\vec{r})$. This is very convenient because it promotes some task simplification (see Chap. 3, Sect. 3.43).

The Green's function $G(\vec{r} - \vec{r}')$ has a *symmetry property;* namely, $G(\vec{r}_1, \vec{r}_2) = G(\vec{r}_2, \vec{r}_1)$. In essence, this is representative of the reciprocity theorem (compare Sect. 1.15, the formula (1.73)): the field in the $\vec{r}_1$ point, which is excited by the source (for example, by the elementary dipole), located in the $\vec{r}_2$ point, is equal to the field in the $\vec{r}_2$ point, excited by the source located in the $\vec{r}_1$ point.

Let us illustrate application of the Green's function $G(\vec{r} - \vec{r}')$ to the solution of Eq. (3.17). For this, we need to multiply Eq. (3.17) by $u(\vec{r})$, and Eq. (3.18) by $G(\vec{r}, \vec{r}')$. Then, we should form their difference and integrate over the region consisting of the source in the $\vec{r}'$ point (Fig. 3.4). As a result, we come to a rather important relation in electrodynamics, which represents the required field at exactly the same $\vec{r}'$ point:

$$u(\vec{r}') = \int F(\vec{r}) G(\vec{r}, \vec{r}') dV + \oint \left(u \frac{\partial G}{\partial n} - G \frac{\partial u}{\partial n} \right) dS. \qquad (3.19)$$

Equation (3.19) is widely used in theoretical, engineering, and computational electrodynamics, and in practice. This is really useful if, for instance, the functions u and G on the S surface simultaneously meet, for example, the zero boundary condition (1.39) (the Dirichlet condition). In this case, the surface integral in Eq. (3.19) disappears, and the required field u is completely defined by the volume integral over the V region. The same result is obtained if the S surface is moved to the infinite distance from the region, which is occupied by sources (owing to radiation condition fulfillment). When we wrote Eq. (3.19), we considered that sources (the $F(\vec{r})$ function) are continuously distributed in the V_1 region. If the field source represents the formation of the separate (discrete, point) sources, then the volume integral in Eq. (3.19) (for $\oint \ldots dS = 0$) should be replaced by the sum distributed onto all point field sources.

Hence, Eq. (3.19) represents the formulation of the *superposition (imposition) principle* of the specific solutions of Eqs. (3.12 and 3.13). The conclusion is evident: knowing the Green's function $G(\vec{r}, \vec{r}')$, it is easy to construct the required solution of Eq. (3.19). However, the thing is that the Green's function is usually unknown for the given specific problem, and in this case, Eq. (3.19) cannot help us to achieve the required solution.

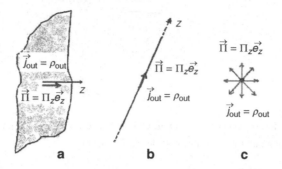

Fig. 3.5 Classification of "ideal" source distribution for the determination of the Green's function. (**a**) The plane. (**b**) The wire. (**c**) The point source

In such a way, the Green's function $G(\vec{r} - \vec{r}')$ is the same auxiliary function, which is suitable, as we see in the future, for the examination of specific tasks, for instance, when reducing the boundary problem to the one type of *integral equations*, for transition to *variational principles* and many others [18, 19, 54, 75].

3.4 The Green's Function and Model Problems in Electrodynamics

Usually, the infinite uniformly charged plane, the infinite length wire (charged or with a current), or the point source (Fig. 3.5) are used as the models of field sources. Accordingly, we should use one-dimensional, two-dimensional, or three-dimensional geometry. In statics, this is the charged plane or a wire, a point charge (or the dipole); in dynamics, this is the plane and the wire with the direct (DC) or alternating (AC) current, in addition to the dipole. At a high enough frequency, in the free space (or, more precisely, in a vacuum: $\varepsilon = \mu = 1$) for a large distance from the source ($kr \gg 1$) and at fulfillment of the radiation condition of the Green's function, we obtain:

$$
\begin{aligned}
G(z, z') &\approx \ldots e^{ik|z - z'|}, \\
G(\vec{r}, \vec{r}') &\approx \ldots \frac{1}{\sqrt{k|\vec{r} - \vec{r}'|}} e^{ik|\vec{r} - \vec{r}'|}, \\
G(\vec{r} - \vec{r}') &\approx \ldots \frac{1}{k|\vec{r} - \vec{r}'|} e^{ik|\vec{r} - \vec{r}'|}.
\end{aligned}
\tag{3.20}
$$

Equations (3.20) accordingly represent the plane (one-dimension $\vec{r} \equiv z\vec{z}^0$, $\vec{z}^0$ is the unit vector of the z-axis), cylindrical (two-dimensional; $\vec{r} = \vec{r}(r, \varphi)$) and spherical (three-dimensional; $\vec{r} = \vec{r}(r, \varphi, \vartheta)$) waves (processes). These situations are conditionally shown in Fig. 3.5. It is obvious that if we are speaking about the uniform homogeneous medium with the refraction index $n = \sqrt{\varepsilon\mu}$, then in Eq. (3.20), the wave number k must be replaced by $k\sqrt{\varepsilon\mu}$.

Of course, the cases of the known Green's function are not limited by Eq. (3.20). These are rather functions for the free space, which are in fact widely used in the theory and practice of radio wave propagation and for calculations of the antenna radiation fields. Really, in the simplest case, when the scattering body is absent in the space (for instance, in Fig. 3.4b the V_1 volume is absent), but the field sources $\vec{J}_{\text{out}}, \rho_{\text{out}}$ are situated inside the "transparent" (or partially "transparent") region $S = S_1 + S_2$, then the field at any point will be determined by the volume integral in Eq. (3.19). If the radiating surface represents the segment of the plane, the volume integral in Eq. (3.19) is replaced by the integral over this segment (for example, over the antenna aperture).

And, finally, if the thin wire with a current radiates ($ka \ll 1$, a is the wire diameter), then the volume integral in Eq. (3.19) should be replaced by the linear integral over the wire.

However, application of the Green's function of the type (3.20) in waveguides or cavities leads to a very complicated procedure in the realization and understanding of the physical sense. Running a few steps ahead, we note that in the tasks of waveguide and cavity excitation by outside sources $\vec{J}_{\text{out}}, \rho_{\text{out}}$, *Kisun'ko* (and independently *Feld*) offered [118] to use the so-called *contra-directional wave* (the *Kisun'ko–Feld* formula, Sect. 12.1) as the Green's function. The required amplitude of the wave, which propagates from the region occupied by sources $\vec{J}_{out}, \rho_{out}$ in the waveguide (the cavity), can be written in the form of the integral (as they say, reduced to quadratures) over the volume (the surface segment or the surface of this wire), which is occupied by sources $\vec{J}_{out}, \rho_{out}$. In the future description, we will become acquainted with the *Kisun'ko–Feld formula* and its applications in more detail.

3.5 The Huygens–Fresnel–Kirchhoff Principle in the Theory of Electromagnetic Wave Diffraction

Equation (3.19) obtained in Sect. 3.3 has great significance for understanding wave processes and for the calculation of diffraction fields from various obstructions, holes in shields, antenna apertures etc. Usually, in Eq. (3.19) the integral is retained over the S surface only, and the spherical wave is chosen as the Green's function (the third line in Eq. (3.19)) $G(r) = \exp \{ikr\}/r$. This reflects the sense of the *Huygens–Fresnel–Kirchhoff* principle, according to which the wave propagation is caused by the action of the secondary sources. Say, if the source region V_1 consists of one point source, then each point in the simplest case on the spherical S surface (or on the cylindrical surface from the hole in the shield as in Fig. 2.10a, b) is the secondary source, which radiates the elementary spherical wave. The enveloping of these elementary spherical waves is the sphere as well. And then, in the simplest scalar case, the *Huygens–Fresnel–Kirchhoff* principle, with accuracy up to an insignificant constant coefficient, can be written as

$$u(r') = \oint_S \left[\frac{\partial}{\partial n} \left(\frac{\exp\{ikr\}}{r} \right) u_S - \left(\frac{\exp\{ikr\}}{r} \right) \frac{\partial u_S}{\partial n} \right] dS. \tag{3.21}$$

Thus, Eq. (3.21) represents the required field value in the r' point (outside the S surface in Fig. 3.4) through the field value and its normal derivative on the S surface.

It is evident that if the exact true field values and its derivative on the S surface are known, Eq. (3.21) gives the exact field values at any point in the space (outside S). However, in practice, the exact field values on the S surface are unknown, and we must replace them with some approximate values. For instance, to replace them with the field of the incident wave, as in Fig. 2.9a, b. Thus, the field in the far zone ($z > 0$, $kz \gg 1$) is defined accurately enough.

In addition, when should we know the field near the body surface (in the near zone)? In the overwhelming majority of books on antenna devices, the clear connection between the conductance current and the radiation field $\vec{E}, \vec{H}$ intensity is absent, which leads in some cases to incorrect physical results. The details of the true approach should be studied according to publications by Prof. *V.A. Neganov* and his colleagues [17–19].

Equation (3.19) can be generalized when the point source is not the elementary radiator, but some source with its directivity pattern.

3.6 Diffraction and the Parabolic Equation Method

3.6.1 Concept of Wave Diffraction

Any deviation of the propagating wave from the laws of *geometric optics* is understood (in the very wide sense) as the wave diffraction phenomenon. The simple example of wave i diffraction on the infinite slot is shown in Fig. 2.9a, b. At the bottom, this scheme presents the transformation of the plane wave i into cylindrical waves in the front ($z > 0$) or back ($z < 0$) half-spaces. In the case of a hole, the spherical waves radiate from it.

In practice, free radio paths are relatively rare. Usually, radio waves, during propagation, encounter some obstacles. The radio wave, which "strikes" against an obstacle during its propagation in the uniform medium, changes its amplitude and phase and penetrates the *shadow* region, as if deviating from the rectilinear path. This phenomenon is called radio wave *diffraction*. As an example, Fig. 3.6 shows a picture of wave incidence onto the half-plane (Fig. 3.6a) and on the hole in the infinite nontransparent shield (Fig. 3.6b). Such a clear picture of the wave diffraction on the obstacle was offered by Russian scientist *G.D. Maliuzhinets* [68]. It is important that after the obstacle, the field in the illuminated region gradually comes into the shadow region (Fig. 3.6a).

Fig. 3.6 Radio wave diffraction on the obstacle [68]. (**a**) On the half-plane. (**b**) On the hole (slot) in the plane shield. (**c**) Geometric–optical representation of the wave incidence on the half-plane

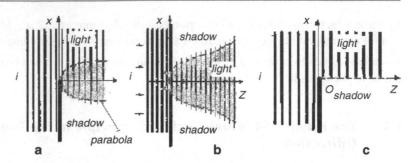

This transition region is called the *diffusion region* or the *half-shadow region*. The yOz plane in Fig. 3.6 is the light–shadow boundary in the *geometric–optical* picture ($\lambda \to 0$) of this phenomenon, which is clearly seen from Fig. 3.6c. From analysis of Fig. 3.6a, b, we see that owing to diffraction, at the edge of the half-plane, the wave really penetrates into the shadow region, i.e., as if it turns around the obstacle. In reality, obstacles may have an arbitrary form and be both nontransparent and half-transparent for radio waves.

Thus, taking diffraction into consideration, we see an infringement of light propagation straightforwardness, i.e., there is a deviation from the *geometric optics* laws (Fig. 3.6a, b; compare Fig. 3.6c).

Determination of the field behind the obstacle can be performed according to Eq. (3.21) or more generally, to Eq. (3.19). In Eq. (3.21), the Green's function of the free space is used in the form of the spherical wave (third line in Eq. (3.20)). Such a procedure is well enough worked through, for instance, in the antenna theory, and it is effective when there are no scattering bodies near the antenna (see Fig. 2.8b; the V_2 volume). In the whelming majority of practical tasks, the Green's function is unknown, and Eqs. (3.19 and 3.18) cannot help us in the task solution. The Kisun'ko–Feld method is an exclusion in the theory of waveguide (cavity) excitation, when the eigenwave of the waveguide (cavity) that is known in advance is used as the Green's function (see Chap. 12). Application of the Green's function as the point source function for the waveguide cavity tasks leads to some difficulties [133].

The way out was found at attentive analysis of the physical picture of electromagnetic field behavior near the obstacle. This peculiarity consists in the sharply different velocity of field variation along the transverse (x) and longitudinal (z) directions. It is obvious that the field along (z) varies more than that along x, i.e., $\partial/\partial z \gg \partial/\partial x$. This allows transfer from the wave equation of type (3.1), (3.2) to the *heat conductivity equation*:

$$\partial \vec{E} / \partial z = (i\lambda/4\pi)\partial^2 \vec{E} / \partial x^2. \tag{3.22}$$

Then, there are two possibilities: for Eq. (3.22) we can write the *Green's function* in the form of the Fresnel integral or solve it numerically, for example using the *finite difference method*. Taking into consideration the general direction of our lecture course, we pause at the numerical solution of Eq. (3.8), allowing its effective implementation into the MathCAD medium, as was offered in Kaganov [3]. There are excellent books on MathCAD [56, 129].

3.6.2   Diffraction on the Half-Plane

At the finite-difference approach, the Eq. (3.22) has the form

$$E_{j+1,m} = E_{j,m} + \text{Im}\lfloor iR\left(E_{j,m-1} - 2E_{j,m} + E_{j,m+1}\right)\rfloor, \tag{3.23}$$

where $R = \frac{1}{4\pi}\frac{M^2}{N}\frac{\lambda}{L}\frac{S}{L}$ is the "step" parameter providing a stable solution at $R \leq 1/2$; N is a number of sample points in the coordinate z, M is the same as the coordinate x, $S = \Delta z\, N$ is the path length in coordinate z, $L = \Delta x M$ is the same as x.

Diffraction calculation of the plane wave i of unitary amplitude, which falls along the normal on the half-plane (Fig. 3.6a) can be performed according to Program P.3.4, Appendix B (http://extras.springer.com/2019/978-3-319-90847-2). Some results of the calculations are given as well. The field directly in the half-plane continuation (Fig. 3.6a) varies from the unit on the light–shadow boundary down to zero in the deep shadow (compare Fig. 3.6c). As far as removal from the shield occurs, the field amplitude varies from 1 on the diffusion zone boundary (the parabola in Fig. 3.6a) to 0 on the lower diffusion zone boundary. On the light–shadow geometric–optical boundary, the field amplitude is equals to 1/2. Thus, at plane wave diffraction on the half-plane, within the limits of the effective diffusion zone, the smooth transformation of the field "magnitude" from 1 (the light region) to 0 (the shadow) occurs. We would like to remind to the reader that we deal with the plane wave, which propagates along the z-axis without variation of the initial amplitude outside (above) the effective diffusion zone. It is clear that the picture is different if the cylindrical (spherical) wave falls on the obstacle.

The physical side of the task, according to data (figures) of Program P.3.4, consists in representation of the initial exciting field on the half-plane $z = 0$, $x \in (0, \infty)$, i.e., on the obstacle continuation in the form of the incident field wave i. This is a traditional approach within the framework of the *physical diffraction theory*, which in many practical cases gives acceptable accuracy. In other words, this is an example of diffusion of diffraction field amplitude over the *plane wave front*. More accurate results are obtained when the amplitude diffusion is calculated according to the "natural" front of the cylindrical i (or spherical in the three-dimensional case) wave. In the front half-space ($z > 0$) the interference picture between incident i and scattered wave is observed at the same distance $z = L$ as if it is done by the shield edge by the cylindrical wave. At $x < 0$, there is no such a picture in this approximation.

At the end of this subsection, we note that the diffraction task of the plane wave (equally as for cylindrical or spherical waves) on the ideally conducting half-plane (or on the wedge) has a *strict solution*. For the first time, its solution was obtained by A. *Sommerfeld* in 1905 and later it was repeated many times using other methods, taking into account the very different boundary conditions of the half-plane in addition to an account of the environmental properties. The strict solution allows estimation of various approximate results and, in particular, the data of the *parabolic equation method* used here [57, 58, 71, 76].

3.6.3 Diffraction on the Slot in the Infinite Plane Shield

The electromagnetic field diffraction on the slot in the infinite shield, i.e., on two half-planes (and also on the band, see later) is one of the classic tasks of electrodynamics and optics. Strictly speaking, on the contrast from the diffraction on the half-plane (Scct. 3.6.2), which, as we already mentioned, has a *strict solution*, the diffraction on the slot is concerned with another class; namely, the class of diffraction on the *finite structure* (the slot – Fig. 3.6b, etc.; see also Fig. 2.4). In this case, the solution can be obtained in the approximate form only (though, with any accuracy given in advance; see, for example, Neganov et al. [17–19]). In the parabolic equation approximation considered by us, the field satisfies Eq. (3.22) and zero boundary conditions on both the half-planes spaced by the distance b between them. The relation $\alpha = \lambda/b$ is the main parameter of the task. The electromagnetic field in the space behind the slot ($z > 0$) at the same distance z is determined according to Program P.3.5 (volumetric picture is shown in Fig. 3.7). In this figure, $\alpha = \lambda/b$, b is the slot width.

The results presented are suitable at the same distance $z \gg \lambda$ as the circuit offered as the diffraction field exciter uses the approximate value of the field on the slot, which can be chosen equal to the incident wave field at this place ($z = 0$; Fig. 3.7). In this circuit, we do not take into account the interaction of slot edges with each other, i.e., the field on the slot is far from true. The further we move from the slot, the effective diffusion zone gradually increases, becomes wider, and the diffraction field gradually decreases.

The considered picture of the field scattered on the slot in the shield (Fig. 3.7; or more generally, on the hole) may serve as the peculiar Green's function (see Sect. 3.4) on consideration of processes in the diaphragm transmission line (Fig. 6.6b; the analysis in Sect. 9.3.3).

3.7 Wave Equations of Classic Electrodynamics and Plane Waves: Properties and Characteristics of Simplest Electromagnetic Waves

The plane waves are the simplest solution of the wave equation (or the Helmholtz equation; (3.1 and 3.2) for the infinite uniform linear medium without losses. Relations for the Green's functions in the form of plane, cylindrical, and spherical waves (Eq. 3.20) show that the Green's function itself is a suitable auxiliary model and the waves from Eq. (3.20) are models only, because in reality they do not exist. Indeed, if, say, the outside sources $\vec{j}_{out}, \rho_{out}$, which are uniformly distributed over the whole infinite plane, are the plane wave source, then, evidently, the total power radiated by this plane is infinitely large according to Eq. (1.30). This is impossible with the usual physical considerations because it contradicts the *energy conservation law*. However, the plane, cylindrical, and spherical waves turn out to be extremely convenient as physical and mathematical models. Below we shall see, for instance, that the problem of the incidence of the plane electromagnetic

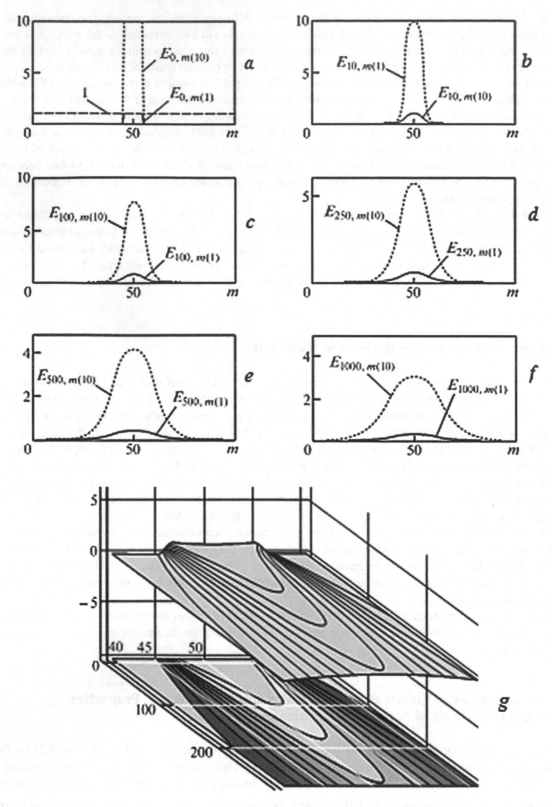

Fig. 3.7 Results of the plane wave diffraction on the slot in the shield for different distances from the slot

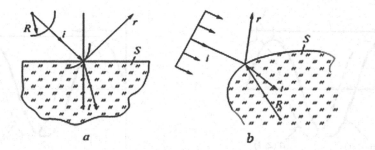

Fig. 3.8 Explanation of the picture of (**a**) the spherical (cylindrical) wave i incidence on the plane boundary S between two media and (**b**) the plane wave incidence i on the spherical (cylindrical) boundary S. In both cases, $kR \gg 1$

wave on the plane boundary of two dielectric media is simple enough. Nevertheless, we cannot consider the task concerning the cylindrical or spherical wave incidence on the same boundary as the simplest task without sufficient substantiation (Fig. 3.8a).

However, on large removal from the region occupied by the field sources $\vec{j}_{\text{out}}, \rho_{\text{out}}$, which are usually in the restricted volume and at a finite distance, though, maybe, at a large enough distance, different simplifications are possible. If so, in the vicinity of the incidence point of each beam, for example, of the cylindrical wave i (Fig. 3.8a) onto the plane boundary S at rather high $R(kR \gg 1)$, we can consider with rather high accuracy (for instance, as $1/kR$) the wave front as a plane in the place of incidence. Similarly, if the plane wave (i) falls onto the nonplane surface (Fig. 3.8b) and the curvature radius of the R surface at the incidence point is of a large value($kR \gg 1$), then the wave reflection at this point is the same as for the plane obstacle surface. Sometimes, the plane wave is called the *transverse electromagnetic wave* and it can be designated as TEM or the T-wave: vectors $\vec{E}$ and $\vec{H}$ are perpendicular and lie in the plane normally located in the direction z of propagation.

Maybe, we should remember here about *sound (acoustic, scalar)* waves, the waves on the water (liquid) surface and, *elastic waves*. Let it be, for example, that two-dimensional ($\partial/\partial y \equiv 0$) sound waves are propagated along the z-axis, and the x-axis is the transverse coordinate. They relate to displacement in the space of regions of gas (air) compression and depression. Each separate air particle oscillates along the z-axis direction and these waves can be referred to as *longitudinal*. We know the longitudinal waves in waveguides (see Chap. 6), in plasma, and in other media. In recent years, *longitudinal electromagnetic waves* have been intensively studied, and they have already found wide application, i.e., waves with explicit longitudinal components of the electric field (see Sect. 4.2) [12, 43, 44].

Explanation of the overwhelming majority of known electromagnetic processes and phenomena is related precisely to the *longitudinal waves*.

Propagation of the plane monochromic wave in any infinite uniform medium without losses can be represented, in the classic variant of the theory, in the form of the function of variables z, t as $a(z, t) = A_m \exp\{i(kz - \omega t)\}$ or, taken the real part from the complex function, as $a(z, t) = A_m \cos(kz - \omega t)$ (Fig. 3.9). Variation of the wave position during some time intervals from the value $a(0, t)$ to $a(z, t)$ is shown in Fig. 3.9a, and with respect to propagation direction from the value $a(z, 0)$ to $a(z, t)$ in Fig. 3.9b. The main characteristics of the wave process, (T is the *oscillation period* and λ is the wavelength corresponding to this period T, are also shown in Fig. 3.9. The plane $x0y$, which is perpendicular to the wave propagation direction z, is called the *wave front* or the *plane of equal phases*. Evidently, at each time moment t, at each point of the z-axis, the wave amplitude has the same value; this is the *plane of equal amplitudes*. We see that in the plane wave propagating in the infinite space without losses, the plane of equivalent phases and the plane of equivalent amplitudes coincide. We will become acquainted with these phenomena in Sect. 4.2.

The solution analysis of wave equation (3.17) in the form $F(x) = \exp\{-x^2/4\}$ by $U_{p,n}(x, t) = U_{p,n}(A, x, v, t, F) = AF$ ($x \mp t$) type was performed according to Program P.3.6 (see Fig. 3.9c–f).

The wave front of the plane wave moves along the z-axis direction with the phase velocity v_{ph}, which is defined as

$$v_{\text{ph}} = dz/dt = \lambda f, \quad f = \omega/2\pi. \tag{3.24}$$

Thus, let the propagation direction of the plane wave coincide with the z-axis of the rectangular coordinate system ($\vec{r} \equiv x, y, z$). The total field of the plane wave is located in the plane $x0y$, perpendicular to the propagation direction z. Therefore, the plane wave field has the components E_x, H_y only, which are nonzero and constant over the whole plane $x0y$. The

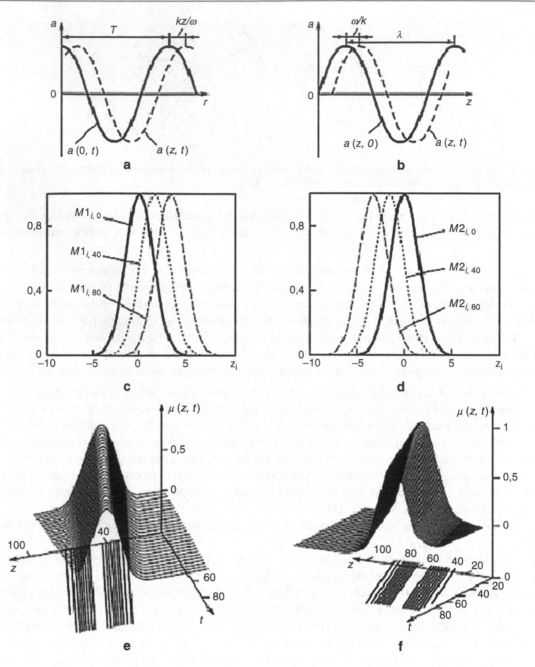

Fig. 3.9 Instantaneous pictures of field distribution of the plane electromagnetic wave in (**a**) time; (**b**) space. Field pictures for the wave propagating along the (**c, e**) positive and (**d, f**) negative directions of the x-axis

longitudinal components are absent:$E_z = H_z \equiv 0$. Thus, the following conditions are fulfilled for the field of the plane's uniform wave, $\partial/\partial x \equiv 0$, $\partial/\partial y \equiv 0$, and hence, the *homogeneous Helmholtz equation* (2.1) for the infinite medium is:

$$d^2 E_x/dz^2 + k^2 \varepsilon \mu E_x = 0. \tag{3.25}$$

Equation (3.25) has the solution

$$E_x = A e^{ik\sqrt{\varepsilon\mu}z} + B e^{-ik\sqrt{\varepsilon\mu}z}. \tag{3.26}$$

In solution (3.26), amplitudes A, B are arbitrary constants that depend on the excitation system, boundary conditions, the medium condition (the absorption presence, obstacles, which do not change; however, the wave character, i.e., which does not transform it, for example, into the cylindrical, spherical or other wave).

Thus, the solution of the homogeneous one-dimensional Helmholtz equation (3.25) is presented in the form of the superposition of two "ideal" plane waves (3.26). The first one is the *forward wave* propagating in the direction of the positive values of the longitudinal coordinate z. The second one propagates in the opposite direction $(-z)$. For this reason, it is sometimes called the *backward wave*, which is not entirely precise (see Chap. 7, Sect. 7.3.7). Apparently, it is more appropriate to call it the *head wave*.

The similar orientation and the parallelism of the $\vec{k}$ (or $\vec{k}\sqrt{\varepsilon\mu}$) wave vector and the Umov–Poynting vector $\vec{U}\ (\vec{r}, t)$ $= [\vec{E}\ \vec{H}]$ (1.37) (see also (1.36)) are significant characteristics of these waves; they are parallel (collinear).

We discuss the definitions and properties of forward and backward waves in Sect. 4.2.

It is important to note that both plane waves from (3.26) propagate with the same speed (compare Sect. 4.2). In Eq. (3.26) we omitted the time factor $\exp\{-i\omega t\}$.. If we add it into the exponent, say, of the forward wave (the first term in (3.26)), the phase factor will have the form $\exp\{i(k\sqrt{\varepsilon\mu}z - \omega t\} \equiv \exp\{i(\gamma z - \omega t\}$. The phase $(v = dz/dt)$ and the *group velocities* of the wave with ω frequency are:

$$v_{ph} = (\gamma/\omega)^{-1} = \omega/k\sqrt{\varepsilon\mu} = c/\sqrt{\varepsilon\mu}, \quad v_{gr} = (d\gamma/d\omega)^{-1}. \tag{3.27}$$

Here c is the light speed in the vacuum, equaling approximately 3.10^8 m/s, $\gamma = \sqrt{\varepsilon\mu}$ is the constant of wave propagation; for waves in media (and in waveguides; see later Chaps. 7, 8, and 9) without losses γ is the real value.

From Maxwell's equations in the complex form (2.37) and from the solution of Eq. (3.25) in the form (3.26), it follows that the second wave field component is:

$$H_y = A\sqrt{\varepsilon/\mu}e^{ik\sqrt{\varepsilon\mu}z} - B\sqrt{\varepsilon/\mu}e^{-ik\sqrt{\varepsilon\mu}z}. \tag{3.28}$$

From plots in Fig. 3.10, we see that these waves are *transverse*: $\vec{E}$ and $\vec{H}$ vectors are always perpendicular. The plane of equal phase and the plane of equal amplitudes represent the *parallel* planes, which are normal for the wave propagation in direction z. Losses in the medium are defined, in general, by its conductivity, so that $\varepsilon = \varepsilon(1 - i\tan\delta)$, where $\tan\delta = \sigma/(\omega\varepsilon)$.

From functions presented in Fig. 3.10a, b, we clearly see the contradiction of the classic theory with usual physical considerations; namely, at some time moments (Fig. 3.10a) or at some distances (Fig. 3.10b) the electromagnetic field becomes equal to zero, and disappears, which does not correspond to views on material life. It would seem that the presence of losses in the propagation medium (Fig. 3.10c, d) saves a situation: here, the field will be nonzero neither in time nor in space, but it propagates with the intensity decrease. Earlier (Sect. 2.5) we saw that for the proof of the unicity theorem, we required

Fig. 3.10 Classic pictures of plane wave propagation. (**a**, **b**) In the medium without losses. (**c**, **d**) In the lossy medium

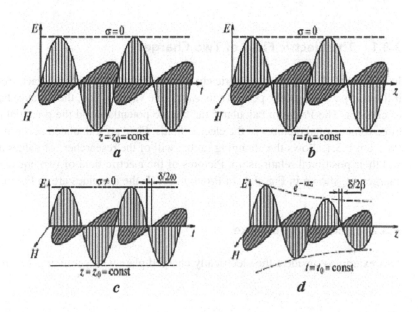

the presence of some, even arbitrarily small losses, and then, based on the analytical properties of the wave equation solution with respect to the wave number $k = k' + ik''$ (or $k\sqrt{\varepsilon\mu}$), it was stated that at $k'' \to 0$ the solution remains true. As we see it, this is wrong; thus, there is the above-mentioned contradiction in the classic theory. The way out of this contradiction, as we shall show in Chap. 4, is given within the limits of macroscopic generalized electrodynamics, when besides "traditional" transverse waves, we take into consideration the longitudinal electromagnetic waves, and the contradiction is eliminated even in the "ideal" case for $k'' = 0$.

Let us look at the solution of the Helmholtz equation (3.25) in detail. Owing to the loss in the wave propagation medium, its propagation constant $\tilde{k} = \omega\sqrt{\varepsilon\mu}$ is the complex quantity: $\tilde{k} = \mathrm{Re}\,\tilde{k} + i \cdot \mathrm{Jm}\,\tilde{k}$ or $\mathrm{Re}\,\tilde{k} + i \cdot \mathrm{Jm}\,\tilde{k} = \omega \cdot \sqrt{\varepsilon\mu(1 - i \cdot \tan\delta)}$. Raising to the second power both parts of this expression and separating real and imaginary parts, we obtain the following system of two algebraic equations:

$$\left(\mathrm{Re}\tilde{k}\right)^2 - \left(\mathrm{Jm}\tilde{k}\right)^2 = \omega^2\varepsilon\mu, \quad 2\left(\mathrm{Re}\tilde{k}\right) \cdot \left(\mathrm{Jm}\tilde{k}\right) = -\omega^2\varepsilon\mu\tan\delta. \tag{3.29}$$

From Eq. (3.29) it directly follows that $2\left(\mathrm{Re}\tilde{k}\right)^2 = \omega^2\varepsilon\mu\left(1 \pm \sqrt{1 + \tan^2\delta}\right)$, and because $\left(\mathrm{Re}\tilde{k}\right)^2 > 0$, we must choose in Eq. (3.29) the "+" sign. If we introduce designations

$$\alpha = \omega\sqrt{(\varepsilon\mu/2)\left(\sqrt{1 \pm \tan^2\delta} - 1\right)}, \tag{3.30}$$

$$\beta = \omega\sqrt{(\varepsilon\mu/2)\left(\sqrt{1 \pm \tan^2\delta} + 1\right)}. \tag{3.31}$$

Thus, $\mathrm{Re}\tilde{k} = \pm\beta$, $\mathrm{Jm}\tilde{k} = \pm\alpha$. Introduced quantities α and β are coefficients of attenuation and the phase variations.

Here, we implicitly pause at the positive value of the root in Eqs. (3.30) and (3.31). However, on transition to the account of longitudinal electromagnetic waves (Sect. 4.2), it is necessary to use both signs of the root.

3.8 Examples of MathCAD Packet Utilization

In the present chapter, several examples of utilization of MathCAD programs for electrodynamics tasks are examined: the electric field of two charges, the electric field of the charged plate, the electric field of the plane capacitor, the diffraction problem on the slot, examination of the wave equation, and an analysis of the forward and backward waves. These examples of computer program utilization are sequentially considered in Appendix B (http://extras.springer.com/2019/978-3-319-90847-2).

3.8.1 The Electric Field of Two Charges

In Program P.3.1 two point electric charges located at the l distance each from other are examined. In this task, we define the following parameters: the permittivity ε_0; charge values q_1, q_2; the distance between charges l; and the geometrical coordinates of charges. The Program calculates the charge potentials and the potential difference, and determines two-dimensional and three-dimensional pictures of the electric fields. Field are calculated according to Program P.3.1 for their potentials and fields. Program P.3.1 allows the changing (at the will of the researcher) of values and signs of charges, the distance between them, and their positional relationship. Pictures of the electric field of two equal-sign (Fig. 3.1b) and two opposite-sign (Fig. 3.1) charges are shown in Fig. 3.1. In Program P.3.1, the ε_0 values are in Far/m, $q_{1,\,2}$ are in Q, and L is in m.

3.8.2 The Charged Plate

This example considers the electrically charged plate with geometry shown in Fig. 3.11.

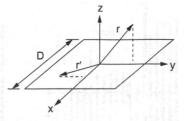

Fig. 3.11 Geometry and coordinate system of the charged plate

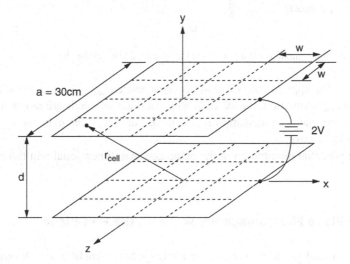

Fig. 3.12 Geometry of the plane capacitor

The Program P.3.2 calculates the electric field components, the electric field distribution above the plate surface, and constructs the two-dimensional and three-dimensional field pictures.

The electric field is defined by expression (compare with Eq. (3.15)):

$$\vec{E}\left(\vec{r}\right) = \frac{\rho_s}{4\pi\varepsilon_0} \int \frac{\vec{r} - \overleftarrow{r'}}{\left|\vec{r} - \overleftarrow{r'}\right|} ds', \tag{3.16}$$

in which the $\vec{r}, \vec{r}'$ is the observation point ($\vec{r} = \vec{e}_y y + \vec{e}_z z$) and the point on the plate surface is $\vec{r}' = \vec{e}_x x + \vec{e}_y y$; values ε_0 and ρ_s are determined in Program P.3.2.

Computation of the integral over the plate surface in (3.16) for y- and z_components of the electric field is fulfilled according to Program P.3.2, allowing determination of the field in the space surrounding the plate. Figure 3.2 shows some results. Using Program P.3.2, we may build the field picture in any sections and have a complete representation of the electric field of the uniformly charged square plate in the surrounding space.

Program P.3.2 allows the representation of the calculations and results for any set of sizes and charge values, which gives enough adequate picture of the active physics of the element under investigation.

3.8.3 The Electric Field of the Plane Capacitor

The general view of the capacitor with plane square plates is presented in Fig. 3.12, where its sizes are shown and the plate fragmentation into partial regions for the application of the method of moments.

Program P.3.3 calculates the distribution of the surface electric charge on plates for different distances (in table form). After that, the field potential is calculated and the electric field intensity between plates. Then, we determine the total charge and the

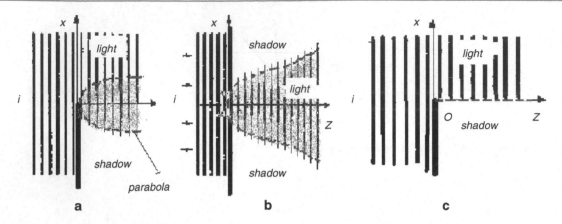

Fig. 3.13 Diffraction of the plane electromagnetic wave on (**a**, **c**) the half-plane and (**b**) on the slot

value of the electric capacitance. The approximate value of the capacitance C_{approx} is obtained without taking into account the edge effects, which are the best approximation, if plate fragmentation sizes into cells are rather small compared with the plate sizes. However, if the sizes of the plate fragmentation into cells are not small enough, the approximate capacitance value C_{approx} can be obtained with a large error.

After that, we calculate the potential distribution in the plane and two-dimensional and three-dimensional pictures of the potential are constructed.

3.8.4   Diffraction of the Plane Electromagnetic Wave on the Half-Plane

This example (Program P.3.4) considers the passage of the plane electromagnetic wave through the segment with vertical shield (Fig. 3.13). As a result, diffraction of the half-plane and diffraction distortion of the wave field occur.

The Program defines the initial data, forms the equation in the finite differences, and solves it. As a result, the three-dimensional picture of field distribution behind the shield is formed in addition to some two-dimensional pictures of dependences on the task parameters.

The physical side of the task, according to data (figures) of Program P.3.4, consists in representation of the initial exciting field on the half-plane $z = 0$, $x \in (0, \infty)$, i.e., on the obstacle continuation in the form of the incident field wave i. This is a traditional approach within the framework of the *physical diffraction theory*, which in many practical cases gives acceptable accuracy. In other words, this is an example of diffusion of the diffraction field amplitude over the *plane wave front*. More accurate results are obtained when the amplitude diffusion is calculated according to the "natural" front of the cylindrical i (or spherical in the three-dimensional case) wave. In the front half-space ($z > 0$) the interference picture between incident i and the scattered wave is observed at the same distance $z = L$, as if it is done from the shield edge by the cylindrical wave. At $x < 0$, there is no such picture in this approximation.

3.8.5   Diffraction of the Plane Electromagnetic Wave on the Slot in the Plane Shield

This example (P.3.5) examines the diffraction of the electromagnetic field on the slot in the plane shield (Fig. 3.14).

The Program P.3.5 defines the initial data, introduces the task smallness parameter, and determines the calculation scheme. As a result, two- and three-dimensional pictures are constructed.

The results presented are suitable the same distance $z \gg \lambda$ because the scheme offered as the diffraction field exciter uses the approximate value of the field on the slot, which can be chosen equal to the incident wave field in this place ($z = 0$; Fig. 3.7). In this scheme, we do not take into account the interaction of slot edges with each other, i.e., the field on the slot is far from true. The further we remove from the slot, the effective diffusion zone gradually increases, becomes wider, and the diffraction field gradually decreases.

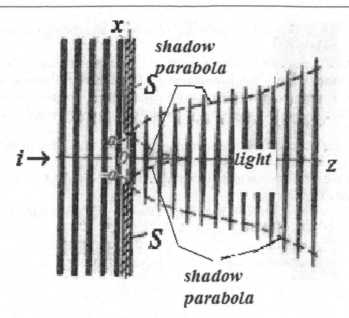

Fig. 3.14 Diffraction of the plane wave on the slot

3.8.6 Wave Equation, Direct and Backward Waves

In the last example of this chapter (P3.6), the total wave equation is analyzed and the parameters of the forward and backward waves are calculated. Two- and three-dimensional pictures are constructed.

3.9 Summary

In this chapter we have considered the rather wide and important issue for general electrodynamics of obtaining the classic wave equations and shown a relation between these equations and acoustic wave equations. Such a connection is important in itself, but it essentially becomes apparent in generalized electrodynamics (Chap. 4). A further step consisted in the introduction of vector and scalar potentials. For elimination of the ambiguity of these potentials, we introduced the Coulomb and Lorentz calibration conditions. These calibration conditions are not required in generalized electrodynamics and significantly widen the activity region of modified Maxwell's equations (Chap. 4). In addition, the clear physical sense is emphasized to the vector potential. Suggested pictures of potential distribution, to a great extent, simplify for the reader the process of acquaintance with the field distributions, for example, in the waveguide transverse sections (Sects. 6.4 and 6.5). It also simplifies the understanding of the excitation essence for waveguides and resonators (Chap. 6).

The Green's function is a widely used concept in physics as a whole and in electromagnetism in particular. Its modification (in the sense of the head wave) finds wide application in practice and in the theory of guiding and resonance structures (the Kisun'ko–Feld theory; Sect. 12.1.2, Chap. 12).

The parabolic equation method is widely used in the physical theory of diffraction, for instance, in the theory of open resonators (Chap. 11).

Checking Questions

1. For which goals do we introduce the electrodynamics potentials $\varphi(\vec{r}, t)$, $\vec{A}(\vec{r}, t)$ and what advantages do they provide?

2. For what reason do we introduce the Lorentz calibration of potentials $\varphi(\vec{r})$, $\vec{A}(\vec{r})$, which has the form $\varepsilon\mu/c^2 \partial\varphi/\partial t + \text{div } \vec{A} = 0$?

3. Why do we introduce the concept of the Green's function?

4. Which physical sense is in the Huygens–Fresnel–Kirchhoff principle and how is it used in the theory of electromagnetic wave diffraction?

5. The wave phase velocity varies according the law $v = v_0\sqrt{1 - \alpha/\omega^2}$. Values of v_0, α do not depend on the frequency ω. How high will the group velocity be?

6. How do we write the vector and scalar D'Alembert's equations for vector and scalar potentials?

7. Using Program P.3.1, calculate the two- and three-dimensional pictures of the electric field of two coupled electric charges located at distance $d = 0.1$ and 0.8 m. Compare the results. Give a qualitative explanation.

8. According to Program P.3.2, construct the field picture of the charged plate, which represents the "gold" rectangle in side sizes $D \times 1,618D$.

9. Using Program P.3.3, calculate the two- and three-dimensional pictures of the electric field of the plane capacitor for $a = 0.15$ and 0.8 cm. Compare the results. Give a qualitative explanation.

10. Using Program P.3.4, examine the diffraction on the half-plane at $R = 0.05$ and 0.3. Construct two- and three-dimensional field pictures. Give physical explanations.

11. According to Program P.3.5, construct the field picture from the rectangular slot with sizes of the "gold" rectangle: $a/b = \Phi = 1,618$, and $a/b = \Phi^2 = 2,618$.

In Chap. 4 of our lecture course we shall become acquainted with the wave equations of macroscopic generalized electrodynamics (Sect. 4.1) and we show their distinctive features compared with the classic case (described in Chap. 3). The consequence of the Stokes–Helmholtz theorem (Sect. 1.3.1) about the fact that an electric field has the vortex and potential components is a significant feature of wave equation formation in generalized electrodynamics. This allows (as was done in Sect. 3.2) equations to obtained for the vector and scalar potentials. At this, the absence of a necessity to use the calibration relationships of Lorentz and Coulomb, respectively, is the essential feature. In this case, the one pair of wave equations obtained defines the transverse electromagnetic waves, whereas the second one defines the longitudinal waves.

The transverse electromagnetic waves are sufficiently well-studied to date, but the longitudinal are far less well-known. Section 4.2 is devoted to the longitudinal class of electromagnetic waves. Nevertheless, we note that longitudinal waves were known earlier, for example, in the solid bodies, in an ionosphere, etc. The generalized theory (Chap. 1) allows explanation of the so-called "zero paradox" at electromagnetic wave propagation, when we need not to introduce "forcedly" a concept of losses at each point of the space. Simultaneously, we show a mechanism of arising and propagation of longitudinal waves on the example of a finite conductor with a current. As a result, we offer a complete pattern of transverse and longitudinal wave propagation in the free space. Separately, we analyze a behavior and the characteristics of electromagnetic waves in the dielectric (Sect. 4.3) and in the lossy medium (Sect. 4.4).

Some of the consequences of generalized macroscopic electrodynamics are included in the special section of this Chapter (Sect. 4.5). It is necessary to do that to emphasize a connection between the generalized theory of electrodynamics and the other disciplines of the physical profile (Sect. 4.5.1). We describe a large and long-standing interest of researchers in the link between electrodynamics and, for instance, mechanics. Moreover, we note that all physical interactions are certainly performed taking into account the participation of a continuum.

We begin our examination from the interaction of an electromagnetic field and the charged particle (Sect. 4.5.2). A phenomenon of the vortex-free electromagnetic induction is shown and proved in experiments. This phenomenon states that the nonstationary scalar magnetic field gives rise to the potential electric field. It is significant that an electric field of the moving charge has a complicated configuration in the reference system. This field represents a superposition of the ellipsoidal Heaviside field and the field of the electric dipole. As a result of this analysis, we can consider the fact that a magnetic field arises at the movement of the charged particle with respect to the vacuum medium, and the magnetic field of the single (solitary) charge has two components: vortex (vector) and potential (scalar).

A connection between a charge and a mass of the charged particle is described in Sect. 4.5.3. At this, the electron model at assumption that it is formed by the process happening in the medium seems to us as important. In other words, an electron is the production of the constitutive vacuum medium, and it is always inside this medium and connected with it inseparably. The suggested model of the solitary electron allows adequate description of its motion in the physical vacuum medium and statement of electromechanical similarities.

The generalized electrodynamics theory allows the connection between the inertia and the gravitation to be set (Sect. 4.5.4). Phenomena of the inertia and the gravitation can be uniformly explained by interaction of charged particles with the physical vacuum at their relative accelerated motion.

The famous experiments of *N. Tesla* have provoked great interest. The one variant of the explanation of these experiments, from the point of view of generalized electrodynamics, is presented in Sect. 4.5.5. For this, in particular, it turns out to be necessary to reveal the principle of the Tesla coil action (Sect. 4.5.6) and to determine the *main resonance frequencies* of the

© Springer International Publishing AG, part of Springer Nature 2019

E. I. Nefyodov, S. M. Smolskiy, *Electromagnetic Fields and Waves*, Textbooks in Telecommunication Engineering,

https://doi.org/10.1007/978-3-319-90847-2_4

Tesla set-up (Sect. 4.5.7). We show that the crucial point in the Tesla experiments is the energy transmission between two ball antennas with the help of energy scalar waves, and this process has a *resonance character*. This fundamental conclusion could be made only within the limits of generalized macroscopic electrodynamics.

The last section on this chapter is devoted to an explanation of longitudinal wave properties from the point of view of microscopic theory.

4.1   Wave Equations

In the heading of this chapter, we emphasize that the material described relates to macroscopic electrodynamics. We once more try to draw the reader's attention to the fact that the theory discussed has some restrictions "at entrance." At the end of this chapter (see Sect. 4.6), we describe the more general field properties, the features of macroscopic theory, and its main results.

Earlier (see Sect. 3.2), we introduced the definition of the electric field intensity $\vec{E}\,(\vec{r},t)$ through potentials $\vec{A}$ and φ in the form (3.5):

$$\vec{E} = -\operatorname{grad}\varphi - \partial\vec{A}/\partial t. \tag{4.1}$$

Taking into consideration that, in accordance with (1.4), the $\vec{E}\,(\vec{r},t)$ field has vortical and potential components, we write (4.1) as

$$\vec{E}_g = -\operatorname{grad}\varphi - \partial\vec{A}/\partial t, \quad \vec{E}_r = -\partial\vec{A}_r/\partial t. \tag{4.2}$$

Acting on the same scheme as at obtaining of the Poisson (3.11) and the Laplace (3.13) equations, we come to two equations for the scalar and vector potentials

$$\Delta\vec{A} - \varepsilon\mu\,\partial^2\vec{A}/\partial^2 t = -\mu\,\vec{j}\,, \tag{4.3}$$

$$\Delta\varphi - \varepsilon\mu\,\partial^2\varphi/\partial^2 t = -\rho/\varepsilon. \tag{4.4}$$

The significant difference of Eqs. (4.3 and 4.4) is that Poisson equations (more precisely, *D'Alembert's* equations) consists in the disuse of calibrations (at their deriving) of the Coulomb (1.21) and Lorentz (1.22) type.

Wave equations for intensities of electric and magnetic fields can be obtained on the basis of the same scheme as for the classic wave equations (3.1 and 3.2) (see Chap. 3).

After simple but bulky manipulations, we obtain two pairs of wave equations of the general type [12]:

$$\Delta\vec{E}_r - \varepsilon\mu\partial^2\vec{E}_r/\partial t^2 = \mu\partial\vec{j}/\partial t, \tag{4.5}$$

$$\Delta\vec{H} - \varepsilon\mu\partial^2\vec{H}/\partial t^2 = -\operatorname{rot}\vec{j}\,, \tag{4.6}$$

$$\Delta\vec{E}_g - \varepsilon\mu\,\partial^2\vec{E}_g/\partial t^2 = (1/\varepsilon)\operatorname{grad}\rho, \tag{4.7}$$

$$\Delta H^* - \varepsilon\mu\partial^2 H^*/\partial t^2 = \partial\rho/\partial t + \operatorname{div}\vec{j}\,. \tag{4.8}$$

The first pair of wave equations (4.5 and 4.6) defines the classic *transverse* electromagnetic field (*TEM* waves); the second pair (4.7 and 4.8) define the longitudinal waves. They are sometimes called "electric scalar waves." Later, we describe the solutions to the equation obtained and perform a physical analysis (Chap. 5).

Now, we note only two interesting facts. The first one was discovered by Poisson in 1828: two types of waves existed in the elastic solid body: the *transverse* waves with the $\sqrt{n/\rho}$ propagation speed, and the *longitudinal* waves propagating with the $\sqrt{(k^2 + (4/3) \cdot n)/\rho}$ speed, where ρ is a density, n is the rigidity constant, and k is the compression modulus [41].

Another fact: longitudinal electromagnetic waves were known a long time ago. They were discovered during the study of electromagnetic wave propagation in an ionosphere (in plasma) in 1942, where they were called *Alfven waves* (*Hannes Alfvén*) [61].

In recent years, it has become clear, for instance, that there is no evident separation between transverse and *longitudinal* oscillations on the liquid surface. The model experience serves as the obvious proof. The sense is that the small ball – a molecule on the surface of the oscillating liquid, on one side, performs the *transverse* oscillations (together with the surface), but on the other side, it is subject to *rotation* motions, i.e., it performs the proper *longitudinal* oscillations. As a whole, the picture reminds us of the pattern of nontraditional screw flow, whose oscillations are described by the Helmholtz equation.[1]

4.2 Longitudinal Electromagnetic Waves

The presence of two types of waves (transverse and longitudinal) in the mutual electromagnetic field is essential for modern science and engineering, keeping within the limits of the macroscopic theory. Transverse waves are adequately studied in detail and widely used in everyday practice. The situation with longitudinal electromagnetic waves is slightly different: they were known a long time ago (almost 100 years), but they have hardly entered into electrodynamics practice, which is related, in the first place, to insufficient substantiation of its presence. We have already mentioned that the classic potential calibrations of Coulomb and Lorentz exclude the longitudinal electromagnetic waves obtained from the classical Maxwell's equations (1.12, 1.13, 1.14, and 1.15). Moreover, they are known in microscopic theory [5, 6].

Meanwhile, confusion with regard to terminology is possible here. In the guiding structures (waveguides, transmission lines, optical guides, etc., see Chaps. 6, 7, 8, 9, 10, 11, and 12), which are adequately described by the classic Maxwell's equations (1.12, 1.13, 1.14 and 1.15), there are so-called longitudinal waves of *TE* and *TM* types, etc. They have longitudinal magnetic and electric components along the propagation direction. The properties of these waves are well studied, confirmed by numerous experiments, and widely used in practice.

To date, in generalized electrodynamics, as we saw in Chaps. 1 and 2, the new concepts of the scalar magnetic field H^* and the potential electric field $\vec{E}_g$ appear (the latter leads to content change in the concept of the total electric field $\vec{E}$). In particular, the new theory allows explanation of the mechanism of electromagnetic wave propagation itself related to the classic paradox, when the presence of losses in the medium is necessary to explain the wave propagation. In a number of books on electrodynamics, we can see the electromagnetic field picture in the plane wave illustrated in Fig. 4.1.

The presence of some points (marked by "0" in Fig. 4.1), in which the electromagnetic field vanishes, is a paradox in this picture ("zero paradox"). In other words, the electromagnetic field disappears at these points, which is in reality is impossible.

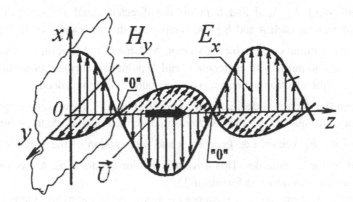

Fig. 4.1 Instantaneous picture of electromagnetic field vector location in the plane wave, which supposedly propagates along the z-axis

---

[1] *Nefyodov E.I., Khublaryan M.G.* Passing of the axis-symmetric screw flow through the channel of given profile (in Russian). Izvestia AS USSR, series Mechan. and Machinery, 1964, No 3, p. 173–176.

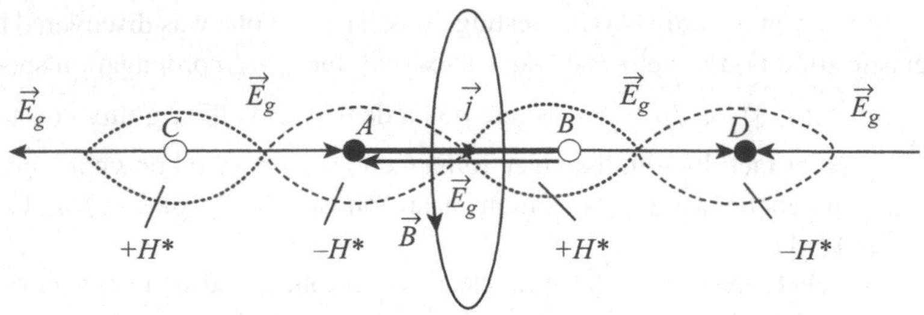

Fig. 4.2 Explanation of the mechanism of longitudinal electromagnetic wave appearance and propagation: the current in the conductor *AB* segment increases

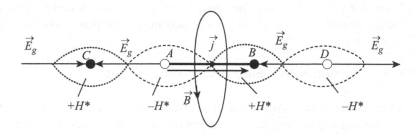

Fig. 4.3 Explanation of the mechanism of longitudinal electromagnetic wave appearance and propagation: the current in the *AB* segment of the conductor decreases

The second side of this paradox in classical interpretation consists in the fact that vectors of the transverse wave (*TEM*) $\vec{E}_r$ and $\vec{B}$ are in-phase. This means that the energy of electric and magnetic fields passes through a maximum and, naturally, *simultaneously* vanish. Thus, no mutual transformation of the vortical electric and magnetic fields takes place, as follows from the main Eqs. (1.12 and 1.13). This is typical both for the temporal (Fig. 3.9) and the spatial (Fig. 3.10) patterns of single plane wave propagation.

The generalized electrodynamics allows elimination of this "long-term" paradox and suggests an explanation for the mechanism of the appearance and propagation of longitudinal electromagnetic waves [12, 41]. Let the magnetic field be created by the rectilinear variable current pulse in a conductor of finite length of the (*AB*)-Hertz vibrator (Fig. 4.2). It is significant that a radiator has a finite length as in this case, it creates the static magnetic field in addition to the vector magnetic field. If this current increases, then the induction of the static magnetic field grows. Let us look at the field near the vibrator ends (*A* and *B*). Near the *A* point, the negative increasing scalar magnetic field arises ($\partial B^*/\partial t < 0$). Thus, at that time, the electric field drain is formed near the *A* point. The positive scalar magnetic field is formed near the *B* point, which increases. Hence, there is an electric field source here. In Fig. 4.2, drains are marked by large black circles, but the sources are marked by white circles. Earlier (see Chap. 1, the text after (1.16)), we mentioned *Lentz's* law. Here, we see its analog, according to which the electric field induced at the *AB* segment is directed against the initial current, and tends to compensate for its growth (Fig. 4.2).

Vectors of potential electric field $\vec{E}_g$ at *A* and *B* points are directed at all sides, i.e., the front of the longitudinal electromagnetic wave propagation near each *A* and *B* point is closed to the spherical one. In Fig. 4.2, at the *A* and *B* points, only $\vec{E}_g$ vectors directed along the $\vec{j}$ initial current line are shown. At some *C* and *D* points, a source and a drain of the electric field are formed respectively. Electric fields with centers at *C* and *D* points are formed with some delay with regard to fields with centers at *A* and *B* points, as the electromagnetic wave propagates with finite speed.

A similar picture arises when the $\vec{j}$ current decreases the modulo (Fig. 4.3). Now, sources and drains change place (with appropriate delay). The main consequence of the mechanism described is the clear picture of electromagnetic wave propagation in the direction of the $\vec{E}_g$ vector, i.e., generation and propagation of the variable electromagnetic field occur, defined by the scalar $B^*(t)$ and vector $\vec{E}_g$ variables. This wave propagation occurs in the $\vec{E}_g$ vector direction, which defines the mechanism of longitudinal electromagnetic wave formation [12].

Owing to the restrictions of this lecture course, we cannot discuss in detail the many interesting examples of longitudinal electromagnetic wave formation, and refer the readers to the publications by Tomilin and by Sacco and Tomilin [12, 35].

We can simplistically represent the longitudinal electromagnetic wave and *TEM* wave propagation in the following two images (Figs. 4.4 and 4.5). The first figure shows two plane waves forming at a sufficient distance from the *O* source. It is obvious that these waves propagate in two mutually perpendicular directions.

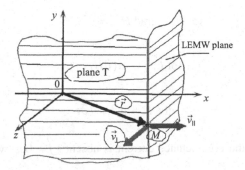

Fig. 4.4 Image of transverse ($\perp$) and longitudinal ($\parallel$) wave propagation far from the source located at the 0 point

Fig. 4.5 A more complete image (than in Fig. 4.4) of transverse and longitudinal electromagnetic wave propagation in the free space

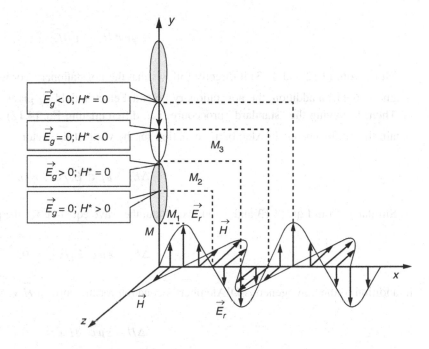

A more complete picture of these two plane waves is shown in Fig. 4.5, from which we can see that when the *TEM* wave achieves a zero value, the longitudinal electromagnetic wave happens to be maximal. This means that transition from one wave type into another expresses the energy exchange between them.

Resolution of the aforementioned paradox of the classic approach consists in the fact that the process of mutual transformation of the plane and longitudinal waves occurs with some "shift" in time; namely, these waves are out-of-phase [12]. If so, then, evidently, the total electromagnetic field of the propagating traveling wave is vanishes. This is clearly shown in Fig. 4.5 (according to A. Tomilin). The usual plane wave propagates along the *x*-axis, but the longitudinal one does so along the *y*-axis. The first completely repeats the image in Fig. 4.1. The main conclusion consists in the fact that transverse and longitudinal waves change according to out-of-phase laws.

At the end of this Chapter (Sect. 4.5), we briefly describe the conclusions of generalized microscopic electrodynamics. These conclusions are of utmost importance, both for the theory and the practice in many issues of modern science and technology.

4.3 Electromagnetic Waves in the Dielectric, Plane Waves

Thus, let the fixed medium fills all the space, in which $\varepsilon' = \text{const}$, $\mu' = \text{const}$ for $\rho = 0$, $\sigma = 0$. Then, from the general system of Eqs. (1.6, 1.7 and 1.8), we have

$$\mathrm{rot}\,\vec{H} + \mathrm{grad}\,H^* = \varepsilon \partial \vec{E}/\partial t, \tag{4.9}$$

$$\mathrm{rot}\,\vec{E_r} = -\mu(\partial \vec{H}/\partial t) \tag{4.10}$$

$$\mathrm{div}\,\vec{E} = \mu\,\partial H^*/\partial t \tag{4.11}$$

Because under problem conditions the conductance current is absent ($\vec{j} = 0$), two independent equations follow from Eq. (4.9):

$$\mathrm{rot}\,\vec{H} = \varepsilon\,\partial \vec{E_r}/\partial t, \tag{4.12}$$

$$\mathrm{grad}\,H^* = \varepsilon\,\partial \vec{E_g}/\partial t \tag{4.13}$$

Thus, from (4.12 and 4.13) it directly follows that the nonstationary vortical electric field $\vec{E_r}$ gives rise to the $\vec{H}$ vortical magnetic field. In addition, the nonstationary potential electric field $\vec{E_g}$ gives rise to the scalar magnetic field H^*.

Then, following the "standard" procedure, i.e., differentiating Eq. (4.12) with respect to time and using Eq. (4.10), we obtain the homogeneous D'Alembert's equation for the vortical $\vec{E_r}$ vector:

$$\Delta\vec{E_r} + \varepsilon\mu\,\partial^2\vec{E_r}/\partial t^2 = 0. \tag{4.14}$$

Similarly, from Eqs. (4.13 and 4.11), we obtain the wave equation for the potential $\vec{E_g}$ vector

$$\Delta\vec{E_g} - \varepsilon\mu\,\partial^2\vec{E_g}/\partial t^2 = 0, \tag{4.15}$$

in addition to the homogeneous D'Alembert's equation for the vortical $\vec{H}$ vector:

$$\Delta\vec{H} - \varepsilon\mu\,\partial^2\,\vec{H}/\partial t^2 = 0 \tag{4.16}$$

and the wave equation for the scalar H^* function:

$$\Delta H^* - \varepsilon\mu\,\partial^2 H^*/\partial t^2 = 0 \tag{4.17}$$

From there, an important conclusion follows: the generalized theory leads to the existence of two electromagnetic processes (waves). The first wave is defined by vortical $\vec{E_r}$ and $\vec{H}$ vectors, and the second by the potential $\vec{E_g}$ vector and the scalar H^* function.

From the form of Eqs. (4.14, 4.15, 4.16, and 4.17), it directly follows that the propagation speed of transverse and longitudinal electromagnetic waves is the same:

$$v_\perp = v_\parallel = 1/\sqrt{\varepsilon\mu} = c/\sqrt{\varepsilon'\mu'}, \tag{4.18}$$

and $c = 1/\sqrt{\varepsilon_0\mu_0}$ is the light speed in the vacuum. From this, the important conclusion follows that transverse and longitudinal waves periodically exchange energy between them during propagation in the fixed space.

In microscopic theory, the situation is much more complicated, and is discussed in Sect. 4.5.

For an adequate distance from the O source (Fig. 4.4), the fronts of both waves at some points can be approximately considered as the plane ones. Thus, the following harmonic processes are the solutions of differential equations (4.14, 4.15, 4.16, and 4.17) (where ω is the radian frequency):

$$\vec{E}_r\,(z,t) = \vec{E}_{0r}\,(z)\exp\{-i\omega t\}, \tag{4.19}$$

$$\vec{H}\,(r,t) = \vec{H}_z\,(z)\exp\{-i\omega t\}, \tag{4.20}$$

$$\vec{E}_g\,(z,t) = \vec{E}_{gx}\,(x)\exp\{-i\omega t + \pi/2\}, \tag{4.21}$$

$$H^*(x,t) = H_x^*(x)\exp\{-i(\omega t + \pi/2)\}, \tag{4.22}$$

Substituting (4.19) into (4.14), we obtain the ordinary differential equation:

$$d^2\vec{E}_{rz}\,(z)/dz^2 + k_\perp^2\,\vec{F}_{rz}\,(z) = 0, \tag{4.23}$$

where $k_\perp = \omega\sqrt{\varepsilon\mu}$ is the wave number of the transverse wave.

Solution (4.23) for the transverse wave propagating in the positive direction of the z-axis is

$$\vec{E}_r\,(\vec{r},t) = \vec{E}_r^0\exp\{i(-\omega t + k_\perp z)\} = \vec{E}_r^0\exp\{i(-\omega t + k_\perp \vec{z}^0 \cdot \vec{r})\}, \tag{4.24}$$

where $\vec{E}_r^0$ is the amplitude of vortical electric field intensity, $\vec{r}$ is a radius vector determining the position of the M point (Fig. 4.4), and $\vec{z}^0$ is the unit vector of the z-axis.

Similarly, using Eq. (4.20) for Eq. (4.16), we have for the vortical magnetic field:

$$\vec{H}\,(\vec{r},t) = \vec{H}^0\exp\{i(-\omega t + k_\perp z)\} = \vec{H}^0\exp\{i(-\omega t + k_\perp \vec{z}^0 \cdot \vec{r})\} \tag{4.25}$$

Substituting Eq. (4.21) into Eq. (4.15), we obtain the ordinary differential equation

$$d^2\vec{E}_{gx}\,(x)/dx^2 + k_\perp^2\vec{E}_{gx}(z) = 0, \tag{4.26}$$

where $k_\| = \omega\sqrt{\varepsilon\mu}$ is the wave number of the transverse electromagnetic wave.

For the intensity of the potential electric field, from Eq. (4.26) we get

$$\vec{E}_g(\vec{r},t) = \vec{E}_g^0\exp\{i(-\omega t - \pi/2 + k_\| x)\} = \vec{E}_g^0\exp\{i(-\omega t - \pi/2 + k_\| \vec{x}^0 \cdot \vec{r})\}. \tag{4.27}$$

Using Eqs. (4.22 and 4.17), we obtain for potential magnetic field intensity

$$H^*(\vec{r},t) = H^{*0}\exp\{i(-\omega t - \pi/2 + k_\| x)\} = H^{*0}\exp\{i(-\omega t - \pi/2 + k_\| \vec{x}^0 \cdot \vec{r})\}. \tag{4.28}$$

Now, more about the directions of the unified electromagnetic field components. For the vortical electric field, $\mathrm{div}\,\vec{E}_r = 0$ and substituting (4.24), we obtain $\mathrm{div}\,\vec{E}_r = ik(\vec{z}^0 \cdot \vec{E}_r) = 0$, and, thus, $\vec{E}_r \perp \vec{z}^0$.

Similarly, for the vector of magnetic field intensity, on the basis of the solution of Eq. (4.25), we get $\mathrm{div}\,\vec{H} = ik(\vec{z}^0 \cdot \vec{H}) = 0$, and thus, $\vec{H} \perp \vec{z}^0$.

4.4   Electromagnetic Waves in the Lossy Medium

Let it be the fixed homogeneous infinite medium, in which $\sigma = \mathrm{const} \neq 0$, $\varepsilon' = \mathrm{const}$, $\mu' = \mathrm{const}$, $\rho = 0$. Then, the equation system (1.6, 1.7, and 1.8) can be written in the form

$$\mathrm{rot}\,\vec{H} + \mathrm{grad}\,H^* = \sigma\vec{E} + \varepsilon(\partial\vec{E}/\partial t), \tag{4.29}$$

$$\mathrm{rot}\,\vec{E}_r = -\mu(\partial\vec{H}/\partial t), \tag{4.30}$$

$$\mathrm{div}\,\vec{E}_g = \mu\,\partial H^*/\partial t, \tag{4.31}$$

Now, similar to (4.24, 4.25, 4.26, 4.27, and 4.28), we have

$$\vec{E}_r(\vec{r},t) = \vec{E}\,^0_r \exp\left\{i\left(-\omega t + \vec{K}_\perp \cdot \vec{r}\right)\right\}, \tag{4.32}$$

$$\vec{H}(\vec{r},t) = \vec{H}^0 \exp\left\{i\left(-\omega t + \vec{K}_\perp \cdot \vec{r}\right)\right\}, \tag{4.33}$$

$$\vec{E}_g(\vec{r},t) = \vec{E}\,^0_g \exp\left\{i\left(-\omega t - \pi/2 + \vec{K}_\| \cdot \vec{r}\right)\right\}, \tag{4.34}$$

$$H^*(\vec{r},t) = H^{*0} \exp\left\{i\left(-\omega t - \pi/2 + \vec{K}_\|\vec{x}^0 \cdot \vec{r}\right)\right\}. \tag{4.35}$$

Substituting (4.32, 4.33, 4.34, and 4.35) into (4.29, 4.30, and 4.31), we obtain the same dispersion complex equation in which the vortical and potential parts can be separated

$$K_\perp^2 = \omega^2\mu'(\varepsilon' - i\sigma/\omega), \tag{4.36}$$

$$K_\|^2 = \omega^2\mu'(\varepsilon' - i\sigma/\omega), \tag{4.37}$$

Thus, the modulus of the wave vectors in transverse and longitudinal directions are equal. In the conducting medium, they are complex:

$$\vec{K}_\perp = \vec{k}_\perp - i\vec{s}_\perp,\quad \vec{K}_\| = \vec{k}_\perp - i\vec{s}_\|, \tag{4.38}$$

when we designate $\vec{s}_\perp = \omega\varepsilon'\mu'\sigma \cdot \vec{z}^0$, $\vec{s}_\| = \omega\varepsilon'\mu'\sigma \cdot \vec{x}^0$

Substituting Eq. (4.38) into Eqs. (4.36 and 4.37), we obtain two biquadratic equations for each wave type respectively:

$$k_\perp^4 - \omega^2\varepsilon'\mu'k_\perp^2 - (1/4)(\omega\mu'\sigma)^2 = 0, \tag{4.39}$$

$$s_\perp^4 + \omega^2\varepsilon'\mu's_\perp^2 - (1/4)(\omega\mu'\sigma)^2 = 0, \tag{4.40}$$

$$k_\|^4 - \omega^2\varepsilon'\mu'k_\|^2 - (1/4)(\omega\mu'\sigma)^2 = 0, \tag{4.41}$$

$$s_\|^4 + \omega^2\varepsilon'\mu's_\|^2 - (1/4)(\omega\mu'\sigma)^2 = 0, \tag{4.42}$$

In "traditional" electrodynamics, at the solution of Eqs. (4.39 and 4.40), we take into account the real roots only, which correspond to the task's physical sense. The positive real roots correspond to the transverse wave propagating in the positive direction along the z-axis (Fig. 4.4):

$$k_\perp = \omega \sqrt{(1/2)\varepsilon'\mu'\left[\sqrt{1 + (\sigma/\varepsilon')^2} + 1\right]}, \tag{4.43}$$

$$s_\perp = \omega \sqrt{(1/2)\varepsilon'\mu'\left[\sqrt{1 + (\sigma/\varepsilon')^2} + 1\right]}. \tag{4.44}$$

Thus, for longitudinal electromagnetic waves, the previous classic result remains:

$$\vec{E}_r(\vec{r},t) = \vec{E}_r^0 \exp\left\{\vec{s}_\perp \cdot \vec{r}\right\} \exp\left\{i(\vec{k}_\perp \cdot \vec{r} - \omega t)\right\}, \tag{4.45}$$

$$\vec{H}(\vec{r},t) = \vec{H}^0 \exp\left\{\vec{s}_\perp \cdot \vec{r}\right\} \exp\left\{i(\vec{k}_\perp \cdot \vec{r} - \omega t)\right\}. \tag{4.46}$$

Solutions of written Eqs. (4.41 and 4.42) have two root pairs: positive and negative. Thus, positive roots correspond to transverse waves, and the negative roots are of special interest:

$$k_\parallel = -i\omega \sqrt{(1/2)\varepsilon'\mu'\left[\sqrt{1 + (\sigma/\varepsilon')^2} + 1\right]}, \tag{4.47}$$

$$s_\parallel = -i\omega \sqrt{(1/2)\varepsilon'\mu'\left[\sqrt{1 + (\sigma/\varepsilon')^2} + 1\right]}. \tag{4.48}$$

For the direct longitudinal electromagnetic wave in the conducting medium, we have:

$$\vec{E}_r(\vec{r},t) = \vec{E}_r^0 \exp\left\{\vec{k}_\parallel \cdot \vec{r}\right\} \exp\left\{i\left(\vec{s}_\parallel \cdot \vec{r} - \omega t(\pi/2)\right)\right\}, \tag{4.49}$$

$$H^*(\vec{r},t) = H^{*0} \exp\left\{\vec{k}_\parallel \cdot \vec{r}\right\} \exp\left\{i\left(\vec{s}_\parallel \cdot \vec{r} - \omega t(\pi/2)\right)\right\}. \tag{4.50}$$

From expressions (4.45 and 4.46) we see that the transverse electromagnetic wave dampens in the conduction medium, which is usual for the traditional representations of classics and experience. Nevertheless, expressions (4.49 and 4.50) state that the longitudinal electromagnetic wave is "amplified" under the same conditions. At $1/k_\parallel$ distance, it is amplified by e times. We would like to remind the reader that the transverse wave, depending on penetration into the absorbing medium, dampens exactly according to the law $1/k_\perp$ Thus, we have the possibility of transmitting the electromagnetic signal in the conducting medium (for example, in the sea water) by distances that essentially exceed the possibilities of usual transverse electromagnetic waves. Numerous experiments confirm this conclusion [43, 44]. Potential possibilities of the longitudinal electromagnetic waves have the great applied application.

4.5   Some Consequences of Generalized Microscopic Electrodynamics

4.5.1   Introduction

The undoubted successes of classic electrodynamics in the various scientific and especially technological applications, to a great extent, weaken the interest in studying interactions and analogies between electrodynamics and other disciplines, for example, between mechanics, hydro-mechanics, etc. At the same time, creation of the elementary charged particle model, which permits the relation between its charge and mass, is relevant. On the other hand, the generalized electromagnetic theory allows the solution of the well-known problem of "4/3", i.e., eliminate the discrepancy between the kinetic energy of the electron and the full energy of its magnetic field. Evidently, there are other interesting problems of interactions between mechanics and electrodynamics that require accurate consideration. Generally speaking, we would like to note that interaction of the classic electrodynamics with other disciplines has had a the long and sometimes difficult history. The limited volume of this book does not allow us to pay adequate attention to this point. Therefore, we are limited to some brief information and references to other publications. Probably, the first Russian researcher who paid serious attention to this problem was a professor of Lomonosov Moscow State University, *N.P. Kasterin* (1869–1947) [8, 78].

The material described in this Section, was taken, to a large extent, from Sacco and Tomilin [35].

According to presentations of the electromagnetism founders, all physical interactions are performed just at indispensable participation of the continuum that surrounds interacting centers [79]. These analogies are used in full measure in the aforementioned work of N.P. Kasterin [78, 79], and in the publication by P.A. Zhilin [55]. It was shown that in general, the magnetic field has both vortical and potential components. The same conclusions were made by *K.J. van Vlaenderen* [81], *D.A. Woodside* [82], *I.A. Arbab, Z.A. Satti* [83], *D.V. Podgainy*, and *O.A. Zaimidoroga* [84]. The results of these investigations were the basis of the generalized electrodynamics theory [12, 43, 44].

4.5.2   Electromagnetic Field of the Charged Particle

We already mentioned that in accordance with the "golden rule of electrodynamics," we should not (if possible) separate the electromagnetic field from its sources, which are charges and currents. From this, far-reaching conclusions can be drawn: for example, that inertia and gravitation phenomena can be uniformly explained by the interaction of charged particles with the physical vacuum at their relatively accelerated motion. But we shall speak about this a little below.

We begin from the charged particle phenomenon and its electromagnetic field. As follows from the results in Tomilin et al. [7], mechanical phenomena and variables are related to electrodynamics, which describe the state and evolution of the vacuum medium. Let us address electrodynamics quantities with the aim of specifying their characteristics and possible interpretations.

As is well-known, the electric field of the conditionally fixed charged particle is spherically symmetric and completely defined by the scalar potential φ. If the charged particle moves in the chosen coordinate system, then the magnetic field is formed around it as well. The magnetic field is characterized by the vector electrodynamics potential $\vec{A}$ The four-vector[2] $(\vec{A}, \varphi)$ is accepted as the main characteristic of the electromagnetic field of the moving charge.

According to the Stokes–Helmholtz theorem [88], which we already discussed in Sect. 1.3, any physical field, unlimited in space, has two components: potential and vortical. It can usually be considered that the potential component of the electromagnetic field is completely defined by the scalar potential φ; therefore, the vector potential is considered as particularly vortical. In the general theory [12], it is shown that such an approach leads to the loss of a physically substantial field part of the moving charged particle. In general, the electrodynamics potential of the vector should be represented in the form of the superposition of two components (see also Eqs. (1.4 and 1.5)):

---

[2] In usual physics, a four-vector is a combination of the three-dimensional space and the fourth coordinate of time. In our consideration of electrodynamics, a four-vector is a combination of the three-dimensional vector potential $\vec{A}$ and the one-dimensional scalar potential φ. This interpretation helps us to describe generalized electrodynamics more precisely. The term "four-vector" was proposed by A. Sommerfeld in 1910. Four-vectors were first considered by A. Poincaré (1905) and then by H. Minkovski.

$$\vec{A} = \vec{A}_r + \vec{A}_g, \tag{4.51}$$

where $\vec{A}_r = \vec{A}_{rot}$ is the vortical (solenoidal) component, and $\vec{A}_r = \vec{A}_{grad}$ is the potential component.

Accordingly, we may write:

$$\nabla \times \vec{A} = \nabla \times \vec{A}_r = \vec{B}^*, \tag{4.52}$$

$$\nabla \cdot \vec{A} = \nabla \cdot \vec{A}_g = -B^*, \tag{4.53}$$

where $\vec{B}$ is the induction vector of the vortical magnetic field, B^* is the scalar function characterizing the potential component of the magnetic field – the scalar magnetic field. It is usually eliminated with the help of the Coulomb and Lorentz calibrations (see Eqs. (1.21 and 1.22)). The theory, which takes into consideration both components of the magnetic field, is called generalized electrodynamics [12].

In the coordinate system K_0, accompanying the charged particle, there is only an electric field: $\vec{E}_0 \neq 0$, $\vec{B}_0 = 0$, $B^* = 0$. For potential, relatively, we have: $\varphi_0 \neq 0$, $\vec{A}_0 = 0$

We should determine the components of the four-potential and characteristics of the electromagnetic field in conditionally fixed coordinate system K, relative to which the particle moves on straight lines with the $\vec{v}$ velocity. We use the following designations in the coordinate system K: $\vec{E} \neq 0$, $\vec{B} \neq 0$, $B^* \neq 0$. и $\varphi \neq 0$, $\vec{A} \neq 0$

Using the Lorentz transform in SI (4.29) for the positive particle, we have:

$$\varphi = \gamma(\varphi_0 - \vec{v} \cdot \vec{A}), \tag{4.54}$$

$$\vec{A} = \vec{A}_0 - \gamma(v/c^2)\varphi_0 + (\gamma - 1)\left[(\vec{v}/v^2)(\vec{v} \cdot \vec{A}_0)\right], \tag{4.55}$$

where $\gamma = (1 - v^2/c^2)^{-2}$

In the case of rectilinear and uniform particle motion, its field in K has the following characteristics:

$$\vec{E} = \gamma\left[\vec{E}_0 - \vec{v} \cdot \nabla \vec{A} - \vec{v} \times (\nabla \times \vec{A})\right] = \gamma(\vec{E}_0 + \vec{v} \cdot \vec{B}^* - \vec{v} \times \vec{B}), \tag{4.56}$$

$$\vec{B} = (\gamma/c^2)\,\vec{v} \times \vec{E}, \tag{4.57}$$

$$B^* = (\gamma/c^2)\,\vec{v} \cdot \vec{E}, \tag{4.58}$$

At low velocities, we use the Galilean transform:

$$\vec{E} = \vec{E}_0 + \vec{v} \cdot \vec{B}^* - \vec{v} \times \vec{B}, \tag{4.59}$$

$$\vec{B} = (1/c^2)\,\vec{v} \times \vec{E}_0 = \frac{\mu_0 q}{4\pi}\frac{\vec{v} \times \vec{r}}{r^3}, \tag{4.60}$$

$$B^* = (1/c^2)\,\vec{v} \cdot \vec{E}_0 = (\mu_0 q/4\pi)(\vec{v} \cdot \vec{r}/r^3). \tag{4.61}$$

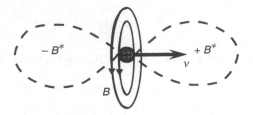

Fig. 4.6 Schematic presentation of the magnetic field of the positively charged moving particle

The formula (4.60) expresses the Bio–Savart–Laplace law, and Eq. (4.61) represents an analog of the same law for the scalar magnetic field. Distribution of vector and scalar magnetic fields of the moving particle is defined by formulas (4.60 and 4.61).

Figure 4.6 schematically shows the generalized magnetic field of the positively charged moving particle (compare with Figs. 4.2 and 4.3).

If the positively charged particle moves along the Ox-axis ($v \ll c$), the vortical magnetic field is represented by the following function in the spherical coordinate system:

$$B(r, \varphi, \theta, t) = \sqrt{\sin^2\theta \sin^2\varphi + \cos^2\theta}. \tag{4.62}$$

Distribution of the scalar magnetic field occurs according to the law:

$$B^*(r, \varphi, \theta, t) = \sin\theta \cos\varphi. \tag{4.63}$$

Here, $r = r(t)$ is the distance from the moving particle's center to the point of the space, in which the field is under determination. Angles θ and φ are also the time functions. Therefore, the magnetic field of the separate moving charged particle is always nonstationary.

From Eqs. (4.61 and 4.63), it follows that in the plane, passing through the particle center orthogonal to its velocity vector, we have $B^* \mid r, \pi, \theta, t \mid = 0$ In front of the moving positive particle, the function B^* has a positive sign, whereas behind the particle the scalar magnetic field has a negative sign (Fig. 4.6). For the negative particle, the scalar magnetic field polarity is opposite.

Time derivative has a dimension of the charge density.

In the generalized electrodynamics, the phenomenon of the vortex-free electromagnetic induction is substantiated both theoretically and experimentally: nonstationary scalar magnetic field gives rise to a potential electric field. The appropriate law can be written as: $\nabla \cdot \vec{D} = \partial B^* / \partial t$ where $\vec{D}$ is the vector of electric induction. In other words, the space point, in which the nonstationary scalar magnetic field is created in the chosen coordinate system, is similar to the point's electrical charge – the quasi-charge.

The moving charged particle happens to obtain additional properties of the electric dipole. In front of the moving positive particle (in a forward way), the positive quasi-charge arises while behind it – the negative one.

For the negative particle, we should change the velocity sign in Eqs. (4.54, 4.55, 4.56, 4.57, 4.58, 4.59, 4.60, and 4.61) by the opposite. At that, the dipole quasi-charge signs change to the opposite as well.

Thus, the electric field of the moving charge in the K coordinate system has a complicated configuration. It represents a superposition of the Heaviside ellipsoidal field (Sect. 4.1) and the field of electric dipole. As quasi-charges forming the dipole have different signs, the potential energy of their interaction is negative. From this, it follows that we must assign the negative sign to the scalar magnetic field energy.

When considering the moving charge interaction, if we worked with electric fields of such a complex configuration, the mathematical expressions would be bulky. Therefore, this task is usually exposed to structuring: one can consider a superposition of the spherically symmetric (Coulomb) electric field and the additional electric field, which has no spherical symmetry. This latter electric field component is called the magnetic field. We know that it depends on the choice of coordinate system.

Let the electron be calm, as before in the accompanied coordinate system K_0. But a trial charge is located in the coordinate system K, linked with the local flow of the physical vacuum. It is acceptable to call such a coordinate system "inertial." If K_0

moves with respect to K in translating and in a uniform manner, it also called inertial. The trial charge does not feel the "vacuum wind"; its electric field is not deformed. In this case, in either coordinate system K_0 or K, the interaction of the moving charge with the fixed one occurs. The trial charge has a spherically symmetric (Coulomb) field, and the electric field of the electron is deformed by the "vacuum wind." The interaction force of these charges differs from the Coulomb one. Hence, in this experiment, we can determine the interaction energy of the electron with a physical vacuum. This is the energy of the magnetic field. From the given mental experiments, it becomes clear that the magnetic field can be revealed only under the condition of relative movement of the main and trial charges. Therefore, both charges must be linked to inertial coordinate systems.

Thus, the magnetic field phenomenon itself proves the existence of the physical vacuum. If we represent the charged particle, which moves in the absolute vacuum, it is impossible to indicate a factor leading to the above-described distortion of its electric field. The relativistic effect in accordance with reduction $l = l_0\sqrt{1 - v^2/c^2}$ for the separate electron, of course, can be manifested, although, for the draft speed of several millimeters per second, it is extremely small. The integral expression of this effect for a current in the conductor can present the vortical magnetic flow only. The second component of the magnetic field (the potential component) cannot be described by the relativistic effect.

Hence:

1. The magnetic field arises at charge particle motion with respect to the vacuum medium.
2. The magnetic field of the solitary charge has two components: vortical (vectorial) and potential (scalar).

4.5.3 Relation Between a Charge and a Mass for Elementary Particle

Scientific discussion concerning the electron structure has remained of importance for many decades [89–92]. A sufficiently complete review of models for the microcosm structure can be found in the paper by A. Kir'yako [80]. The electron structure and associated phenomena can be explained by exclusively internal processes. The Poincaré model has achieved wide spread occurrence. Nevertheless, it has a priority advantage: refusal from point idealization and use of the definite electron's radius r_e. We note that it differs from the well-known classic electron radius $R_e = 2.81 \cdot 10^{-15}$ m, which defines the size of its effective electrical field. It is clear that in this model, the following relation should be true: $r_e < R_e$.

Another model represents an electron by the electromagnetic process occurring in the region with fuzzy boundaries [80]. At this approach, there is no necessity to introduce into consideration the forces of non-electromagnetic nature – this is its advantage. Beyond this region, the electrical field arises. This model can be developed and applied for the study of the process occurring inside the particle, with the purpose of explaining the nature of the elementary electric charge. The difficulty in utilizing this model concludes with the absence of a clear region boundary, in which the charge is generated [80]. Besides, the question about environment and its properties remains unanswered.

In Tomilin et al. [7], the authors used the energy method, on which the hybrid electron model is based. Let its charge be generated by the electromagnetic process localized in a sphere of definite radius r_e. The r_e value is a subject for determination. Thus, an electron is represented by the localized particle of spherical form with a clear boundary. The particle has a charge and a mass. Beyond the spherical particle, the proper electric field arises.

A question about the sense of the electromagnetic field was touched on above repeatedly. It relates to conceptual fundamentals of natural sciences. The materialistic conception of near-action contradicts a vacuum as the absolute emptiness. The empty space, which has no physical properties, cannot be used in the description of physical interaction, even as abstraction. Therefore, physicists use the "physical vacuum" – the material continuum with known electromagnetic properties. The electron model offered in Tomalin et al. [7] assumes that the electron is formed by the process that occurs in this medium. Therefore, the electron is a production of the material vacuum medium, always exists inside it, and is inseparably linked to it.

Within the framework of such a concept, the electromagnetic field is represented in the form of this medium's perturbations: flows, deformations, and waves. We do not know all the medium's properties; moreover, even its nature remains unknown. Therefore, the offered model does not pretend to be complete.

Let us consider the charged particle with r_e radius, which moves in straight lines and uniformly with the velocity v. For distinctness, we consider it as the positive charge. Let us place the observer at some conditionally fixed point. The call charge q of the particle passes by the observer during time equal to the ratio of its longitudinal size and the motion speed: $t = l/v$.

In general, the linear size of the particle is defined with a relativistic reduction account:

$$l = l_0 \sqrt{1 - v^2/c^2} \tag{4.64}$$

where $l_0 = 2r_e$.

The observer determines the local current: $I = qv/l$. The energy

$$W = LI^2/2 \tag{4.65}$$

corresponds to this current.

Usually, the inductance L is assigned to the conductor, but here it should be ascribed to the charged particle. Such a characteristic is known as the "kinetic inductance" [44]. Let us determine the value of the kinetic inductance for the spherical particle with the r_e radius.

It is surprising, but a formula for the calculation of spherical conductor inductance is unknown; however, there is a formula for the cylindrical conductor:

$$L = \frac{\mu_0 h}{8\pi} + \frac{\mu_0 h}{2\pi} \left(\ln \frac{2h}{r} \right), \tag{4.66}$$

where r is the cylinder radius, h is its height.

In the above-mentioned considerations, we analyzed the particle motion on the segment with $l = 2r_e$ length. Hence, in Eq. (4.66) $h = 2r_e$ In other words, we need to consider and compare the inductance of the sphere with $2r_e$ diameter and the cylinder with the same height. Thus, we consider a cylinder and a sphere inscribed into it. In this case, the relation $r = \sqrt{\frac{2}{3}} r_e$ takes place. At that, the kinetic inductance of the spherical particle can be calculated with great accuracy by the formula:

$$L = \mu_0 l/4\pi = \mu_0 r_e/2\pi. \tag{4.67}$$

We note that at particle modeling by the material point, we lose its kinetic inductance and the possibility of determining its electromagnetic energy disappears.

Having written Eq. (4.65), taking into account (4.67), we obtain an expression for the current energy:

$$W = \mu_0 q^2/8\pi. \tag{4.68}$$

On the other hand, the moving particle with the m mass has the kinetic energy:

$$K = mv^2/2 \tag{4.69}$$

Variation of each of these energies represents a work of forces, which causes the acceleration (deceleration) of the particle. In essence, formulas (4.68 and 4.69) express the same quantity; therefore, we can equate them. From this, we obtain the expression linking a charge and a mass:

$$m = \mu_0 q^2/4\pi l. \tag{4.70}$$

Neglecting the relativistic effect, we assume $l = l_0 = 2r_e$ and obtain the value that it is acceptable to call the "rest mass" of the charged particle:

$$m_0 = (\mu_0/8\pi)(q^2/r_e). \tag{4.71}$$

From Eqs. (4.70 and 4.71) we see that the particle mass does not depend on its charge sign. In particular, these formulas can be applied to an electron. Taking into consideration the known rest mass of the electron, we obtain its own radius:

$$r_e = (\mu_0/8\pi)(q^2/m_0) = 1.4 \cdot 10^{-15} \text{m}, \tag{4.72}$$

where $m = 9.1 \cdot 10^{-31}$ kg is the electron rest mass, $q = 1.5 \cdot 10^{-19}$ Q/m; q is the elementary charge, and $\mu_0 = 1,256 \cdot 10^{-6}$ is the magnetic constant.

The electron radius obtained happens to be half its classic radius

$$R_e = 2.81 \cdot 10^{-15} \text{ m.}$$

We note that the limited measurement accuracy of geometrical sizes is restricted by the Planck length $l_p = 1.6 \cdot 10^{-35}$m.

Evidently, these electron quantum properties do not essentially influence the accuracy of the determination of their own radius (4.72).

Taking into consideration the approach to mass determination (4.70), we can call it the inertial mass. However, the question arises about the gravitational (heavy) mass and about the gravitation phenomenon itself. We discuss this question, in addition to the problem of equivalence of inertial and gravitational mass, in a later section.

Let us discuss the result obtained and its authenticity. We considered the acceleration process of the separate electron and equate the work of accelerating force with the kinetic energy, which the electron acquires. At electron acceleration, obviously, the energy of its interaction with the physical vacuum varies, and the magnetic field arises. The question about electromagnetic field energy: is it necessary to add more of the kinetic electron energy to its field energy? This question assumes the hypothetic separation of the motion process. First, we analyze the electron moving in the empty space and determine its kinetic energy. Then, we introduce the external medium and determine its energy perturbation due to the electron motion. Such an approach is used, for instance, taking into consideration the body motion in the viscous medium. However, the known expression for kinetic energy is obtained for the material object motion in the physical vacuum, not in the empty space. Therefore, it takes into account the energy variation for object connection with this medium. In our case, the kinetic energy of the electron and its field energy are the same. Formulas of mechanics and electrodynamics look different, but express the same sense. Therefore, these energies in our task do not summarize but equate.

The particle mass, as we know, depends on its motion speed:

$$m = m_0/\sqrt{1 - v^2/c^2}. \tag{4.73}$$

The charge quantity (in contrast to the mass) does not depend on the particle motion speed. The charge is the relativistic invariant. The relation (4.73) revealed corresponds to these properties of a charge and a mass. Relativistic mass growth is linked to particle size reduction in accordance with Eq. (4.64).

In a similar way, another elementary particle can be considered – the proton.

The material discussed allows us to draw the following conclusions:

1. The offered model of the solitary electron allows adequate description of its motion in the physical vacuum medium and determination of electromechanical analogies.
2. The charged particle mass is unambiguously defined by its charge and its own size, which is calculated in the chosen coordinate system.
3. The mass of the moving charged particle grows in the chosen coordinate system depending on the speed, exclusively due to the relativistic reduction of its size in the direction of motion.

4.5.4   Inertia and Gravitation

Let us consider the case of accelerated motion of the charged particle. The convection current is not already constant; hence, its time derivation is nonzero:

$$I' = \partial I/\partial t = (\partial/\partial t)(qv/2r_e)a, \tag{4.74}$$

where $a = \partial v/\partial t$ is the particle acceleration.

According to the electromagnetic induction law, such a current causes the electromagnetic field of self-induction, which contradicts the variation of its generating current: $U = -L\dot{I}$ taking Eq. (4.67) into consideration, we can write:

$$U = -\frac{\mu_0 r_e}{2\pi} \frac{q}{2r_e} a = \frac{\mu_0}{4\pi} a. \tag{4.75}$$

The following work is done at charge movement:

$$U_q = (\mu_0 q^2 / 4\pi) a. \tag{4.76}$$

At positive acceleration, the work has the negative sign; on the contrary, at particle deceleration the work done is positive. Let us determine the force, which performs this work for movement $2r_e$:

$$F = \frac{U_a}{2r_e} = \frac{\mu_0 q^2}{4\pi} \frac{1}{2r_e} a = -ma. \tag{4.77}$$

The force (4.77) is directed opposite to acceleration irrespective of the charge sign; hence, it is the inertia force. Thus, the origin of the inertia force can be explained by the electrodynamics. We know that the forces arise as a result of material object interaction. The inertia forces should not be excluded. The charged particle is one of the participants in the interaction. The question of the second interaction object arises. The model used by us assumes that the particle moves not in the absolute empty space, but in the material medium with physical properties. Different versions of this concept has been used in physics for a long time.

The inertia forces happen to form as a result of body interaction with the vacuum medium. Within the limits of such a scientific concept, inertia forces cease to be the "peculiar" class of forces, for which the law of "action–counteraction" is inapplicable.

From Eq. (4.78), it follows that the inertia becomes apparent only at accelerated motion of the charged particle with respect to the physical vacuum. At uniform and rectilinear particle motion with respect to the physical vacuum, inertia is not demonstrated. This corresponds to the Newton's law of inertia.

Let us discuss the issue of the possibility of utilizing the coordinate system, which relates to the physical vacuum. Because the physical vacuum is represented by the continuum, in which there can be "flows" and "deformations," then it is clear that it is impossible to link it to the unified coordinate system and take it as the absolute one. However, we may always introduce and use the conditionally fixed "local" coordinate system, in which adequate volume of the physical vacuum remains practically fixed, at least along one of the directions.

The presentation on mass described above defines its inertia properties only. However, the concept of the physical vacuum opens up a way to explain gravitation as well. In the book by Stokes [85], the author shows that the same physical process: the interaction of charged particles with accelerated flows of the physical vacuum is the reason for inertia and gravitation. Thus, it does not matter which object of interaction is considered to be moveable and which is considered as conditionally fixed. If we assume that near the stars and planets, there exist accelerated flows of the physical vacuum, then the reason for gravitation becomes evident. When using this concept, a question about the equivalence of the inertial and heavy mass has an explanation: they are equivalent because they are caused by the same physical phenomenon.

Thus, the phenomena of inertia and gravitation are equally explained by the interaction of charged particles with the physical vacuum at their relative accelerated motion.

4.5.5   Sense Explanation of the Famous Tesla Experiments

Detailed investigation of the Tesla–Meyl experiments was performed in the very circumstantial publication by Sacco and Tomilin [35]. The scheme of experiments based on Tesla's ideas and later (by a hundred years!) by K. Meyl is shown in Fig. 4.7. The receiver and the transmitter are fully symmetrical. As the "load" at the receiver output, two light-emitting diodes were used, which are the same as in the transmitter path. The original resonator from the sphere and the Tesla induction coil is the main element of the system. In contrast to usual cylindrical coils, the Tesla inductance represented the plane construction. The main idea of the Tesla–Meyl set-up consists in implementation of some resonant phenomena, which were an essence of experiments. They remind us of the dipole antennas operating on the different types of oscillations.

Fig. 4.7 The scheme of the
Tesla–Meyl experiments

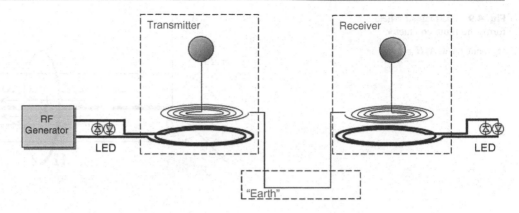

Fig. 4.8 Diagrams for
representation of (**a**) azimuth and
(**b**) radial currents in the Tesla coil

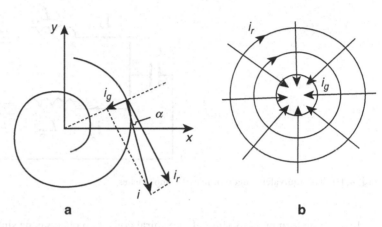

4.5.6   Operation Principle of the Tesla Coil

First of all, we clarify: how do the set-ups of Tesla and Meyl differ from the usual radio system? Let us take into consideration the spiral Tesla coils – they differ from the solenoidal winding of the usual transformers. In the usual transformer, we use the phenomenon of vortical electromagnetic induction, i.e., joint penetration of the vortical magnetic fields occurs. In the windings of the usual transformer, the circular (vortical) currents flow (Fig. 4.8).

The Tesla transformer is constructed in such a way that there are two components of current: tangential (vortical) $\vec{j}_r$ and radial (irrotational or nonvortical) $\vec{j}_0$: $\vec{j} = \vec{j}_r + \vec{j}_0$.

Hence, the electric field in the coil can be represented as the superposition of vortical (solenoidal) and potential (irrotational) components: $\vec{E} = \vec{E}_r + \vec{E}_0$.

Equation (1.6) for the description of processes existing in the spiral coil, can be decomposed into two independent differential equations:

$$\operatorname{rot} \vec{H} = \vec{j}_r; \quad \operatorname{grad} H^* = \vec{j}_g.$$

At the stationary process, the relation between the amplitude values of these current components and the electric fields depends on spiral coil construction and is defined by the α angle:

$$\tan \alpha = j_g/j_r = E_g/E_r. \tag{4.78}$$

However, in the nonstationary case, owing to induction phenomena, this relation becomes incorrect, and the components $\vec{E}_g$ and $\vec{E}_r$ become independent. Figure 4.9 shows (separately) the tangential and radial current components in the Tesla coil at the given time instant.

Fig. 4.9 The magnetic field
forms the finite conductor

segment $(A \equiv \vec{A}; H = \vec{H})$

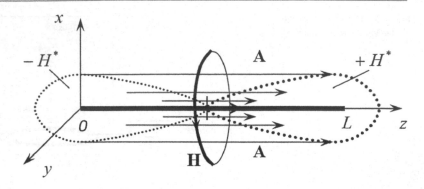

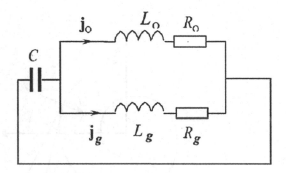

Fig. 4.10 The equivalent circuit of reactive impedances

In the transformer, consisting of two spiral coils, two phenomena simultaneously occur: vortical electromagnetic induction and irrotational electromagnetic induction.

The vortical electromagnetic induction, as we know, leads to vortical current transformation. Owing to irrotational electromagnetic induction, the transformation of the radial current takes place. The radial current leads to the separation of electric charges in the radial direction between the center and periphery of the secondary coil. On the spherical transmitting antenna connected with the center, the nonstationary electric charge arises. This allows creation of the strong electric field around the ball. It is nonstationary, has a radial structure and is characterized by the $\vec{E}_g\left(\vec{r}, t - \dfrac{r}{c}\right)$; $\vec{r} = x, y, z$ vector.

Let us write laws for the vortical and irrotational processes respectively:

$$U_r = -L_r \frac{dJ_r}{dt}; \quad U_g = -L_g \frac{dJ_j}{dt}. \tag{4.79}$$

Thus, the inductive properties of the spiral coil are characterized by two different coefficients: L_g and L_r. Hence, there are two types of inductive impedance:

$$\text{circular } X_{L(r)} \quad \text{and} \quad \text{radial } X_{L(g)}. \tag{4.80}$$

The reactive impedances in tangential and radial directions are obviously different:

$$X_r = \omega L_r - 1/\omega C, \quad X_g = \omega L_g - 1/\omega C. \tag{4.81}$$

Currents $\vec{J}_r$ and $\vec{J}_g$ form as though they are two parallel branches. The equivalent circuit is presented in Fig. 4.10. At operation on the frequency f_1 (and also f_{01}), the current passes onto the upper branch as its reactive impedance is equal to practically zero. On frequency f_2 (and f_{02}), on the contrary, the upper branch is closed, and the reactive impedance of the lower branch is equal to zero. Thus, we must distinguish two types of active resistance: R_r is the resistance to vortical currents $\vec{J}_r$ and R_g is the resistance to irrotational (radial) currents $\vec{J}_g$.

As the inductive elements with coefficients L_g and L_r are connected in parallel, their total inductance (under conditions: $f \neq f_{01}$ and $f \neq f_{02}$ can be determined by the formula:

$$L = \frac{L_g L_r}{L_g + L_r} \tag{4.81}$$

4.5.7 Main Resonance Frequencies

Starting from these considerations, we can calculate three resonant frequencies:

$$\omega_1 = 2\pi f_1 = \sqrt{\frac{1}{L_r C}}, \quad \omega_2 = 2\pi f_2 = \sqrt{\frac{1}{L_g C}}, \quad \omega_3 = 2\pi f_3 = \sqrt{\frac{1}{LC}} \tag{4.82}$$

For $\omega_1 = \omega_2$, the azimuth (circular) current $\vec{J}_r$ is amplified in the spiral coil. At $\omega = \omega_2$, the picture essentially changes. The scalar magnetic field arises, whose gradient is directed along the coil radius. Thanks to this, the resistance in the radial direction is significantly reduced; therefore, the radial current $\vec{J}_g$ increases. This results in creation of strong AC scalar magnetic field in the center of the secondary coil; hence, an effective charge of large amplitude occurs. We may say: "*the charge resonance*" takes place. Thus, the total current strength increases insignificantly, because the azimuth resistance remains high. On the third resonant frequency ω_3, the total current $\vec{J}$, which flows in the spiral coil, is amplified as the reactive impedance is minimal in the direction of the total current.

4.5.8 Electromagnetic Processes in the Tesla Coil

Let us look at the mechanism of radial current transformation. The merits of the case can be explained by the picture of two segments of the conductor located along the same line (Fig. 4.11a). The left segment is the modeling of the element of the radial current flowing in the primary coil. The right segment is the modeling of the secondary coil element. We assume that the condition of each current closure is fulfilled. Let the primary current obey the law:

$$\vec{J}_1 = \vec{J}_{01} \sin \omega t. \tag{4.83}$$

From the general analysis, it follows that in the space between conductors, the J_1 current creates the scalar magnetic field (Fig. 4.11a):

$$B_1^* = B_{01}^* \sin \omega t \tag{4.84}$$

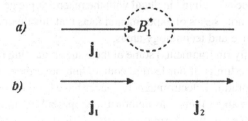

Fig. 4.11 The diagram of the transformation mechanism of radial currents in the Tesla coil $\left(J_{1,2} \equiv \vec{J}_{1,2} \right)$. (**a**) Currents out-of-phase. (**b**) Currents in-phase

Hence, in this region, a nonstationary effective charge arises:

$$\rho_{\text{eff}} = \frac{\partial B_1^*}{\partial t} = B_0^* \omega \cos \omega t. \tag{4.85}$$

In accordance with the continuity equation (1.19), in this region, the current source arises:

$$\operatorname{div} \vec{j}_2 = -\varepsilon' \varepsilon_0 \partial^2 B_1^* / \partial t^2 = \varepsilon' \varepsilon_0 B_{01}^* \omega^2 \sin \omega t. \tag{4.86}$$

Thus, in the secondary coil, the J_2 current is induced, which is directed in-phase with the primary current J_1 (Fig. 4.11b). For these considerations, we neglected the delay, assuming that the conductors were located at a distance that was significantly less than the appropriate wavelength.

Naturally, in the first conductor, the countercurrent arises (an analog of the Lenz's law). In other words, the primary current is slightly reduced and its energy decreased. We may say that in this process, the energy transfer occurs from the primary current to the secondary one. In a similar manner, the energy transfer takes place between radial currents flowing in the adjacent winds of the spiral coil. This means that the radial currents are not blinkered by tangential currents, and the relation between them is not determined by the spiral coil construction in accordance with the formula $\tan \alpha = \frac{J_g}{J_r} = \frac{E_g}{E_r}$. We can face the case when radial currents has a larger amplitude whereas the azimuth currents are very small.

The above-mentioned theory explains the operation principle of the Tesla transformer, which consists of two spiral coils. This theory allows adequate description of the "three-humped" resonance.

The calculation of the main resonant frequencies $f_{1, 2, 3}$ was performed in Sacco and Tomilin [35]. The calculated data were confirmed experimentally with high accuracy. The restricted volume of this book did not allow us to examine the results in detail [35]. Nevertheless, the following interesting conclusions could be drawn:

- It is stated that signal transmission between the transmitter and the receiver is performed with the help of a wave electromagnetic process, which happens between spherical antennas.
- It is shown that in this system, the electromagnetic waves differ in their properties from the transverse Hertz waves.
- Experimental results are explained based on the generalized (four-dimensional) electromagnetic theory, which unites vortical and potential electrodynamics processes.
- It is proved that the signal transmission between spherical antennas happens by means of electric scalar waves.
- The operation principle of the Tesla transformer, consisting of two spiral coils, is theoretically described.

Thus, the main points in Tesla's experiments (and later in Meyl's experiments) proves that energy transfer between two ball antennas is carried out with the help of electric scalar waves, and that this process has a *resonance character*. This fundamental conclusion could be drawn within the limits of generalized macroscopic electrodynamics.

4.6   Features of Longitudinal Electromagnetic Waves, Conclusions from the Microscopic Theory

As we already mentioned, during recent years, the interest in so-called longitudinal electromagnetic waves has grown significantly. To a great extent, this is caused by a whole series of discoveries in medicine, biology, ecology, and in other applied areas [6]. We had already become acquainted in detail with the macroscopic generalized electrodynamics (Chaps. 1, 2, 3 and 4). However, we can see from a whole series of experimental facts that this theory, though not in full measure, meets the requirements of the wide circle of science and technology areas.

In this section, we describe very briefly (in fragments) some of the issues regarding this important problem. Any reasonable discussion of this problem is beyond the limits of our lecture course, but, nevertheless, we must describe it briefly.

First, we would like to note that longitudinal electromagnetic waves have been known for a long time, approximately since the 1920s, and the brightest information about them was demonstrated, probably, during study of radio wave propagation in the plasma and in the ionosphere. Waves in the plasma are distinguished by their volumetric character and a variety of properties. Using an expansion into the Fourier (Laplace) series, any small perturbation in the medium (plasma or ionosphere) can be represented as a set of simplest waves of the sinusoidal form. Each such a wave (monochromatic) is characterized by the definite frequency ω, the wavelength λ, and the so-called phase velocity of propagation. Besides, waves may differ by polarization, i.e., by direction of the electric field vector in the wave. If this field is directed along the propagation velocity, this wave is called *longitudinal*, whereas if it is directed across, it is called *transverse*. Three types of waves are possible in the plasma (without a magnetic field): the longitudinal Langmuir waves with ω_0 frequency, the longitudinal sound waves (more precisely, ion sound waves), and the transverse electromagnetic waves. The transverse waves can have two polarizations and can propagate in the plasma without a magnetic field if only the frequency ω exceeds the plasma frequency ω_0. Otherwise, in $\omega < \omega_0$, the refraction index becomes imaginary, and transverse waves could not propagate inside the plasma, but they reflect off of its surface. Exactly, therefore, radio waves with $\lambda > \sim 20$ m are reflected by the ionosphere, which provides the possibility of long-range radio communication in the Earth.

However, in the presence of a magnetic field, the transverse waves, which have resonances with ions and electrons on their cyclotron frequencies, can propagate inside the plasma for $\omega > \omega_0$. This means the appearance of two more waves in the plasma, which are called *Alfvén* waves, or *fast magneto-sound* waves. The Alfvén wave represents the transverse perturbation, which propagates along the magnetic field. In addition, the slow magneto-sound wave may propagate in the plasma and it represents the usual sound wave with characteristics that are slightly changed by the varied magnetic field.

The electromagnetic field is capable of propagating in the form of the transverse, transverse–longitudinal, and longitudinal waves. As we know, the transverse–longitudinal wave is the wave for which the highly observable vector quantity, which characterizes the wave, lies in the plane perpendicular to the direction of its propagation. Such a method for classification remains the known arbitrary rule for formal identification of the wave process. If we choose the longitudinal Hertz vector $\vec{U}$ (the polarization vector potential of the electromagnetic field) as the determining quantities instead of the intensity of the electric $\vec{E}$ and magnetic $\vec{H}$ fields, then this wave, by definition, should be called longitudinal.

In electrodynamics, at electromagnetic wave classification as the transverse wave or the longitudinal wave, it is usually assumed (if we have no other recommendations) that we use $\vec{E}$ and $\vec{H}$ quantities as parameters characterizing the wave process. The generalized equations of electrodynamics allow the existence both transverse and longitudinal electromagnetic waves. In longitudinal electromagnetic waves, vectors $\vec{E}$ and $\vec{H}$ are oriented in the direction of wave propagation (see Fig. 4.6).

It is known that the nonzero projections of $\vec{E}$ and $\vec{H}$ in the direction of electromagnetic wave motion are typical for the near-zone of the electromagnetic radiator, and they are present in *TE* and *TM* waves in the waveguide (see Sect. 11.5). Such electromagnetic waves are not purely transverse and according to the classification used they have the mixed transverse–longitudinal type.

Propagation of longitudinal electromagnetic waves in the plasma relate to the presence of the long-range action of Coulomb forces in the plasma. When the electron group in the plasma is displaced from the equilibrium position, it is subject to the influence of the electrostatic recovery force, which leads to oscillations. Such nonpropagated oscillations (standing waves) and propagated oscillations have the vector $\vec{E}$, which is collinear to the direction of the wave movement (Fig. 4.5), i.e., they are longitudinal Langmuirian waves.

The existence of longitudinal electromagnetic waves was discovered in crystals (D.W. Berreman, 1963); peculiarities of longitudinal electromagnetic wave propagation are investigated in the conducting media and dielectrics [5, 12], in alive media and organisms [6, 35], etc.

The longitudinal electromagnetic waves propagating in the free space differ because of their high penetrating ability and may serve as the carriers of the perceptive informational channel [5, 6]. The main characteristics of longitudinal electromagnetic waves are obtained during analysis of the hypothetical process of the resonant interaction of electrons in the substance and in the physical vacuum [6], and are listed in Table 4.1.

It is necessary to note that wave processes with the group velocities u_2, u_3, and u_4 proceed into the deep layers of the physical vacuum, causing a response in the observing (immediately preceding the substance) layer of the hierarchical structure of the physical vacuum in the form of longitudinal electromagnetic waves $(H_0, \vec{E})$, $(E_{01}, \vec{H})$, and $(E_{0t}, \vec{H})$ with velocities u_2, u_3, and u_4. However, in this case, with respect to directly observing the structural physical vacuum level, velocities u_2, u_3, and u_4 have the sense of phase velocities.

The Umov–Poynting vector, in contrast to Eq. (1.33), is defined for longitudinal electromagnetic waves by Eq. (1.36), in which the permanent scalar fields are designated as E_0 and H_0. Without going into detail (about something we already discussed in this chapter and earlier in Chaps. 1 and 2), we can consider them, for instance, as permanent fields of the Earth, Space, etc.

Table 4.1 Parameters of electromagnetic waves

Wave type	Electric constant, Far/m	Magnetic constant, Hn/m	Velocity, m/s	Action quantum, J·s
Transverse $(\vec{E}, \vec{H})$	ε_0, $8.8541878 \times$ $\times 10^{-12}$	μ_0, $1.25663706 \times$ $\times 10^{-6}$	c, $2.99792458 \times$ $\times 10^8$	h, $6.6260755 \times$ $\times 10^{-34}$
Longitudinal $(H_0, \vec{E})$	ε_0, $8.8541878 \times$ $\times 10^{-12}$	$\mu_{0e} = \alpha^{-2}\mu_0$, $2.3598219 \times$ $\times 10^{-2}$	$u_2 = \alpha c$, $2.1876912 \times$ $\times 10^6$	$h_2 = \alpha h$, $4.835281 \times$ $\times 10^{-36}$
Longitudinal $(E_{01}, \vec{H})$	$\varepsilon_{01} = \alpha^4 \varepsilon_0$, $2.5107891 \times$ 10^{-20}	μ_0, $1.25663706 \times$ $\times 10^{-6}$	$u_3 = \alpha^{-2} c$, $5.6297608 \times$ $\times 10^{12}$	$h_3 = \alpha^{-2} h$, $1.2443015 \times$ $\times 10^{-29}$
Longitudinal $(E_{0t}, \vec{H})$	$\varepsilon_{0t} = \alpha^2 \varepsilon_0$, $4.7149758 \times$ $\times 10^{-16}$	μ_0, $1.25663706 \times$ $\times 10^{-6}$	$u_4 = \alpha^{-1} c$, $4.1082355 \times$ $\times 10^{10}$	$h_4 = \alpha^{-1} h$, $9.0810079 \times$ $\times 10^{-32}$

$\alpha = 1/137,0359895(61)$ –fine-structure constant

It is significant that the listed velocity values u_3 and u_4 do not contradict the results of astronomic observations concerning the registration (at practically to the same time as the visual picture) of the true and future positions of the star source of irreversible processes with the help of physical and biological sensors, located in the telescope focus [93]).

From the data listed in Table 4.1, it follows that the main peculiarities of longitudinal electromagnetic waves:

1. Velocity is 4–8 orders greater than the speed of light.
2. The quantum of longitudinal electromagnetic waves is 4–5 orders greater than the quantum of the transverse waves.
3. Longitudinal electromagnetic waves have a high penetration ability, in particular, when passing through the good conduction media.

At the same time, we should note that serious investigations of properties and opportunities of ultra-fast systems of information processes based on longitudinal electromagnetic waves, in essence, are just beginning, and are extremely interesting and promising for future theory and practice.

4.7   Summary

In this chapter, we have become acquainted with absolutely new opinions (for the educational textbooks) on macroscopic generalized electrodynamics. A significant point is the uselessness of the Coulomb and Lorentz calibration relationships and the introduction of the new concept of the scalar magnetic field. This gives a clear physical interpretation of the vector potential concept and solves the age-old problem of "zeros" in the physical picture of the simplest plane wave in the free space. The material in this chapter material is all-encompassing and does not directly influence the other chapters of our lecture course. Nevertheless, we introduce a complete and unambiguous definition of the longitudinal electromagnetic wave. We will repeatedly encounter similar (sounding) concepts in the theory and practice of waveguides and resonators (Sects. 6.4, 6.5, 6.6, 6.7, 6.8, 6.9, 6.10, and 6.11). Having considered the materials of Chap. 6, the reader must accurately use with terminology. In essence, the term longitudinal electromagnetic wave itself has long been encountered in radio wave propagation.

In a certain sense, the material of this chapter has a self-consistent character: the reader could be sure to become acquainted with new physical properties of the presented generalized theory. In particular, we showed that the magnetic field arises at charged particle movement with respect to the vacuum medium and that this field of the solitary charge has two components: vortex (vector) and potential (scalar).

The problem of connection between a charge and a mass of the elementary particle is that the charged particle mass is unambiguously defined by its charge and its own size.

Finally, the generalized theory gives a transparent and clear explanation of the essence of Tesla's famous experiments.

Checking Questions

1. What is the goal of the introduction of the vector $\vec{A}$ and scalar φ potentials into classic electrodynamics?
2. How can we determine the strengths of electric $\vec{E}$ and magnetic $\vec{H}$ fields through the vector $\vec{A}$ and scalar φ potentials?
3. What is the difference between the wave equations in classic and generalized electrodynamics?
4. What is the difference between longitudinal electromagnetic waves in classic and generalized electrodynamics?
5. How is Lentz's rule formulated in generalized electrodynamics?
6. How can the paradox of zero be eliminated in generalized electrodynamics?
7. How can we explain the possibility of long-range propagation of the longitudinal electromagnetic waves in sea water compared with the transverse waves?
8. What is the difference between the Tesla coil and the usual cylindrical coil?
9. What is the peculiarity of the electromagnetic field of the Tesla coil?

Part II

Electromagnetic Waves in Infinite Space Wave Phenomena on the Media Boundary

Plane, Cylindrical, and Spherical Electromagnetic Waves

5

We are prepared now to study the wave processes both in the infinite space, which was already mentioned earlier (Sect. 3.7), and with an account of the interface presence of some media. We begin our consideration with the wave properties in the uniform infinite medium (Sect. 5.1). Plane waves are the simplest representatives of this class of waves (Sect. 5.1.1).

A concept of the wave impedance, which is a ratio of transverse components of electric and magnetic fields in the plane wave (Sect. 5.1.2), has found wide application in practical electrodynamics. We must be especially accurate with the definition of wave impedance in the case, for instance, of two different types of transmission line connections (Chap. 7). There are many such types of examples, for instance, the creation of three-dimensional microwave integral circuits.

We introduce concepts of a phase plane (Sect. 5.1.3) and dispersion (Sect. 5.2). Dispersing dependences of the permeability and the permittivity for both lossless and lossy media are discussed.

A polarization phenomenon in the plane electromagnetic waves both in natural (Sect. 5.3.1) and in artificial (Sect. 5.3.2) media is described. Waves with linear, circular, and elliptic polarization are considered. Waves in the chiral media are described for artificial media.

A concept of "twisted" electromagnetic waves (Sect. 5.3.3) is introduced. Patterns of twisted electromagnetic waves of the right and left polarizations are presented, together with the shape of the classical parabolic antenna in addition its modified shape for twisted wave transmission. The forms of multi-lobe antennas for twisted waves are shown.

Section 5.4 is devoted to the plane wave incidence onto the plane interface of two half-infinite media. The Fresnel formulas are obtained for reflection and transmission coefficients of plane waves through the plane interface of two uniform media. The account of the presence of loss in the plane wave propagation medium is performed in Sect. 5.5.

The process of the plane wave reflection from the plane with anisotropic impedance is studied in Sect. 5.6. On the plane, the field satisfies the already-known anisotropic impedance boundary conditions (Sect. 2.1.4). As a result, the generalized Fresnel formulas can be derived. At the end of this section, we give some information about the effects on nonlinear media (Sect. 5.6.2). In particular, reflections from the half-space with the nonlinear medium are examined.

Features of other wave equation solutions, namely, cylindrical and elliptical waves, are considered in Sects. 5.7 and 5.8 respectively.

The last section (Sect. 5.9) of this chapter is devoted to waves in gyrotropic media. First, we discuss Maxwell's equations for media with the permeability tensor (Sect. 5.9.1). Then, we study the transverse propagation of waves in magnetized ferrite (Sect. 5.9.2). As a consequence, we obtain the solution in ordinary and extraordinary waves; for the latter, the Cotton–Mouton effect is typical. In optics, this effect is known as the effect of double refraction. The longitudinal propagation of waves in the magnetized ferrite characterized, in particular, by the well-known Faraday effect, is considered at the end of this chapter.

To reinforce the material studied, examples of MathCAD packet application are presented (Sect. 5.10). In particular, the phase plane method (Sect. 5.10.1), the animation method with the phase plane (Sect. 5.10.2), the method of reflection from the plane boundary for a modulus (Sect. 5.10.3), and phase (Sect. 5.10.4) of the first polarization. In further programs, we may see the modulus of wave transmission coefficient for the first polarization (Sect. 5.10.5), and the reflection coefficient modulus (Sect. 5.10.6) and phase (Sect. 5.10.7) for the second polarization. Programs (Sects. 5.10.8 and 5.10.9) allow investigation of the behavior of the modulus of the transmission coefficient and phase for waves of the second polarization. Finally, Program (5.10.10) allows the beam trajectories in the medium with the varying refraction index to be drawn.

© Springer International Publishing AG, part of Springer Nature 2019
E. I. Nefyodov, S. M. Smolskiy, *Electromagnetic Fields and Waves*, Textbooks in Telecommunication Engineering,
https://doi.org/10.1007/978-3-319-90847-2_5

5.1   Definition, Waves in the Homogeneous Infinite Medium

5.1.1   Note About Plane Electromagnetic Waves

We already became acquainted with some definitions of plane waves in Sect. 3.7, and we noted that expressions for the Green's functions in the form of the plane, cylindrical, and spherical waves (3.20) show that both the Green's function itself is a suitable model and waves from Eq. (3.20) are the more or less convenient physical models, and nothing more. In this chapter, we continue the acquaintance with plane, cylindrical, and spherical waves. The fact is that the plane, cylindrical, and spherical waves turn out to be extremely convenient and common as physical and mathematical models.

5.1.2   Wave Impedance

In electrodynamics and especially in engineering practice, the ratio of transverse components of electric and magnetic fields in the plane wave traditionally has great significance. It is acceptable to express this relation as follows:

$$E_x = wH_y, \tag{5.1}$$

where the value w, which we have already encountered (compare (2.13)), is:

$$w = \sqrt{\mu_{\text{abs}}/\varepsilon_{\text{abs}}}; w = \sqrt{\mu/\varepsilon} \times w_0, w_0 = \sqrt{\mu_0/\varepsilon_0} = 120\pi\,\Omega \tag{5.2}$$

The w quantity characterizes the medium and is called the *characteristic impedance*; w_0 is the wave impedance of the free space. Thus, these rather *conditional* names connect with rather dimension (w is measured in units of impedance) than with some deep sense. In essence, this is simply the ratio of field quantities in the plane wave (and not only), as written in Eq. (5.1). Later, we extend (within some limits) the concept of wave impedance (Sect. 6.13). But we must always keep in the mind that the value of w depends on the structure of the field $\vec{E}$, $\vec{H}$ at each space point, the transmission line, the cavity, etc. The question about w definition is especially important for microwave and millimeter-wave three-dimensional integrated circuit (3D-IC), in which, as we already mentioned, transmission lines with completely different field structures must be joined, and the identity of the w numerical value for both joining transmission lines does not guarantee the absence of reflection from their joining point.

If the medium of plane wave propagation has losses, i.e., the medium ε, μ characteristics are complex quantities, then its refraction index n may be written as $n = \sqrt{\varepsilon\mu} = n' + in''$.. We should draw the reader's attention to the fact that the square root in definition of the refraction index n must be extracted according to the rule:

$$n = \sqrt{\varepsilon\mu} = \sqrt{|\,\varepsilon\,| \cdot |\,\mu\,|}\exp\{(i/2)(\delta + \Delta)\}, \tag{5.3}$$

Then, the wavelength is determined as: $\lambda = (2\pi/k)/\,\text{Re}\,n$. Hence, during propagation in the lossy medium, the plane wave amplitude decreases according to the exponential law: $|\vec{E}\,|$, $|\vec{H}\,| \approx \exp\{-kn''z\}$. This is clearly seen from Figs. 3.11, and 3.12.

The δ and Δ angles in Eq. (5.3) are called the angles of electric and magnetic losses respectively. In particular, $\tan\,\delta = \sigma/\omega\varepsilon$, where σ is the medium conductivity.

5.1.3   Concept of the Phase Plane

Sometimes, the process of decrease (increase) in the ordinary differential equation solution of the second order (of type (3.25)) is described for better obviousness on the so-called *phase plane*. In general, the phase space is n-dimension, and for the equation system of the second order it reduces into the phase plane. We know that the two-order equation corresponds to two equations of the first order:

$$dy_1/dt = \Phi_1(t, y_1, y_2), \quad dy_2/dt = \Phi_2(t, y_1, y_2). \tag{5.4}$$

Functions $y_1(t)$, $y_2(t)$ are solutions to the equation system (5.4). Excluding the t-time from these solutions, we shall have the same function $y_2 = \Psi(y_1)$. If we draw the plot $y_2(y_1)$ on the y_1, y_2 plane, then it will be the *phase trajectory*, which reflects the totality of all possible system (5.4) states. Especially often, an approach of the phase plane is used in microwave electronics [94, 125]. In the simplest case, when $y_2(t) = y_1(t)/dt$, the phase trajectory is the function derivative with regard to the function itself (for example, voltage–current). The equation system (5.4) allows the creation of various pictures with phase trajectories on the phase plane. Creation of phase trajectories (including animation) is illustrated in Program P.5.1 (Appendix B – http://extras.springer.com/2019/978-3-319-90847-2).

5.2 Concept of Dispersion

Under *wave dispersion*, we understand the dependence of its phase and group velocities upon the frequency. The dispersion is defined by physical properties of the medium or the guiding system in which waves are propagating. Evidently, a more complicated case is possible, when the guiding system is completely or partially filled by the medium, which has parameter dispersion, but on the other hand, the transmission line itself has structural frequency dispersion (see Chaps. 6, 7, 8, and 9). We know that in a vacuum, the electromagnetic waves propagate without dispersion, whereas in the material medium, even in such a rarefied medium as the Earth's ionosphere (see Kuzelev and Rukhadze, and Gluschenko and Zakharchenko [61, 95]), the wave dispersion occurs. The presence of dispersion leads to distortion of the waveform at propagation in the medium or in the guiding system. It is explained by the fact that monochromic waves of various frequencies, which can be used as a carrier signal, propagate with different velocities. From secondary school, we know well the phenomenon of light dispersion at its passage through a transparent prism, which leads to expansion of white light into the spectrum (for example, a rainbow).

For the plane waves, as we see from Eq. (5.4), material dispersion is only possible, when $\varepsilon = \varepsilon(\omega)$ and/or $\mu = \mu(\omega)$. Typical functions $\omega_\varepsilon(\omega)$ and $\omega_\mu(\omega)$ with respect to ω frequency are shown in Fig. 5.1 [96, 125] for a medium without losses. It is assumed that the medium has electric ω_ε (Fig. 5.1a) or magnetic ω_μ (Fig. 5.1b) *resonances*. Media with dipole particles (molecule or electromagnetic particles of an artificial medium) are such media.

The picture of the electromagnetic field for the lossy medium, when $\varepsilon = \varepsilon' + i\varepsilon''$ and/or $\mu = \mu' + i\mu''$, is, naturally, more complicated. Thus, the real parts of the dispersion dependences on permittivity $\varepsilon(\omega)$ (Fig. 5.2a) and permeability $\mu(\omega)$ (Fig. 5.2b) already have no breaks, as in Fig. 5.1. But their derivatives have both positive and negative values (the latter case is typical of the area with resonance frequencies ω_ε, ω_μ).

In the case of small losses $[\varepsilon''] \ll [\varepsilon']$, $[\mu''] \ll [\mu']$,, and thus, $[\gamma''] \ll [\gamma']$, which results in the following estimations [96]:

$$\frac{\gamma'}{\omega} = \begin{cases} \left[(\varepsilon'\mu')^{1/2}\right] & \text{for } \varepsilon', \mu' > 0 \\ -\left[(\varepsilon'\mu')^{1/2}\right] & \text{for } \varepsilon', \mu' < 0 \end{cases}. \tag{5.5}$$

For waves in media and in transmission lines without losses, we can also use the concept of energy speed

Fig. 5.1 Dispersion functions of (**a**) permittivity and (**b**) permeability for the dipole electromagnetic model of the medium without losses. (Taken from Nefyodov and Kliuev [73], p. 109)

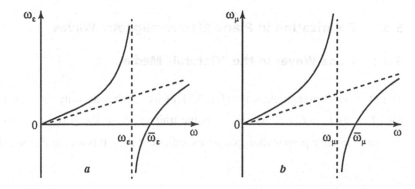

Fig. 5.2 Dispersion functions of (**a**) permittivity and (**b**) permeability for the lossy medium. (Taken from Nefyodov and Kliuev [73], p. 110)

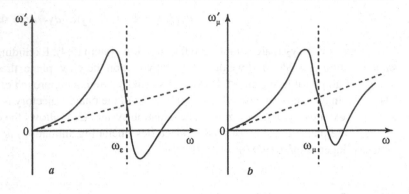

Fig. 5.3 Dispersion dependence of propagation coefficients of (**a**) the direct and (**b**) the backward waves, guided by a metallic waveguide with the internal dielectric rod and the plane waveguide with a negative guiding layer. *1* the backward wave, *2* the direct wave

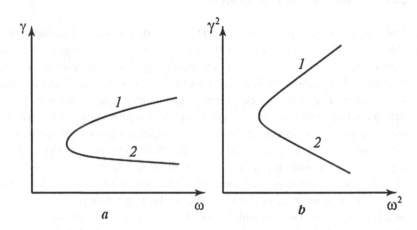

$$v_{\mathrm{en}} = P/W, \tag{5.6}$$

in which for the plane wave P is the power flow through the unit area perpendicular to the flow direction $P = (1/2)\mathrm{Re}\left(\vec{E} \times \vec{H}\right)\vec{z}$, and W is the wave energy density $W = (1/4)\left(\dfrac{d\omega\varepsilon}{d\omega}\left|\vec{E}\right|^2 + \dfrac{d\omega\mu}{d\omega}\left|\vec{H}\right|^2\right)$.

Expressions for $v_{\mathrm{en}} = P$, W justify that they are true for media both with positive (ε, $\mu > 0$) parameters, and with negative (ε, $\mu < 0$) parameters. Thus, the wave, which has the phase and group velocities directed in an opposite manner, namely, the phase incursion and the energy flow have opposite directions, is the *backward* wave (see, for instance, Shevchenko [96]). The dispersion dependences of the propagation coefficients of direct and backward waves, guided by the metallic waveguide with a dielectric rod and the plane waveguide with a negative guiding layer, are shown in Fig. 5.3.

At the same time, we would like to note that there is a different point of view, the essence of which consists in the non-acceptance of the dispersion concept, i.e., the frequency dependence of the plasma and dielectric permittivity [50, 61]. Maybe there is something new in it, but the authors are not yet ready to discuss it, or at least, not within the limits of this lecture course.

5.3   Polarization in Plane Electromagnetic Waves

5.3.1   Plane Waves in the "Natural" Media

The plane electromagnetic wave (Eq. 5.3), as we see, has the only component of the electric field $\vec{E}_x$, so that the vector of the total field is $\vec{E} = E_x \vec{e}_x$, where $\vec{e}_x$ is the unit vector of the x-axis of the rectangular coordinate system $\left(\vec{r} \equiv x, y, z\right)$. The direction of wave propagation coincides with the z-axis. It is acceptable to call the plane $x0z$, in which the vector $\vec{E}$ is located,

Fig. 5.4 (a) Left and (b) right polarization of the plane electromagnetic wave. (Taken from Nefyodov and Kliuev [73], p.112)

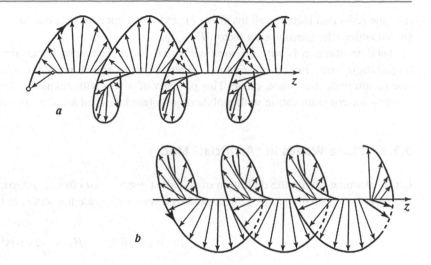

the *polarization plane* of the plane electromagnetic wave. Thus, the uniform plane electromagnetic wave (5.3), (5.5) with the definite polarization plane $x0z$ is the wave with the *linear polarization*.

Evidently, in general, the polarization plane may be arbitrary. Indeed, let the total field represent a sum of two plane waves from the only source; therefore, they have the same frequency of ω, given by their source. Then, various situations are possible. For instance, let the one of these waves be polarized in the $x0z$ plane, and the other waves are in the $y0z$ plane. Thus, the total field of two waves, under the condition of this phase equality (*in-phase*), is:

$$E_x = A_1 \cos \omega \ t, \quad E_y = A_2 \cos \omega \ t, \tag{5.7}$$

It is clear that the $\vec{E}$ vector of the total field moves along the rectangle diagonal to sides $2A_1$ and $2A_2$. The polarization plane of the total field forms the angle $\varphi = arc \tan (A_2/A_1)$ with the x-axis.

Another case: let the phase shift of E_y in (5.7) be 90°, i.e.,

$$E_x = A_1 \cos \omega \ t, \quad E_y = A_2 \sin \omega \ t. \tag{5.8}$$

In this case, the $\vec{E}$ vector of the total field rotates, naturally, with the same ω frequency and depicts in the space by not already the line along the rectangle diagonal $2A_1 \times 2A_2$, as at the linear polarization, but an ellipse (inscribed into the same rectangle $2A_1 \times 2A_2$), and it is a wave with *elliptic polarization*.

Thus, two cases are possible. In the first, the $\vec{E}$ vector formed from waves (5.8) rotates counterclockwise (if looking from its end; Fig. 5.4a), and this wave is called *left-polarized*. In the second case, if, say, we change the sign of the E_y component from Eq. (5.8) to the opposite, i.e., if we write $E_y = - A_2 \sin \omega t$, then this wave is *right-polarized* (Fig. 5.4b). It is essential that both of these waves (left- and right-polarized) propagate along the z-axis with a *similar velocity* to the propagation factor $\gamma = k\sqrt{\varepsilon\mu}$ (compare Sect. 4.2).

We should note the important case of circular polarization, which is significant for practical applications, for instance, in antenna systems. In this case, the *left-* and *right-polarized* waves have equal amplitude, i.e., in (5.12) $A_1 = A_2 = A$ (see Fig. 5.3).

From a practical point of view, the wave with linear polarization, say, at $\vec{E} = E_x \vec{e}_x$, induces in the linear, for example, stub antenna the current $j_x = \sigma E_x$ whereas the current from the orthogonally polarized wave $\vec{E} = E_y \vec{e}_y$ will be equal to or close to zero: $j_y = \sigma E_y = 0$ Really, this circumstance allows construction of reception antennas for two mutually perpendicular and independent polarizations. The similar systems are possible, as we shall see, for instance, in the square waveguide on waves H_{10} and H_{01} (Sects. 6.4 and 6.5), in the light-guides (Sect. 9.2), etc. We can widely use these properties of media and structures on implementation of polarization strained screens, band-pass, and polarization filters, etc. In general, here is a simple rule for frequently periodic wire lattices (see, for example, Katsenelenbaum et al. and Kuzelev and Rukhadze [50, 61]): the wave with the direction of electric field strength $\vec{E}$ vector, which is parallel to the lattice wires, is reflected (almost completely) from it (i.e., the reflection factor modulus $|R| \to 1$), and if it is not perpendicular to them, the wave almost completely passes through such a lattice (the transmission factor $|T| \to 1$).

Another situation is with *circular polarized* waves. The same linear antenna located normally at the axis of the wave propagation with circular polarization, gives the radio signal at output with the constant amplitude independently on its orientation in the transverse plane. This property of waves with circular polarization is widely used in practice, especially in systems for communication with mobile objects (airplanes, missiles, transport, etc.).

5.3.2 Plane Waves in "Artificial" Media

Let us examine briefly the definition of so-called *generalized circular polarization* [50, 60, 97]. If the medium is not chiral ($\aleph = 0$ in Eq. (1.48)), then two linear polarized waves (5.3) are the waves in the infinite medium:

$$E_x = \exp\{ik^c z\}, \quad H_y = \frac{1}{w}\exp\{ik^c z\}, \tag{5.9}$$

$$E_x = \exp\{ik^c z\}, \quad H_y = -\frac{1}{w}\exp\{ik^c z\}, \tag{5.10}$$

where $k^c = kn$, $n = \sqrt{\varepsilon\mu}$, $w = \sqrt{\mu/\varepsilon}$. The plane waves (5.9 and 5.10) have the same propagation factors k^c. Any of their linear combinations is an eigenwave.

Another case in the chiral medium when Eq. (1.68) $\aleph = 0$. Waves (5.9 and 5.10) already cannot exist separately, and only two of their linear combinations are eigenwaves:

$$E_x = \exp\{ik^c_+ z\}, \quad E_y = -\exp\{ik^c_+ z\}, \tag{5.11}$$

$$H_x = \frac{i}{w}\exp\{ik^c_+ z\}, \quad H_y = \frac{1}{w}\exp\{ik^c_+ z\}, \tag{5.12}$$

$$E_x = \exp\{ik^c_- z\}, \quad E_y = \exp\{ik^c_- z\}, \tag{5.13}$$

$$H_x = -\frac{i}{w}\exp\{ik^c_- z\}, \quad H_y = \frac{1}{w}\exp\{ik^c_- z\}. \tag{5.14}$$

It is significant that, as we see from Eqs. (5.11 and 5.12), and also from Eqs. (5.13 and 5.14), the propagation factors of these two waves are *different*:

$$k^c_\pm = k(n \pm \aleph). \tag{5.15}$$

Thus, the wave (5.11 and 5.12) is the wave of *left circular polarization*, whereas the wave (5.13 and 5.14) is the wave of *right circular polarization*. Fields of these waves are connected by relations:

$$\vec{H}_\pm = \pm(1/w)\vec{E}_\pm. \tag{5.16}$$

In the chiral medium, any fields $\vec{E}_+, \vec{H}_+$ and $\vec{E}_-, \vec{H}_-$ may exist *independently*. Precisely they are called the fields with *generalized circular polarization*. The upper sign in Eq. (5.16) corresponds to the wave with left circular polarization, whereas the lower sign corresponds to right circular polarization.

Equations presented in this Section are the main ones and they permit the reader to study further the physics of devices based on chiral media and elements (for details, see Katsenelenbaum et al. and Neganov and Osipov [50, 51]).

Fig. 5.5 Pictures of twisted electromagnetic waves.
(**a**) Right and (**b**) left polarization.
(**c**) The classic parabolic antenna.
(**d**) The modified "parabolic" antenna for transmission of the twisted wave

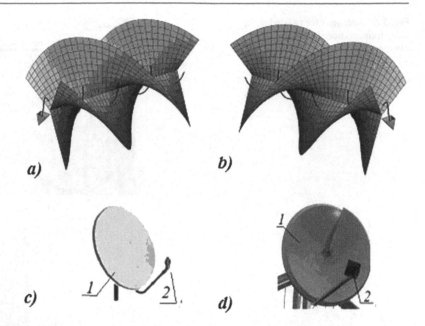

5.3.3 Concept of "Twisted" Electromagnetic Waves

The classic electromagnetic wave has amplitude, a frequency, and a phase. In recent years, the information appears about so-called "twisted" electromagnetic waves, which in addition to the aforementioned characteristics has the orbital corner moments (the same coefficient) (Fig. 5.5). The reality of such a type of waves was confirmed experimentally. According to preliminary calculations, up to 55 messages can be transmitted at each frequency.

The twisting of the electromagnetic wave can differ, not only by direction (clockwise or counterclockwise), but by the degree of twisting (the ratio between the spiral step and the wavelength). At first glance, the twisted electromagnetic wave reminds us of the plane wave with circular polarization (Fig. 5.4). However, there is an essential difference between these processes: the Umov–Poynting vector is directed along the z-axis of wave propagation for the plane wave with circular polarization, whereas for the twisted wave the Umov–Poynting vector is perpendicular to the wave-front surface and its direction differs from z. In Fig. 5.5a, its direction is conditionally shown by oscillating arrows. The twisted wave reminds us of the surface wave, whose intensity dampens in the propagation direction, which is perpendicular to the axis. We shall see a similar picture for waves in dielectric waveguides (optical guides; see Chap. 9). In Fig. 5.5, the classic parabolic antenna is shown (Fig. 5.5c), and the modified "parabolic" antenna is shown for transmission of the twisted waves (Fig. 5.5d).

The other new direction for the generation of twisted waves is the multi-lobed antennas (Fig. 5.6), developed by *M.V. Smelov* [33].

Evidently, this is the very first information on the new class of electromagnetic processes and on new radiators. Nevertheless, we understand just now that this direction has a great and interesting future, which requires serious research and development activity, and it will make a significant contribution to technology.

5.4 Plane Wave Incidence on the Plane Boundary of Two Half-Infinite Media, Fresnel Formulas

The classic task on the incidence of the plane (*TEM-* or *T-*) wave on the same plane boundary $x0y$ also has great significance in applied electrodynamics. It even has its own name; namely, the *Snell–Descartes–Fresnel task*. This task is the *key task* (or model task; compare Fig. 1.17) to a definite degree (see also Nikolsky and Nikolskaya, and Schwinger [53, 90]) in the consideration of many more complicated structures, which have a plane boundary (or only part of it) between two homogeneous media. There are various boundaries between media in waveguides and resonators, dielectric and waveguide–dielectric resonators, 3D-IC of microwave and millimeter-wave ranges, in optical guides, in optics, acoustics, etc.

Thus, let the plane wave of unity amplitude (*i* in Fig. 5.7) in the form of $\exp\{ikz'\}$ fall onto the plane boundary $x0y$ of two media ($j=1$) and ($j=2$) under the some arbitrary angle φ. The medium (j) is characterized, in general, by the complex refraction index $n_j^2 = \varepsilon_j \mu_j$.

Fig. 5.6 Antennas for twisted wave transmission. (**a**) Three-, (**b**) five-, and (**c**) 15-lobed antennas

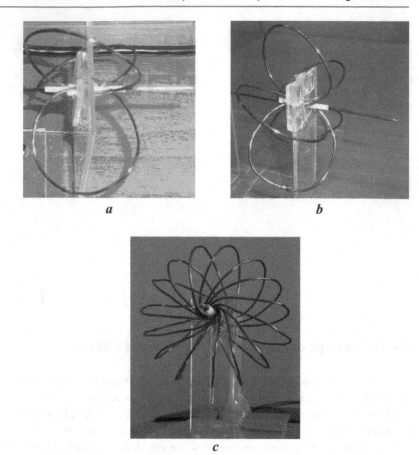

a

b

c

Fig. 5.7 The diagram of slanted incidence of the plane wave onto the plane boundary *x0z* of two media (*1*) and (*2*). (**a**) The dynamic process picture is presented in Program P.5.4. (**b**) The diagram of the normal incidence of the *i* pulse of Gaussian type on the boundary $z = 0$ and after passage (*t*) – reflection (*r*). (**c**) The dynamic picture is presented in Program P.5.4

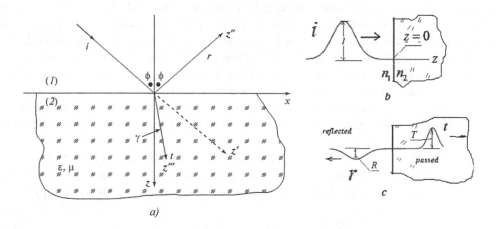

It is acceptable to call the plane *x0z* in Fig. 5.7 the *incidence plane*, and we distinguish two polarizations of the falling wave *i* with respect to this plane:

1. The *first polarization*. In the practice of radio wave propagation, this polarization is called *vertical* according to electrical field orientation with respect to the Earth's surface and designated by the sign "⊥." The magnetic field of the incident wave is perpendicular to the plane of incidence *x0y* and, thus, it is parallel to the boundary *x0y* and has only one component H_y. The electric field has components E_x and E_z and is completely located in the incidence plane *x0y*.
2. The *second polarization* (*horizontal*; "⫠"). The electric field is perpendicular to the incidence plane *x0y* and, hence, it is parallel to the boundary *x0y* and has only one component E_x. The magnetic field has components H_x and H_z, and is parallel to the boundary *x0y*. Because in waves of both polarizations, the field does not depend on the *y* coordinate ($\partial/\partial y \equiv 0$), the task is *two-dimensional*.

It is clear that the case of arbitrary wave falling can be formed from these specific examples, as the system under consideration is linear, and the medium (2) is homogeneous and isotropic. For simplicity, we assume that the medium (1) is the vacuum ($\varepsilon = \mu = 1$).

So to *first polarization*. The magnetic field of the falling (along the z'-axis) plane wave is as follows:

$$H_y^{(i)} = e^{ikz'}. \tag{5.17}$$

Let the incident angle be φ, then $z' = z \cos \varphi + x \sin \varphi$ and the field component H_y is written for the wave passed into the medium (2) as:

$$H_y^{(i)} = T_1 e^{ik(z \cos \varphi + x \sin \varphi)}. \tag{5.18}$$

Here, T_1 is the *transmission coefficient* for the wave into the medium (2).

From sufficiently evident physical consideration and from previous experience (school, university course in physics), we assume that the picture of falling of the plane wave (5.17 and 5.18) onto the plane boundary $x0y$ of two half-infinite media (1)–(2) become completely exhausted by the set of incident (*i*), reflected (*r*), and passed (*t*) *plane waves* into the medium (2). Obviously, it is difficult to imagine that reflected or passed waves may be, for instance, cylindrical, spherical, or some other (besides plane waves). There is no wave from the region $z = +\infty$ owing to the emission condition (Sect. 2.5).

Then, we can write the expression for the total field in the medium (1), which contains the sum of incident (*i*) and reflected (*r*) waves (reminding us of the *first Snell's rule* that the falling angle must be equal to the reflected angle):

$$H_y = \left(e^{ikz \cos \varphi} + R_1 e^{-ikz \cos \varphi}\right) e^{ikx \sin \varphi}, \quad z < 0. \tag{5.19}$$

Here, R_1 is the *reflection coefficient*.

To find out the unknown coefficients R_1 and T_1, it is necessary on the boundary $x0y$ to fulfill the continuity conditions (2.9) of tangent components of H_y and E_x for $z = 0$. From Maxwell's equations in the complex form (2.37), we can write $E_x = (1/ik \, \varepsilon)\partial H_y/\partial z$ for the component E_x in the medium (2). From this, the boundary condition for H_y follows:

$$\partial H_y / \partial z \big|_{z=-0} = (1/\varepsilon) \partial H_y / \partial z \big|_{z=+0}. \tag{5.20}$$

The boundary condition for the E_x component is:

$$E_x \big|_{z=-0} = E_x \big|_{z=+0}. \tag{5.21}$$

Substituting expressions (5.17 and 5.18) into (5.20 and 5.21), we obtain (at $z = 0$) the equation system (where $v = \sqrt{n^2 - \sin^2 \varphi}$, $n^2 = \varepsilon \mu$,):

$$1 + R_1 = T_1, \quad (1 - R_1)\varepsilon \cos \varphi = v T_1,$$

whose solution gives the following for the *first polarization*:

$$R_1 = \frac{\varepsilon \cos \varphi - v}{\varepsilon \cos \varphi + v} \quad T_1 = \frac{2\varepsilon \cos \varphi}{\varepsilon \cos \varphi + v}. \tag{5.22}$$

It is acceptable to state that coefficients R_1 and T_1 are coefficients of reflection and transmission *on the magnetic field*.

Second polarization is investigated in a similar manner, but we should write $E_y^{(i)} = \exp\{ikz'\} = \exp\{ik(z \cos \varphi + x \sin \varphi)\}$ for the incident wave (*i*) and fulfill the same transformations as for the first polarization. As a result, we obtain for R_2 and T_2 coefficients of reflection and transmission:

$$R_2 = \frac{\mu \cos \varphi - v}{\mu \cos \varphi + v}, \quad T_2 = \frac{2\mu \cos \varphi}{\mu \cos \varphi \phi + v} \tag{5.23}$$

These coefficients R_2 and T_2 are reflection and transmission coefficients *on the electric field* (the component $E_y^{(i)}$ is the basing component).

Expressions (5.22 and 5.23) have great importance in various areas of science and technology (physics, optics, radio wave propagation, acoustics, etc.) and they are referred to as *Fresnel formulas*. Some general dependences of reflection coefficients R_1, R_2 upon the incidence angle φ are presented in Fig. 5.8 (see also Nikolsky and Nikolskaya [53]).

Reflection coefficients for waves of both polarization can be calculated according to Programs P.5.3 and P.5.4 as functions of the incidence angle φ and material constants ε, μ in accordance with Eqs. (5.27 and 5.28). In the aforementioned images, the appropriate equations are presented, which allow determination of calculation accuracy for the required $R_{1,2}$, $T_{1,2}$ coefficients. It is easy to check that they are satisfied with great accuracy, i.e., they are satisfied according to the energy conservation law.

The represented Fresnel formulas have a "rich" physical sense. We note some important specific cases of these formulas. First of all, the well-known *reflection* and *refraction laws* are contained in them. Thus, the propagation direction of the reflected wave (z''-axis in Fig. 5.7) forms with the z-axis the same angle φ as the incident wave (see Eq. (5.22)). The refracted beam (t; the passed wave) is deflected from the z-axis by the angle ϕ, which satisfies the *second Snell's law* (the first medium is a vacuum):

$$\sin \varphi / \sin \phi = n = \sqrt{\varepsilon \mu}. \tag{5.24}$$

The medium (2; Fig. 5.7) completely (more exact, almost completely) reflects the incident wave (i), if μ is finite, and $\varepsilon \to \infty$. Then, at the limit, ($\varepsilon \to \infty$) $R_1 \to 1$, $R_2 \to -1$. If, in contrast, ε is finite, but $\mu \to \infty$, then $R_1 \to -1$, $R_2 \to 1$. However, there are not yet any substances (with $\mu \to \infty$) in nature, although materials with a large μ are known.

Full wave reflection from the plane boundary is interesting. Let us designate the incident angles φ, at which the full reflection for both polarizations occurs, as φ_1, φ_2 respectively. Then (see also Sivukhin, and Weinstein [99, 125]), we have:

$$\cos \varphi_1 = \sqrt{(\varepsilon \mu - 1)/(\varepsilon^2 - 1)}, \ \cos \varphi_2 = \sqrt{(\varepsilon \mu - 1)/(\mu^2 - 1)}. \tag{5.25}$$

If $\varepsilon = \mu$ from (5.25) it follows: $\cos \varphi_1 = \cos \varphi_2 = 1$, i.e. $\varphi_1 = \varphi_2 = 0$. This half-space $z > 0$ does not reflect the wave, and is fully transparent. On the other hand, at $\mu = 1, n = \sqrt{\varepsilon}$ from Eq. (5.25) we see that

$$\cos \varphi_1 = (\varepsilon + 1)^{-1/2}, \ \cos \varphi_2 = \infty. \tag{5.26}$$

It is clear that waves of second polarization always reflect, whereas waves of first polarization do not reflect at incidence for the so-called *Brewster's angle*, which is equal to

$$\varphi_1 = \arctan n. \tag{5.27}$$

This can be seen from data comparison on Programs P.5.3 and P.5.4.

The full internal reflection $|R_1| = |R_2| = 1$ occurs when $\varepsilon_1 \mu_1 > \varepsilon_2 \mu_2$ and the following condition is satisfied

$$n^2 \leq \sin^2 \varphi. \tag{5.28}$$

The last inequality is, as we shall see in the future, the basis for the development of a large number of devices, among which there are *dielectric waveguides, dielectric resonators, optical guides, dielectric antennas*, etc.

In conclusion, we consider the case of normal incidence, i.e., $\varphi = 0$. For this, from Eqs. (5.27 and 5.28) it directly follows:

$$R = -R_1 = R_2, \ T = wT_1 = T_2, \ w = \sqrt{\mu / \varepsilon}. \tag{5.29}$$

Thus, in this example, the polarization properties of the incident wave are not manifested.

Some general dependences of the Fresnel coefficients for both polarizations and in the wide range of relative permittivity of adjacent media are shown in Fig. 5.8. As we see, these coefficients behave differently.

Numerical data presented in Programs P.5.3, P.5.4, and many other relations between parameters of media *1* and *2* (including the presence of loss in media) allow the creation of an adequate picture of classic electrodynamics. The reader can use these (and similar) Programs for investigation of a great number of practical cases for the tasks of electromagnetic wave reflection with different, arbitrarily "exotic" relations between task parameters. Here, we are limited to some "standard" examples.

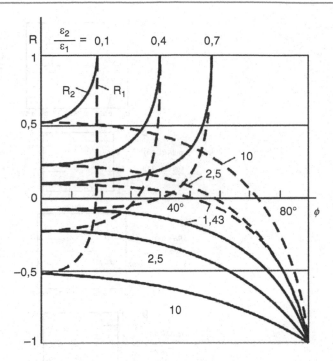

Fig. 5.8 Reflection coefficients from the plane boundary of two homogeneous half-infinite dielectric media for waves of both polarizations depending on the incidence angle φ and on the ratio of half-space permeability. *Dotted lines* the first polarization, *solid lines* the second polarization

Fig. 5.9 Pictures of field lines in the plane wave. (**a**) The "ideal" wave (from the source, which is uniformly distributed in the infinite plane). (**b**) the field, which has a finite cross section. (Taken from Nefyodov and Kliuev [73], p. 122)

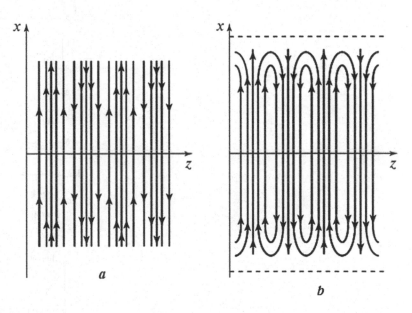

5.5 Losses, Classic Generalized Plane Waves

Under the concept of *classic generalized* plane waves we shall understand in the future the solutions of classic wave Eqs. (2.1), (2.3) in their homogeneous version (5.2), when the field sources are absent, $\vec{j}_{\text{out}} = 0, \rho_{\text{out}} = 0$, and solutions are expressed in the form of waves (5.3), (5.5). The *very* generalized plane waves were briefly considered in Chap. 4.

The suitable and simple wave model with a front of infinite length should be supplemented by, at least, two circumstances. First, the wave front may be restricted, as, for instance, in the coaxial waveguide (Fig. 6.1b), in the symmetric stripline (Fig. 6.1f, g), the plane waveguide (Fig. 6.1d), etc.

The field absence outside the line volume (the cross section) is typical of these systems, i.e., the field in the line has the finite cross section (along the x-axis). Twos such situations are shown in Fig. 5.9 [52]. Another picture is observed, if, say, the

Fig. 5.10 (a–j) Coefficients of reflection and transmission $T_{1,2}$ of the plane wave from the plane boundary of two homogeneous media in accordance with Fresnel formulas (5.22 and 5.23). In the case of **i** and **j**, the permittivity of the second media is $\varepsilon = 9.7(1 + i)$

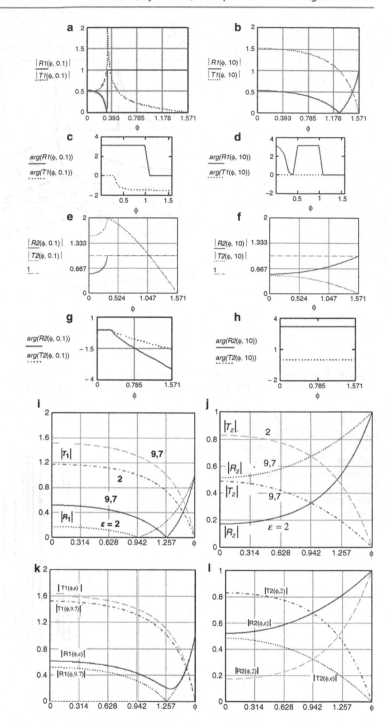

medium (2) (Fig. 5.7a) has losses: $\varepsilon = \varepsilon' + i\varepsilon''$, $\mu = \mu' + i\mu''$, $(n = n' + in'')$. Distribution of the instantaneous values of the wave process in the infinite medium with losses is shown in Fig. 5.10.

Reduction of the wave amplitude from the selected line beginning ($z = 0$) occurs according to the exponential law: $a(z, t) = A_0 \exp\{-n''z\}$, where A_0 is the wave amplitude at $z = 0$.

Now, we go into detail about plane wave falling onto the half-infinite ($z > 0$) absorbing medium (Fig. 5.11b). If in the medium (2) is loss-free ($\varepsilon'' = \mu'' = 0$), then the *plane of equal amplitudes* and the *plane of equal phases* coincide (Fig. 5.11a). Another situation occurs, when the medium (2) has losses (Fig. 5.11b): in this case, planes of equal amplitudes and equal phases form between them at the same angle (γ). This is clear because the wave, which passed some distance from the boundary, at any place in the plane of equal amplitudes dampens in a similar manner ($\exp\{-n''z\}$) and the plane of equal amplitudes turns out to be parallel to the boundary plane $x0y$. Obviously, in the case of normal incidence ($\varphi = 0$), there are no

Fig. 5.11 Pictures of the plane wave (*i*) incidence on the plane boundary (*x0y*) of two media for three cases. (**a**) The second medium is homogeneous without losses. (**b**) The second medium is homogeneous with losses. (**c**) The second medium is heterogeneous along the *x*-axis without losses: $n = n(x)$. (Taken from Nefyodov and Kliuev [73], p. 123)

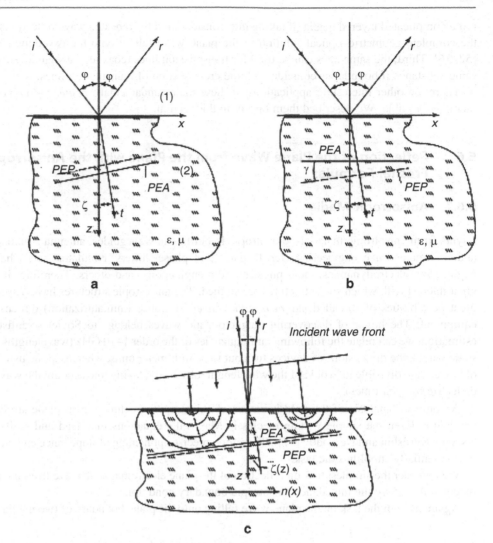

differences in the position of planes of equal amplitudes and equal phases regardless of the fact whether the medium (*2*) has losses or not.

The case is interesting when losses in the medium *2* are large ($|n| >> 1$). Then, the angle γ (Fig. 5.11b) between planes of equal amplitudes and equal phases can be arbitrarily small ($\gamma \ll 1$) and propagation direction of the passed wave practically coincides with the *z*-axis and weakly depends on the incident angle φ, i.e. $\varsigma \ll 1$. Thus, the ratio of electric and magnetic fields in the passed wave corresponds to the *Schukin condition* (2.13).

The case is also interesting when the lower medium is heterogeneous, say, along the *x*-axis, as shown in Fig. 5.11c. Then, because the more the wave velocity, the higher the refraction index $n = n(x)$, the wave front in the lower medium (tangent to the cylindrical and spherical fronts) deviates from the *z*-axis direction.

Evidently, this is only an approximate picture, and it must be specified by solution of the appropriate wave equations. In the simplest case, to construct the trajectory for heterogeneous media, we should use Program P.5.5, where there is a calculation formula and the corresponding plot is presented.

To conclude this section, we would like to make some notes [18].

The above-considered phenomena, which occur when a wave falls onto the boundary of two media, have a much wider area of application than it may seem at the first glance. At first, the reflection and refraction laws are used in optics, quasi-optics, radio wave propagation, and in cases when electromagnetic waves and the boundary surfaces are not planes (at calculation of the optical guides of optical systems, transmission lines in millimeter-range containing lens and mirrors; lines containing the refractive prisms). We may show that such a utilization of refraction and reflection laws must lead to the correct results if the curvature radii of both the boundary surface and the wave front surface are much greater than the wavelength (see Fig. 3.10). Second, the same laws can be used for calculation of wave transmission through the plates and more complicated layered media (if taking into consideration the repeated wave reflections). Thus, for instance, we may offer the complete geometric–optical solution to the plane wave falling onto the two-dimensional ($\partial/\partial z \equiv 0$) dielectric wedge [52, 76]. Third, the same laws can be used for transmission lines (coaxial, waveguide, strip), which are partially filled by the same substance if boundaries coincide with the cross-section of transmission lines.

There are other interesting applications of these main fundamental *Fresnel formulas* and appropriate conclusions from them are possible. We described them briefly in this section.

5.6   Reflection of the Plane Wave from the Plane with the Anisotropic Impedance, Generalized Fresnel Formulas

5.6.1   Anisotropic Media

Layers from anisotropic materials, anisotropic impurities in waveguides, antenna structures, basing elements of microwave and millimeter-wave engineering (ferrite gates and phase-shifters, circulators, etc.) have a huge area of application in engineering electrodynamics, radio physics, radio engineering, and others. Therefore, it is quite understandable to see the great interest with which such structures are studied. The anisotropic structures have a special affinity in connection with the great possibilities of so-called *magneto-static waves* to reduce (miniaturization) dimensions and the weight of electronic equipment. The honor of discovering magneto-static waves belongs to Soviet scientist *V.V. Nikolsky* (1926–1993). For estimation, we can relate the following: at frequencies of the order 1–10 GHz (wavelengths are a few microns for the magneto-static wave), the dimensions of devices turn out to be millimeter units, whereas at the usual transverse waves, the dimensions of devices are divisible to ¼ or ½ of the wavelength (obviously, taking into account the wavelength shortening in the magneto-dielectric by $\sqrt{\varepsilon\mu}$ times).

A complete analysis of the field behavior of anisotropic structures (even plane structures) is a complicated and bulky procedure. Even the simple writing of classic Maxwell's equations may lead and really does lead alas to errors (see, for instance Kurushin and Nefyodov [57]). Therefore, precise application of impedance anisotropic boundary conditions of (2.14) type essentially simplifies the task.

We consider the simplest task on reflection of the plane electromagnetic wave from the plane $x0y$ (Fig. 5.12), on which the field should satisfy the impedance anisotropic boundary conditions (2.14).

Again, as with the task on the plane wave falling onto the plane boundary of two media (Sect. 4.2), we examine two cases:

1. When the $\vec{H}$ vector is perpendicular to the incidence plane $x0z$ (the first polarization, *s*-case)

2. When the $\vec{E}$ vector is perpendicular to this plane (the second polarization, *p*-case).

However, if at plane wave reflection from the plane $x0y$, which is the boundary of two half-infinite dielectric media (Fig. 4.1), we had only one reflected wave, then now, owing to the impedance anisotropic boundary conditions of (2.14) type:

$$E_x = -\rho_0\left(-a_{11}H_x + a_{12}H_y\right), \quad E_y = -\rho_0\left(-a_{21}H_x + a_{22}H_y\right) \tag{5.30}$$

and we have a more complicated picture. Let us start from the first polarization.

The first polarization (*s*-case). Let the incident plane wave falling onto the plane $x0y$ have the form:

Fig. 5.12 Derivation of the generalized Fresnel formulas. (**a**) The first polarization (*s*-). (**b**) The second polarization (*p*-)

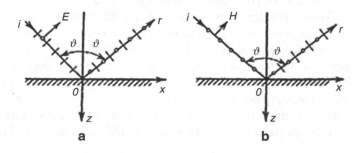

$$\vec{E}^{(i)}(x,z) = \vec{E}\, e^{ik(x\sin\varphi + z\cos\varphi)}, \quad |\vec{E}| = 1. \tag{5.31}$$

Now, taking into account our assumption regarding two reflected waves (of different polarizations), we present the reflected field as:

$$\begin{Bmatrix} E_s \\ E_p \end{Bmatrix} = \begin{Bmatrix} R_s^{(1)} \\ R_p^{(1)} \end{Bmatrix} e^{ik(x\sin\varphi - z\cos\varphi)}. \tag{5.32}$$

The upper index (*1*) for the reflection coefficient $R_{s,p}^{(1)}$ means that the incident wave (*2*) has the *s*-polarization.

Again, as earlier (Sect. 5.2), we determine the magnetic field of the incident wave from Maxwell's equations (1.12...1.13) as:

$$H_y = \rho_0^{-1} e^{ik(x\sin\varphi + z\cos\varphi)}, \tag{5.33}$$

and for reflected waves it is as follows:

$$\begin{Bmatrix} H_y \\ H_x \end{Bmatrix} = \rho_0^{-1} \begin{Bmatrix} R_s^{(1)} \\ R_p^{(1)}\cos\varphi \end{Bmatrix} e^{ik(x\sin\varphi - z\cos\varphi)}. \tag{5.34}$$

We assume that fields (5.32, 5.33, and 5.34) satisfy the boundary conditions (5.30) at $z = 0$, and we obtain the system of algebraic equations, from which the required reflection coefficients are determined for the *s*-polarized incident wave (*2*):

$$\begin{Bmatrix} R_s^{(1)} \\ R_p^{(1)} \end{Bmatrix} = \begin{Bmatrix} (\cos\varphi - a_{12})(1 + a_{21}\cos\varphi) + a_{11}a_{22}\cos\varphi \\ 2a_{22}\cos\varphi \end{Bmatrix}$$
$$\times \frac{1}{(\cos\varphi + a_{12})(1 + a_{21}\cos\varphi) - a_{11}a_{22}\cos\varphi}. \tag{5.35}$$

From these equations, it follows that at $a_{22} = 0$ the orthogonal polarization does not arise, and formulas (5.35) transfer to known *Fresnel formulas* (5.22 and 5.23) for the reflection coefficient of the plane wave from the plane boundary of two isotropic media [99, 125].

The second polarization (p-case). Acting in a similar way, we obtain:

$$\begin{Bmatrix} R_p^{(2)} \\ R_s^{(2)} \end{Bmatrix} = \begin{Bmatrix} (1 - a_{21}\cos\varphi)(\cos\varphi + a_{12}) + a_{11}a_{22}\cos\varphi \\ 2a_{11}\cos\varphi \end{Bmatrix}$$
$$\times \frac{1}{(1 + a_{21}\cos\varphi)(\cos\varphi + a_{12}) - a_{11}a_{22}\cos\varphi}. \tag{5.36}$$

At $a_{11} = 0$ we obtain $R_s^{(2)} = 0$, i.e., the orthogonally polarized wave does not arise. Equations (5.36) also transfer to known Fresnel formulas for the reflection coefficients of the plane wave from the plane boundary of two isotropic media.

Thus, the presented results (5.35 and 5.36) are the generalization of known Fresnel formulas (compare (5.22 and 5.23)).

From the general considerations according to the unicity theorem fulfillment (see Sect. 1.10) in the lower half-space ($z > 0$), the wave propagating to the boundary ($z = 0$) should not exist. The fulfillment of this condition applies the definite conditions on coefficients a_{ij}. They can exist in two versions (details may be found in Kurushin and Nefyodov [57]):

$$\text{Re}\,a_{12} \le 0, \text{Re}\,a_{21} \le 0, \quad \text{Re}\,a_{12}\text{Re}\,a_{21} \ge 0.25|a_{11} + \bar{a}_{22}|^2. \tag{5.37}$$

Here, $\bar{a}_{22}$ is the complex-conjugated quantity.

In the second case

$$R_s^2 + R_p^2 \leq 1. \tag{5.38}$$

The reader can perform a whole series of experiments on studying electromagnetic wave reflection from the anisotropic impedance surface using Eq. (5.36). We can track, in particular, the process of another polarization arising depending on the properties of the anisotropic reflecting surface.

In a similar manner, we can obtain the whole series of expressions of generalized Fresnel formulas. For instance, we can study the reflection of the plane wave from the half-space with chiral properties (see Sect. 1.5), but we do not present these expressions because of their bulky nature.

5.6.2 Note on Effects in Nonlinear Media

Our lecture course is predominantly oriented toward examination of linear media. However, in the theory and practice of radio engineering and radio physics, often phenomena are encountered, about which it is impossible to pass over in silence, although, of course, they should be considered in separate lecture courses [61, 102].

We examine the reflection of the plane electromagnetic wave from, say, in the simplest case, the plane half-space boundary with the nonlinear medium. The typical geometric–optical picture of conversion of the wave (*i*), which falls under angle φ with frequency ω onto this boundary, is presented in Mittra, and Kugushev et al. [70, 102]. The simplest case of the occurrence of the second harmonic only with frequency 2ω in the system of two media (linear *1* and nonlinear *2*) is presented.

If the media *1* is a vacuum, the generalized *Snell's formulas* (compare with (5.27)) have the form

$$\sin \varphi(2\omega) = \frac{\sin \varphi}{\sqrt{\varepsilon(2\omega)}}, \qquad \sin \varphi_p(2\omega) = \frac{\sin \varphi}{\sqrt{\varepsilon_2(\omega)}}, \qquad \sin \varphi_0(2\omega) = \sin \varphi, \tag{5.39}$$

where the *p* index designates the wave of nonlinear polarization $\vec{P}_{(2)}(2\omega)$.

Thus, the "reflected" wave of the second harmonic $\vec{E}_0(2\omega)$ in a vacuum propagates in the same direction as the reflected wave of fundamental frequency ω. The wave $\vec{E}_0(2\omega)$ in the second medium propagates in the same direction as the refracted wave of fundamental frequency field $\vec{E}_{(2)}(2\omega)$. At the same time, the wave propagation direction of the field second harmonic $\vec{E}_{(2)}(2\omega)$ does not coincide with the propagation direction of $\vec{E}_{(2)}(2\omega)$ if $\varepsilon_2(2\omega) \neq \varepsilon_2(\omega)$. The propagation directions of these waves coincide in two cases only when $\varepsilon_2(2\omega) \neq \varepsilon_2(\omega)$ and for the normal incidence ($\varphi_0 = 0$).

More detailed description of these problems may be found in Kurushin and Nefyodov, and Kugushev et al. [57, 102]. The generalized Fresnel formulas for nonlinear cases are presented, in particular, in Kurushin and Nefyodov [57].

5.7 Cylindrical Waves

We already encountered in Sect. 3.3 with the general representations of cylindrical and spherical waves generated by the linear wire and the point source, respectively (the Green's functions). The main Eqs. (3.1 and 3.2) were presented in the same place for fields $\vec{E}$, $\vec{H}$ and potentials φ, $\vec{A}$ (3.9) of these waves. The asymptotic representation for the Green's function are given in the second and third lines in Eq. (3.20).

We underline once more that both plane and cylindrical waves and equally as the spherical wave (see Chap. 5) are only suitable mathematical models. Nevertheless, in combination, say, with the Huygens–Fresnel–Kirchhoff principle, they turn out to be suitable for discovering fields of antenna emission, in tasks of radio wave propagation in different media, etc.

The wave equation (more precisely, the Helmholtz equation for excluded time function $\exp\{-i\omega t\}$) for the case of the absence of an outside source $\left(\vec{j}_{\text{out}} = \rho_{\text{out}} \equiv 0\right)$ in the cylindrical coordinate system (r, φ, z) has the form:

$$\frac{\partial^2 u}{\partial r^2} + \frac{1}{r}\frac{\partial u}{\partial r} + \frac{1}{r^2}\frac{\partial^2 u}{\partial \varphi^2} + \frac{\partial^2 u}{\partial z^2} + k^2 \varepsilon \mu \, u(r, \varphi, z) = 0. \tag{5.40}$$

Here, under $u = u(r, \varphi, z)$ we understand the electric $\vec{E}$ or magnetic $\vec{H}$ field, and media permittivity ε, μ may be complex.

Different representations of Eq. (5.40) solutions are described in detail in Chap. 6. Usually, Eq. (5.40) is solved by the variable separation method. In electrodynamics, it is acceptable to call the solutions of Eq. (5.40) *cylindrical waves*, if they do not depend on the z-coordinate, i.e., when $u(r, \varphi) = R(r)\Phi(\varphi)$. For functions $R(r)$, $\Phi(\varphi)$, each of which depends on the one "own" coordinate r or φ respectively, we obtain two equations in ordinary derivatives after their substitution in Eq. (5.40). In turn, their solutions are the following azimuth trigonometric functions

$$\Phi(\varphi) = \left\{ \begin{array}{c} \sin \\ \cos \end{array} \right\} m\varphi \text{ or } \Phi(\varphi) = \exp\{\pm im \ \varphi\} \tag{5.41}$$

and radial cylindrical functions

$$J_m(x), \quad N_m(x), \quad x = kr. \tag{5.42}$$

In Eqs. (5.41 and 5.42) we designated via m the positive real number, which may be a non-integer. Functions $J_m(x)$ are called the Bessel functions, and $N_m(x)$ are Neumann functions; in it m is the index, and $x = kr$ is an argument.

The first representation in Eq. (5.41) (on the left) is called the *standing wave*; on the right in Eq. (5.41) are the *traveling waves* (along the azimuth coordinate φ).

Solutions to Eq. (5.42) form the linear-independent solution, i.e.,

$$R(x) = AJ_m(x) + BN_m(x), \tag{5.43}$$

in which A, B are arbitrary constants determined from the excitation task (see Chap. 12).

The significant feature of the presented solutions is the presence of unlimited growth of the $N_m(x)$ function for $x \to 0$. It is well seen, for instance, from approximate expression for the $N_m(x)$ function at small values of its argument:

$$N_0(x) = -(2/\pi) \ln (2/\gamma x), \quad N_m(x) = -[(m-1)!/\pi](2/x)^m, m = 1, 2, \ldots, \tag{5.44}$$

where $\gamma = 1.7811\ldots$ is the Euler constant. From Eq. (5.49) we really see that at $x \to 0$ the function $N_m(x)$ has a singularity. On the contrary, Bessel functions at small values of the argument have a form of power expansion

$$J_0(x) \approx 1 - x^2/4 + \ldots, \quad J_1(x) \approx x/2 - x^3/16 - \ldots \tag{5.45}$$

At large values of the argument $(x \gg m^2)$, the functions $J_m(x)$ and $N_m(x)$ have the following asymptotic representation:

$$J_m(x) \approx \sqrt{2/\pi x} \cos \xi, \quad N_m(x) \approx \sqrt{2/\pi x} \sin \xi, \quad \xi = x - (2m + 1)\pi/4. \tag{5.46}$$

Expressions presented in Eq. (5.46) show that Bessel functions $N_m(x)$ and Neumann functions $N_m(x)$ determine the standing waves, which have an infinite number of zeros or nodal circles and the amplitude, which dampens in the radial direction of the waves as $1/\sqrt{kr}$, i.e., represents the waves divergent in the radial direction.

The other cylindrical functions, namely, the Hankel functions of the first and second kind, have found wide application in engineering electrodynamics:

$$H_m^{(1)}(x) = J_m(x) + iN_m(x), \quad H_m^{(2)} = J_m(x) - iN_m(x). \tag{5.47}$$

At real values of the x argument, Hankel functions become infinite at $x = 0$. The behavior of these functions at large values of the argument is a more interesting case. It directly follows from Eq. (5.47):

$$H_m^{(1)} \approx \sqrt{2/\pi x}\exp\{i\xi\}, \quad H_m^{(2)} \approx \sqrt{2/\pi x}\exp\{-i\xi\}. \tag{5.48}$$

From Eq. (5.48) we clearly see that Hankel functions represent the *traveling waves*. Thus, if the Hankel function of the first kind describes the cylindrical wave, which *diverges* from the axis of the coordinate system, then the Hankel function of the second kind, on the contrary, describes the *convergent* wave.

Cylindrical functions of Bessel, Neumann, and Hankel have found very wide application in physics and microwave and millimeter-wave range. In essence, the electromagnetic field of all axis-symmetric transmission lines and waveguides, for instance, the coaxial and circular waveguides (see Sects. 6.9, 6.10), the dielectric waveguides (the optical guides; see Chap. 9.2), quasi-optical transmission lines (Sect. 9.3) are presented in the form of expansions over cylindrical functions. Naturally, the same ideas concern resonators used on segments of such transmission lines. Through them, we can describe the fields of traveling wave antennas, slotted transmission lines (Chap. 7), etc.

It should be noted that cylindrical functions are very well-studied in mathematics. For them, the various approximations, asymptotic relations are obtained, which essentially simplify fulfillment of simple engineering calculations. In due course, the detailed tables were formed for cylindrical functions. Now, they have lost their original significance, because any CAD programs include the appropriate programs for their calculation, although, as we have emphasized more than once, unthinking computer utilization, as a whole, does not help the deep understanding of the process physics, and before application of this computing instrument, it is necessary to understand the physical essence of the task.

In Appendix B (http://extras.springer.com/2019/978-3-319-90847-2), there are programs for calculation of the Bessel functions $J_n(x)$ and roots of equation $J_n(x) = 0$. (P.6.12), in addition to a program for calculation of the Neumann function $J_n'(x)$ and roots of equation $J_n'(x) = 0$ (P.6.13).

5.8   Spherical Waves

Spherical waves in the engineering electrodynamics have a less vast are of application than the above-considered cylindrical waves. Nevertheless, in a series of practical systems they do play a significant role. First of all, this is the emission from the sources of limited dimensions l (compared with the wavelength λ) and for a large distance r from it, so that $kr \gg 1$. On the other hand, this is the emission from small, say, circular holes with the a diameter at a small distance r from it, so that $ka \ll 1$, $r \leq \lambda$. The latter case turns out to be determining the development of radio equipment for medical–biology application [6, 58].

Spherical waves have a great significance in the study of wave scattering on the Earth and for the model of the multi-layered Earth. Here, we note the most essential moments in the spherical wave theory. Having written the wave Eqs. (3.1) or (3.2) for the case of the absence of a source in the spherical coordinate system r, φ, ϑ and using the variable separation method, i.e., representing the solution as a product:

$$u(r, \varphi, \vartheta) = R(r)\Omega(\varphi, \vartheta), \tag{5.49}$$

we obtain for the radial function $R(r)$ the equation in ordinary derivative form

$$d^2R/dx^2 + \left\{1 - \left[v(1 + v)/x^2\right]\right\}R = 0. \tag{5.50}$$

The following functions are solutions of (5.55):

$$\psi_v = \sqrt{\pi x/2}\, J_{v+1/2}(x) \quad \text{and} \quad \zeta_V(x) = \sqrt{\pi x/2}\, H^{(1)}_{V+1/2}(x). \tag{5.51}$$

Here, the function $J_{v+1/2}(x)$ is the Bessel function and it represents the standing wave, and the Hankel function $H^{(1)}_{V+1/2}(x)$ is the divergent function with singularity at $x = 0$.

The behavior of ψ_v and $\zeta_v(x)$ functions at large argument values is directly determined from the asymptotic formulas (5.51):

$$\psi_v(x) \approx \cos\eta, \quad \zeta_v(x) \approx \exp\{i\eta\}, \quad \eta = i[x - (v + 1)\pi/2]. \tag{5.52}$$

The equation for the angular function $\Omega = \Omega(\vartheta, \varphi)$ has the form $\Delta\Omega + \chi\Omega = 0$ and its solution by the variable separation method gives:

$$\Omega = P_m^\nu(\cos\vartheta)\cos(m\varphi), \quad \Omega = P_m^\nu(\cos\vartheta)\sin(m\varphi). \tag{5.53}$$

Functions $P_m^\nu(x)$ in (5.53) are the *associated Legendre functions*.

It is interesting to note that m, n indices may be exclusively integer numbers at spherical wave study in the free space, when the φ coordinate takes any value, and $\vartheta \in [0, \pi]m$. The definite successes in spherical wave study and their application, for example, at its diffraction on the spherical Earth, belong to the Soviet school of physics. The essential contribution was made by V.A. Fock.

5.9 Waves in the Gyrotropic Medium

5.9.1 Introduction, Maxwell's Equations

The equation system (1.12)...(1.15) in addition to the constitutive equation (1.11) forms к the basis for consideration of electromagnetic wave properties, when the permeability tensor $\|\mu\|$ is

$$\|\mu\| = \begin{Vmatrix} \mu_{xx} & -\mu_{yx} & 0 \\ \mu_{yx} & \mu_{yx} & 0 \\ 0 & 0 & 1 \end{Vmatrix}. \tag{5.54}$$

Here

$$\mu_{xx} = \mu_T = 1 - \frac{\Omega\,|\gamma|\,M_0\mu_0^{-1}}{\omega^2 - \Omega^2}, \quad \mu_{yx} = i\alpha, \quad \alpha = \frac{\omega\,|\gamma|\,M_0\mu_0^{-1}}{\omega^2 - \Omega^2}, \tag{5.55}$$

where $\Omega = |\gamma|\,H_0$ is the proper frequency of precession, and γ is some constant. For instance, for the electron spin $\gamma = -2.21 \cdot 10^5$, $(A/m)^{-1}\text{sec}^{-1}$.. Relations (5.52) are written for the case of absence of loss.

Maxwell's equations for the gyrotropic medium, when there is an absence of outside sources, take the following form in accordance with Eq. (1.11):

$$\text{rot}\ \vec{H} = -i\omega\varepsilon_a\,\vec{E}, \tag{5.56}$$

$$\text{rot}\ \vec{E} = i\omega\mu_0\|\mu\|\,\vec{H}. \tag{5.57}$$

The $\|\mu\|$ tensor from Eq. (5.54) is usually called the *Polder tensor*. The diagonal components of this tensor are real when losses are absent in the medium, whereas the out-of-diagonal components are purely imaginary quantities. Thus, $\mu_{ij} = \mu_{ji}^*$ is always true, and the asterisk sign indicates the complex-conjugated quantity. We already mentioned (Sect. 5.2) that ferrite media have frequency dispersion. The special case for Eq. (5.54) is the frequency of *ferromagnetic resonance*, when components of the Polder tensor break, i.e., they lose continuity, and thus, we can use Eq. (5.54) only at frequencies that are far enough from the ferromagnetic resonance frequency.

The equation system (5.56 and 5.57) with the tensor of (5.54) type is sufficiently complicated for general analysis. Let us proceed to consideration of typical specific cases, which allows us to obtain a full representation of properties of electromagnetic waves propagating in these media. We start with cases of the infinite medium.

5.9.2 Transverse Wave Propagation in Magnetized Ferrite

1. *The Helmholtz equation.* The anisotropy is manifested at the presence of the permanent, so-called, *magnetic bias* field $\vec{H}_0$.

We choose the rectangular coordinate system (x, y, z) so that the $\vec{H}_0$ vector is directed along the x-axis, the wave propagates along the z-axis according to the $\exp\{ihz\}$ law. Here, h is the longitudinal wave number. The electromagnetic field is uniform in planes that are parallel to the $x0y$ plane, i.e., $\partial/\partial x = \partial/\partial y = 0$. Those assumptions tell us that we are dealing with the plane wave propagation – the T wave.

Let us start from the case mentioned in the title of this Section; namely, the case of transverse wave propagation, when the $\vec{H}_0$ vector is directed along the x-axis. We write the vector Eq. (5.56) in the scalar coordinate form:

$$dH_y/dz = i\omega\varepsilon_a E_x, \quad dE_x/dz = -i\omega\mu_0 H_y. \tag{5.58}$$

From Eq. (5.58), it follows that under chosen conditions $E_z = H_z = H_x \equiv 0.$. Having differentiated, for instance, the first equation from Eq. (5.58) over z and substituting in it (from the right) the expression dE_x/dz from the second equation in Eq. (5.63), we obtain the Helmholtz equation (compare with Eq. (2.2))

$$d^2 H_y/dz^2 + \omega^2 \varepsilon_a \mu_0 H_y = 0. \tag{5.59}$$

2. *The ordinary wave.* The solution of the last equation represents a sum of two linearly independent uniform plane waves

$$H_y = Ae^{ihz} + Be^{-ihz}, \tag{5.60}$$

which propagate along the z-axis in opposite directions with arbitrary constant amplitudes and with the same wave number

$$h = \omega\sqrt{\varepsilon_a \mu_0}. \tag{5.61}$$

Each of the waves from Eq. (5.60) is the plane electromagnetic wave and it is acceptable to call it the *ordinary wave*, which propagates in the gyrotropic medium with the permanent magnetic bias field $\vec{H}_0$ directed along the x-axis. We note that in the case of the ordinary wave, the $\vec{H}$ vector is parallel (collinear) to the direction of the $\vec{H}_0$ vector.

3. *The non-ordinary wave. The Cotton–Mouton effect* (discovered in 1907). Let the wave have only one nonzero component E_y of the electric field, i.e., perpendicular to the permanent magnetic field $\vec{H}_0$. Using the coordinate writing of Eq. (5.63) and repeating the same calculation, we come to the Helmholtz equation for the H_x component

$$d^2 H_x/dz^2 + h^2 H_x = 0. \tag{5.62}$$

Two linearly independent waves of (5.60) type are its solutions, for which the propagation constants are not the same:

$$h = \omega\sqrt{\varepsilon_a \mu_0}\sqrt{\mu - \left(\mu'^2/\mu\right)}. \tag{5.63}$$

The essential difference between solutions (5.63 and 5.61) is the presence in it of the longitudinal component of the magnetic field H_z vector together with difference in propagation constants (5.61 and 5.63). Therefore, the non-ordinary wave is the H-wave. Because of this, such a wave is called the *non-ordinary* wave, and thus, the propagation speeds of ordinary and non-ordinary waves are generally different. This leads to the interesting physical picture of wave propagation with an arbitrarily oriented direction of the electric field strength. Let such wave falls onto the unlimited in x, y coordinates plane layer of the ferrite with l thickness. Looking aside at the behavior picture of the natural field near the layer, i.e., not considering the reflection from its forward wall ($z = 0$), in addition to reflection from the second wall ($z = l$), we track only for behavior of the incident wave polarization in the layer and after it. Let the polarization of the electric field $\vec{E}$ vector be arbitrary, i.e., with respect to the magnetic bias field $\vec{H}_0$, the $\vec{E}$ vector has two components: $\vec{E} = E_x \vec{e}_x + E_y \vec{e}_y$. Thus, the E_x component is parallel to the $\vec{H}_0$ vector, whereas the E_y component is perpendicular to it. In the linear layer (its parameters does not depend on the field intensity), these two components excite the ordinary and non-ordinary waves respectively. However, because propagation speeds are different (compare Eqs. (5.66 and 5.68)), i.e., they happen to be shifted in phase, then, overall, they give the uniform plane wave with rotating elliptical polarization. When the phase shift of these waves is $\pi/2$, and the amplitudes are the same, the passed wave polarization is circular. The property of the plane electromagnetic wave in the gyrotropic layer to change the polarization plane obtained the name the *Cotton–Mouton effect*.

4. *The double refraction effect.* In optics, the Cotton–Mouton effect is well known under the name *double refraction* of the light in the isotropic substance, which is placed in the transverse magnetic field (perpendicular to the light beam). This effect was first discovered in colloid solutions by *J. Kerr* and (independently) by Italian physicists *K. Maiorana* (1901). This effect was examined in detail by *E. Cotton* and *A. Mouton* (1907). The scheme of their experiments are: through a sample of a transparent isotropic substance, which is placed between strong magnetic poles, the monochromatic light is passed, which is linearly polarized in the plane constituting the 45° angle with respect to magnetic field direction. In the magnetic field, the substance turns out to be optically anisotropic (its optical axis is parallel to the H magnetic field), and the passed light becomes elliptically polarized because it propagates in the substance in the form of two waves: ordinary and non-ordinary, which have different phase speeds (see Eqs. (5.61 and 5.63)).

Let us consider the falling of the arbitrary polarized wave (i) under some angle φ onto the half-infinite (along z) gyrotropic medium. Let the permanent magnetic field $\vec{H}_0$ be perpendicular to the $y0z$ plane. As the media are linear, we can expand the incident wave i into a sum of two waves of different polarizations, when in one case the $\vec{E}$ vector, and in other case the $\vec{H}$ vector are parallel to the $\vec{H}_0$ vector. Thus, one of these waves is capable of exciting in the gyrotropic medium ($z > 0$) the only ordinary wave, and the other – the only non-ordinary wave. From Snell's law (5.29), it is easy to determine refraction angles for these angles: ϑ_{ord} and $\vartheta_{\mathrm{non\text{-}ord}}$. They are found from expressions

$$\frac{\sin \vartheta_{\mathrm{ord}}}{\sin \varphi} = \frac{n}{n_{\mathrm{ord}}}, \qquad \frac{\sin \vartheta_{\mathrm{non\text{-}ord}}}{\sin \varphi} = \frac{n}{n_{\mathrm{non\text{-}ord}}}. \tag{5.64}$$

In latter equations, the refraction indices n_{ord} and $n_{\mathrm{non\text{-}ord}}$ are determined via propagation coefficients h from Eqs. (5.61 and 5.63) as $h/(\omega/c)$ respectively.

In a similar manner, we can consider the case of the gyrotropic plasma exerting a significant influence on radio wave propagation, for example, in the waveguide the Earth – the ionosphere, in space radio communication, and in other cases [57].

5.9.3   Longitudinal Propagation of Waves in Magnetized Ferrite, Faraday Effect

1. *Waves in the infinite ferrite space.* Under longitudinal propagation of waves in the infinite gyrotropic medium we usually understand the wave propagation along the direction of the magnetic bias field $\vec{H}_0$. As in the previous section, we assume that the dependence of all field components with respect to the longitudinal coordinate z is defined by the $\exp\{\pm ihz\}$ multiplier. We also assume that the waves are uniform in the cross-section planes: $\partial/\partial x = \partial/\partial y \equiv 0$, i.e., $E_z = H_z \equiv 0$
Now, we write the required solution of Eqs. (5.61 and 5.62) in the form of the following linear combination:

$$\vec{E} = \left(\vec{e}_x E_x + \vec{e}_y E_y \right) e^{ihz}, \qquad \vec{H} = \left(\vec{e}_x H_x + \vec{e}_y H_y \right) e^{ihz}, \tag{5.65}$$

where h is the required propagation constant. Substituting Eq. (5.65) in Eqs. (5.61 and 5.62), we obtain the equation for h determination:

$$\left[h^2 - (\omega/c)^2 \varepsilon \mu_T \right]^2 = (\omega/c)^4 (\varepsilon \alpha)^2, \tag{5.66}$$

for which the Polder tensor elements α, μ_T are presented in Eq. (5.55). From Eq. (5.66), it directly follows that

$$h^2 = (\omega/c)^2 \varepsilon (\mu_T \pm \alpha). \tag{5.67}$$

Extracting the square root from both parts of (5.66) (with sign conservation), we obtain values $h = \pm h^+$, $h = \pm h^-$, where

$$h^+ = (\omega/c)\sqrt{\varepsilon(\mu_T + \alpha)}, \quad h^- = (\omega/c)\sqrt{\varepsilon(\mu_T + \alpha)}. \tag{5.68}$$

From this result, we see that in the gyromagnetic medium, at longitudinal magnetization, there are two types of longitudinal waves, which propagate in direct ($+z$) and reverse ($-z$) directions.

Under definite conditions, namely, at $h^2 = (h^+)^2$ and $h^2 = (h^-)^2$, amplitudes of these waves in Eq. (5.70) are connected as

$$H_y = -iH_x, \quad H_y = iH_x. \tag{5.69}$$

The latter means that each of the waves has circular polarization (see also Sect. 5.3). Thus, the wave with the propagation constant h^+, which propagates along the z-axis, corresponds to the right circular polarization, whereas the wave with h^- corresponds to left circular polarization (see Sect. 5.3). For further analysis, we need the knowledge of complex amplitudes of these $\vec{E}$, $\vec{H}$ waves. For waves propagating in the positive direction of the z-axis we have:

$$\vec{H}^{\pm} = C(\vec{e}_x \pm i\vec{e}_y)\exp\{ih^{\pm}z\}, \quad \vec{E}^{\pm} = CW^{\pm}(\pm i\vec{e}_x - \vec{e}_y)\exp\{ih^{\pm}z\}, \tag{5.70}$$

whereas for waves propagating in the negative direction of the z-axis:

$$\vec{H}^{\mp} = C(\vec{e}_x \mp i\vec{e}_y)\exp\{-ih^{\mp}z\}, \quad \vec{E}^{\mp} = CW^{\mp}(\mp i\vec{e}_x + \vec{e}_y)\exp\{-ih^{\mp}z\}, \tag{5.71}$$

where the constant unknown amplitude coefficient C must be determined by excitation conditions (Chap. 12), and the wave impedances are

$$W^{\pm} = \sqrt{\varepsilon_0/\mu_0}\sqrt{(\mu_T + \alpha)/\varepsilon}. \tag{5.72}$$

2. *The Faraday effect.* From general considerations, we proceed now to the physical analysis of the result (5.71) obtained. Let the waves of both polarizations exist simultaneously and have the same amplitudes C. Superposition of two waves from Eq. (5.71), which propagate along the z-axis, leads to the following:

$$\vec{H}(z) = \vec{H}^+(z) + \vec{H}^-(z) = 2C\exp\{i[(h^+ + h^-)/2]z\}$$
$$\times \left\{\vec{e}_x \cos[(h^+ - h^-)/2]z + \vec{e}_y \sin[(h^+ - h^-)/2]z\right\}. \tag{5.73}$$

The analysis of the equation obtained for the two-wave field takes into consideration the behavior of the $\vec{H}$ vector (or more precisely, its orientation) at a distance z variation from some chosen point, for example, from $z = 0$ to the same value, say, $z = l$. So, let $z = 0$. From Eq. (5.73), it directly follows that $\vec{H}(0) = 2C\vec{e}_x$, and for $z = l$ the magnetic field vector is $\vec{H}(l)$. From Eq. (5.73) we also see that the rotation angle of the $\vec{H}(l)$ vector is

$$2\vartheta = (h^+ - h^-)l. \tag{5.74}$$

Thus, when the coordinate z grows, i.e., at propagation of this sum of two waves, the rotation angle continuously rotates. The fact is that $\vec{H}^+$ and $\vec{H}^-$ vectors rotate toward each other. However, the phase speeds of the propagation of these waves are different and equal to ω/h^+ and ω/h^-. The considerations presented are true in the ideal case, when losses are absent. The described rotation of the electromagnetic wave polarization plane forms the essence of the *Faraday effect*.

The half-sums $(h^+ + h^-)/2$ and half-differences $(h^+ - h^-)/2$ of the propagation constants obtained in Eqs. (5.73 and 5.74) have a sense of the propagation constant and the rotation angle (ϑ/l) of the polarization plane per unit of length of the line respectively. The latter characteristic has the name the *Faraday constant*.

We note the important physical consequence of the Faraday effect, which is that it is *irreversible*. This means that the rotation ϑ of the polarization plane in (5.74), which is obtained as a result of wave train passage on the distance along the z-axis, will not be compensated, for instance, at wave reflection from any obstacle.

The basing elements of microwave IC and 3D-IC created on the basis of the Faraday effect find wide application in microwave radio electronics and radio physics 14, 20, 21, 57, 99].

5.10 Examples of MathCAD Packet Utilization

In the present chapter, ten examples of the utilization of MathCAD programs to electrodynamics tasks are examined: the illustration of the phase plane method, the phase plane with animation, electromagnetic field reflection from boundaries (two variants), the field passage through the boundary, another two examples of reflection, another three examples of the electromagnetic field passage, and another example of trajectory in a non-uniform medium. These examples of computer program utilization are considered sequentially in Appendix B (http://extras.springer.com/2019/978-3-319-90847-2).

5.10.1 Phase Plane Method

We introduced two types of sine signals with damping and growth with different initial parameters. Program P.5.1 obtains the equations of the phase trajectories and constructs the phase portraits of these equations. The program can construct the volumetric (three-dimensional phase trajectories) for both cases (damping and growth).

5.10.2 Phase Plane Method with Animation

The same example, but with animation.

5.10.3 Reflection from the Boundary (the Modulus, First Polarization)

First of all, the geometry of the task is defined. We determine the complex permittivity and permeability and introduce angles of reflection and refraction. The Program defines the formula for the reflection coefficient and constructs a three-dimensional picture of the reflection coefficient modulus and a two-dimensional picture for different parameters.

5.10.4 Reflection from the Boundary (the Phase, First Polarization)

The same as for the phase of the reflection coefficient.

5.10.5 Passage of the Plane Wave (First Polarization)

First of all, the geometry of the task is defined. We determine the complex permittivity and permeability and introduce the angles of reflection and refraction. The Program defines the formula for the transmission coefficient and constructs a three-dimensional picture of a transmission coefficient modulus and a two-dimensional picture for different parameters.

5.10.6 Reflection from the Boundary (the Modulus, Second Polarization)

First of all, the geometry of the task is defined. We determine the complex permittivity and permeability and introduce the angles of reflection and refraction. The Program defines the formula for the reflection coefficient for second polarization and constructs a three-dimensional picture of the reflection coefficient modulus and a two-dimensional picture for different parameters.

5.10.7 Reflection from the Boundary (the Phase, Second Polarization)

The same as for the phase of the reflection coefficient for second polarization.

5.10.8 Passage of the Plane Wave (the Modulus, Second Polarization)

First of all, the geometry of the task is defined. We determine the complex permittivity and permeability and introduce the angles of reflection and refraction. The Program defines the formula for the transmission coefficient modulus and constructs a three-dimensional picture of the transmission coefficient modulus and a two-dimensional picture for different parameters.

5.10.9 Passage of the Plane Wave (the Phase, Second Polarization)

The same as for the phase of the transmission coefficient for second polarization.

5.10.10 Beam Trajectory in the Medium with Variable Refraction Index

First of all, the geometry of the task is defined. We determine the refraction index for both media, select the calculation scheme, and construct the two-dimensional picture for different parameters.

5.11   Summary

In this chapter, we examined the important (for an understanding of further materials in our lecture course) problems regarding electromagnetic waves in the infinite space in addition to phenomena on the media interfaces. These problems are determinative for understanding and mastering our lecture course. We introduced the fundamental characteristics of wave processes: wave impedance, phase and group velocities, the concepts of the phase plane, and the dispersion. They will be necessary for us in sections concerning the transmission lines (Chaps. 6 and 7). The dispersion concept is especially important for practice because it determines a quality of dielectric and light-guiding transmission lines (Chap. 9).

The concept of electromagnetic wave polarization is important for practice. It is especially significant in connection with the discovery of new classes of twisted waves, which have a great future. Polarization phenomena are observed in the transmission lines. For instance, we examined the cases of degeneration of different types of waves, which have a different velocity (Sect. 6.10).

In general, tasks about wave reflection and transmission through a two-media interface are the crucial problems of electrodynamics. They become apparent, for example, at expansion of waveguide waves into plane waves (Chaps. 6 and 7).

The well-known Fresnel formulas on plane wave falling on the plain interface of two semi-infinite media are described. Their analogs at plane wave reflection from a plane with anisotropic impedance (generalized Fresnel formulas) are obtained as well.

For all problems in Chap. 5 we use the MathCAD packet programs, whose implementation should help the reader to reinforce the studied material.

Checking Questions

1. What is the frequency of ferromagnetic resonance if ferrite has $\vec{H}_0 = 37.5$ A/m?

2. Let ferrite magnetization be $M = 3,6 \cdot 10^4$ A/m, the magnetic bias field be $\vec{H}_0 = 9 \cdot 10^4$ A/m, and the frequency be $\omega = 1,5 \cdot 10^{10}$ 1/s. What will the values of the Polder tensor components (Eq. 5.59) be?

3. Let the same ferrite (as in the previous point) have $\varepsilon = 10$ and be located in the EMW field with frequency $\omega = 1,7 \cdot 10^{10}$ 1/s. How thick is the ferrite plate l, at which the phase shift between ordinary and non-ordinary waves is $90°$?

4. What will the pictures of electric and magnetic waves be from infinitely lengthy, arbitrarily thin wire with a current at a large distance?

5. What will the pictures of electric and magnetic waves be from two infinitely lengthy, arbitrarily thin, parallel located conductors with a current at a large distance?

6. Which wave – cylindrical or spherical – will dampen faster in the far zone?

7. What will be the Umov vector for two parallel wires in the cases: (1) currents in conductors are in-phase, (2) currents are out-of-phase?

8. Why can we consider the emission of the elementary source as the local plane wave in the far zone only?

9. What kinds of electromagnetic wave polarizations are used in practice?

10. Is the statement true that left- and right-polarized waves along the z-axis propagate with the same speed determined by the propagation constant $k\sqrt{\varepsilon\mu}$ and why?

11. How can we define the γ angle between surfaces of equal amplitude and surfaces of equal phases in the case when the "lower medium" has losses for both polarizations?

12. For which structures is the effect of full internal wave reflection from the boundary essential?

13. What is the behavior of R_1 and R_2 coefficients at $\mu = 1$, $\varepsilon >> 1$ depending on the incidence angle?

14. What physical sense do the generalized Fresnel formulas?

15. Draw the beam trajectories in the heterogeneous media, whose refraction index is: $f(y) = 1 + ay^{1/2}$.

Part III

Electromagnetic Waves in Guiding and Resonance Structures

Hollow Waveguides

6

We proceed to the study of the third part of our lecture course devoted to electromagnetic waves in guiding and resonant structures, and start from the oldest and most developed subject; namely, from hollow waveguides. We give a definition of the waveguide of the transmission line as the main constituent part of the modern systems of ultra-fast information processing. The so-called basing elements and functional blocks of all types of the radio-electronic equipment are built based on them.

A description begins with the classification of transmission lines and the brief characteristic of electromagnetic wave properties, which are propagated in these lines (Sect. 6.2). The classes are: transmission lines with transverse waves (Sect. 6.2.1), transmission lines (waveguides) with transverse-magnetic and transverse-electric waves (Sect. 6.2.2), transmission lines of the open (semi-open) type, and lines with surface waves (Sect. 6.2.3). Quasi-optical transmission lines and some elements of quasi-optics follow (Sect. 6.2.4).

Having become acquainted in the general outline with transmission line classes, we pass to familiarization with the main characteristics of transmission lines (Sect. 6.2.5). Among them, we note the wave type, the electromagnetic field structure, and the critical frequency. With the respect to each specific type of waves, we mark the propagation constant, the phase coefficient, the attenuation coefficient, and phase and group velocities, their dispersion characteristics, in addition to a wavelength in the transmission line. The wave impedance and the maximal throughput are transmission line parameters.

Depending on the extension in z, electrodynamics structures can be infinite, half-infinite, and finite (Sect. 6.2.4).

We are acquainted with the fundamentals of the element base and concepts of electrodynamics, and now we shall have to consider these structures in their specific representation. To that end, the methods of classical electrodynamics are sufficient. We start by studying waveguide eigenwaves and the general method of variable separation (Sect. 6.3). Concepts of the line proper wave, the longitudinal and transverse wave numbers, and the critical wavelength and the wavelength in the waveguide are introduced (Sect. 6.3.1). We give a representation of the complete field of the guiding structure (Sect. 6.3.2).

Electrical (Sect. 6.4) and magnetic (Sect. 6.5) waves of the rectangular metallic waveguide are examined. The Helmholtz equations (Sect. 2.1) with appropriate boundary conditions constitute the basis of the analysis.

As a result, the eigenfunctions (in trigonometric form) and analytical expressions for eigennumbers are obtained. Volumetric pictures of electromagnetic wave distribution for various wave types in the waveguide cross-section are presented. The reader can reproduce these pictures based on the appropriate Programs P.6.1 and P.6.2.

In Sect. (6.6) the rectangular waveguide with electric and magnetic walls are examined together with the plane waveguide model. Pictures of some eigenfunctions are presented for such a waveguide.

Section 6.7 is devoted to the physics investigation of the rectangular waveguide with a dielectric layer. In this case, the total field is presented in the form of expansion over the complete systems of electric (or transverse-magnetic) and magnetic (or transverse-electric) waves. Each reader can create and observe the field pictures of this interesting and class of transmission lines that is important for practice (Programs P.6.6, P.6.7, P.6.8 and P.6.9).

The reader will obtain a deeper understanding of the physical process from pictures in transmission lines after familiarization with an expansion procedure of the waveguide waves onto plane waves (Sect. 6.8).

The coaxial cylindrical waveguide, which is traditionally used on the transverse electromagnetic (TEM) main wave (Sect. 6.9.1) is the "oldest" transmission line. We investigate its higher order types: electric and magnetic waves. Eigennumbers should be obtained from a transcendent characteristic (dispersion) equation, which should be investigated using numerical methods (Sects. 6.9.2 and 6.9.3). Expressions for the fields of electric and magnetic waves (Sects. 6.9.4 and 6.9.5) are

© Springer International Publishing AG, part of Springer Nature 2019
E. I. Nefyodov, S. M. Smolskiy, *Electromagnetic Fields and Waves*, Textbooks in Telecommunication Engineering,
https://doi.org/10.1007/978-3-319-90847-2_6

presented. Separately, the main wave and its properties are studied (Sect. 6.9.6), and a numerical investigation of the characteristic equations is carried out (Sect. 6.9.7; Programs P.6.10 and P.6.11).

The knowledge obtained at painstaking assimilation of Sects. 6.4, 6.5, 6.6, and 6.7 facilitates to a considerable degree the creation of strip-slot structures (Chap. 7), which are important for practice.

In practice, the circular waveguide (Sect. 6.10) has become widespread. We pay special attention to asymmetrical magnetic waves owing to the unique properties of some of them (Sect. 6.9.8). Consideration of coaxial waveguide properties is completed by an interesting class of waves, called "whispering-gallery" waves (Sect. 6.9.9).

Examination of the transmission line class – the circular waveguide (Sect. 6.10) – is performed according to a well-known scheme: the Helmholtz equation with appropriate boundary conditions. We consider electric (Sect. 6.10.2) and magnetic (Sect. 6.10.3) waves. The characteristic equations are derived and their numerical solutions are discussed (Programs P.6.12 and P.6.13). The typical features of H_{01} and H_{1m} magnetic waves are described, which find interesting practical applications. We note features in the expansion of circular waveguide waves into plane ones (Sect. 6.10.4), when the total field is represented by a non-finite number of eigenwaves, as for the rectangular waveguide (Sect. 6.8), but by their infinite totality. Currents in the waveguide walls are discussed (Sect. 6.10.5). Examples of current distribution for several wave types are given.

Examination of the hollow waveguide structures is completed by information about the elliptic waveguide (Sect. 6.11). We note its capability to eliminate the degeneration of polarization of some waves. It is interesting that a value of a ratio of the ellipsoid axes sizes is close to a value of the golden ratio, and thus, the maximal bandwidth of the single-wave mode is provided at a relatively small attenuation.

This chapter is of fundamental importance for the training of physicists and engineers in practical and engineering activity. For this reason, we draw the reader's attention once more to the persistent necessity for the materials of this chapter in development work using the computer programs included.

6.1   Introduction

We begin our consideration from an analysis of those electrodynamics structures, on the basis of which modern radio engineering, radio physics, radar, radio astronomic, and many other systems and devices, units, and modules of microwave and millimeter-waves are built. The uniform (or regular) *transmission lines* (*a waveguide, the waveguiding structure*) with distributed (in length or in cross-section) parameters. The uniform transmission line is characterized by the constant cross-section and its constancy along the whole line length. The *basing elements* are created on the basis of segments of regular or irregular transmission lines, from which, in turn, the more complicated formations and so-called *functional units* are composed of the microwave and/or millimeter-wave module (block) of radio systems.

The transmission line is intended for guiding of electromagnetic energy flow in the necessary direction. Transmission lines serve for the transmission (conveying, transportation) of electromagnetic energy from a source to a consumer, for instance, from a transmitter to an antenna, and also for the connection of separate parts and units (basing elements and functional units) in the whole radio system.

Many devices of microwave and millimeter-wave ranges are constructed on the basis of uniform or non-uniform segments of transmission lines.

The transmission line is referred to as *uniform* if the cross-section is filled by the *uniform medium* and its shape remains unchanged along the length of the transmission line. We can distinguish between open lines and waveguides. The shield restricting the propagation area of electromagnetic energy in the transverse direction is absent in *open transmission lines*. The one or several surfaces restricting the electromagnetic energy propagation in the transverse direction are necessary in the *waveguides* (*shielded* transmission lines).

6.2   Classification of Transmission Lines and Brief Characteristics of Electromagnetic Wave Properties Propagating in Them

We start our examination from uniform (longitudinally uniform, regular) transmission lines, which are the fundament of devices of engineering electrodynamics.

Fig. 6.1 Cross-sections of the main classes of transmission lines with the T-wave. (**a**) The double-wire transmission line. (**b, c**) Coaxial transmission lines. (**d**) The plane waveguide (**e**) The plane waveguide with spasmodic variation of the cross-section. (**f**) The plane waveguide with a conductor. (**g**) The plane waveguide filled by magneto-dielectric with a conductor. (**h**) The "closed" asymmetric stripline. (**i, k**) Transmission lines of the sector type. (Taken from Nefyodov and Kliuev [73], p. 65)

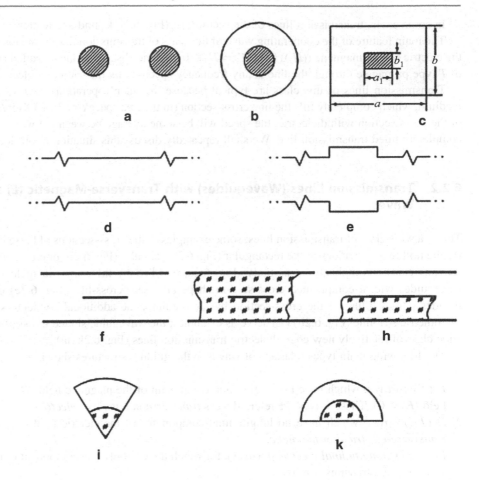

The transmission line classification described below, in the definite regard, is approximate, because it reflects the main (defining) properties only of the aforementioned transmission lines. Frequently, some properties of one class of transmission lines can, probably, to a lesser extent, be manifested in another class of lines.

6.2.1   Transmission Lines with Transverse Waves

Some examples of this class of transmission lines with the transverse (*TEM* or *T*-) wave are shown in Fig. 6.1, where we can see its cross-sections. The one from the first transmission lines was the *double-wire* line (Fig. 6.1a) and the coaxial (Fig. 6.1b) transmission line.

Thus, the shape of the external shield and the internal conductor of the coaxial transmission lines does not play an essential role. This circumstance makes these lines suitable in the constructive connection, because it fits into the necessary structure.

In practice, transmission lines of this type are used when we have instead of two conductors (say, in structures in Fig. 6.1a–f), several conductors. Such a transmission line is called a *multi-conductor*. Other examples of lines when the *T*-wave are lines based on two parallel conducting planes (Fig. 6.1d–g). These are either two planes (Fig. 6.1d), or transmission lines, in which, for example, two rectangular ridges are made in both planes (Fig. 6.1e). There can be several ridges and they may be in both planes or in one of them.

The wide class of transmission lines represent structures based on two parallel planes with a conductor located between them (Fig. 6.1f–h). Thus, the space between the planes can be completely (Fig. 6.1g, h) or partially (Fig. 6.1h) filled by the dielectric of magneto-dielectric.

Transmission lines of this type (together with many others) are the main ones for the *plane* (planar) *integrated circuits* (IC) and the *three-dimensional integrated circuit* (3D-IC)[1] of microwave and millimeter-wave ranges.

---

[1] These integrated circuits of microwave and millimeter-wave ranges are often referred to as 3D-IC.

In some cases, transmission lines of the sector type (Fig. 6.1i, k) find application.

The main feature of the cooperating wave of this class of transmission lines is an absence of the longitudinal component of the electric (E_z) or magnetic ((H_z)) field ($E_z = H_z \equiv 0$) and the absence of the so-called *critical frequency* ($f_{cr} = 0$), i.e., waves of T type propagate through the line at any frequency specified by the source of electromagnetic oscillations.

Transmission lines of this class are typical because the main (operation) T-wave propagates with light speed c in the medium, which completely fills the line cross-section (in the free space $c \approx 3 \cdot 10^{10}$ cm/s) (Fig. 6.1b–g). At incomplete filling of the cross-section with dielectric, the speed will be some average between the wave speed in the hollow (unfilled) and the completely filled transmission line. We shall repeatedly discuss this situation in our lecture course.

6.2.2 Transmission Lines (Waveguides) with Transverse-Magnetic (*E*) and Transverse-Electric (*H*) Waves

This is a wide class of transmission lines; some examples of the cross-sections of these lines are shown in Fig. 6.2. First, these are the hollow waveguides of the rectangular (Fig. 6.2a), circular (Fig. 6.2b) or elliptic (Fig. 6.2c) cross-sections. Their cross-sections can be completely or partially (Fig. 6.2d, e) filled by the magnetic–dielectric medium. In practice, the hollow waveguides with a complicated cross-section shape are used: cross-like (Fig. 6.2e) or U-shaped (Fig. 6.2g) waveguides. Besides partial filling of the cross-section by the medium, the additional conductors can be located there. The screened asymmetric stripline (Fig. 6.2h) can serve as examples, the waveguide slotted transmission line (Fig. 6.2i), in addition to the wide class of relatively new edge-dielectric transmission lines (Fig. 6.2k–m).

The following main types (classes) of waves in the guiding structures shown in Fig. 6.2 are:

1. *E (TM)-waves*, which have no longitudinal component of the magnetic field ($H_z = 0$). They have a longitudinal electric field ($E_z \neq 0$). These waves are referred to as *transverse-magnetic* or *electrical*.
2. *H (TE)-waves*, which have no longitudinal components of the electric field ($E_z = 0$). These waves are referred to as *transverse-electrical* or *magnetic*.
3. *LE-waves* (*longitudinal-electrical waves*), for which there is only one component of the electric field ($E_z \neq 0$) in the cross-section of the transmission line.
4. *LM-waves* (*longitudinal-magnetic waves*), for which there ise only one component of the magnetic ($H_z \neq 0$) in the cross-section of the transmission line.
5. *Hybrid waves of the type HE or EH*, for which the presence of all six components of the electromagnetic field is typical.

Fig. 6.2 Cross-sections of transmission lines with different types of waves. (**a–c**) Rectangular, circular, and elliptical waveguides (*H*- and *E*-waves). (**d, e**) The partially filled rectangular waveguide (*LE*- and *LM*-waves). (**f, g**) The cross-like and U-shaped waveguides (*H*- and *E*-waves). (**h**) The screened asymmetric stripline (hybrid waves). (**i**) The waveguide-slotted line (hybrid waves). (**k–m**) The edge-dielectric lines (hybrid waves). (Taken Nefyodov and Kliuev from [73], p. 63)

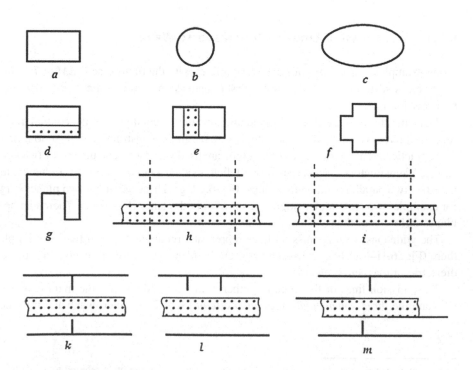

Waves of H- and E-type exist in waveguides with the uniform dielectric filling (Fig. 6.2a, b, e, g). The *critical frequencies* of these waves are nonzero and depend on the shape and dimensions of the cross-section, and on the parameters of the filling dielectric. For waveguides with a *coordinating* shape of the cross-section (rectangular, circle, ellipse, etc.; 12 coordinate systems are known, into which variables in the *wave equation* are divided), the exact formulas exist or the dispersion (usually transcendent) equations for calculation of the critical frequency f_{cr}.

Waves of the LE and LM types are typical for partially filled longitudinal-uniform waveguides (Fig. 6.2d, e). *Critical frequencies* and *propagation constants* for them are defined, as a rule, numerically from the solution of *dispersion equations* (usually, transcendent equations). The *analytical* (a priori) *approximate results* are possible. Sometimes, the *approximation estimations* are known. In our lecture course, there are corresponding *examples* from the theory of strip-slotted structures.

6.2.3 Transmission Lines of the Open (Semi-Open) Types, Surface Waves

Some examples of transmission lines of this class are shown in Figs. 6.3 and 6.4. Transmission lines based on dielectric (or magnetic–dielectric) structures are shown in Fig. 6.3. First of all, this is a dielectric layer in the free space (Fig. 6.3a) and the layer on the conducting substrate (Fig. 6.3b).

The electromagnetic field of the wave beyond the layer on the vertical coordinate decreases according to the exponential law, i.e., quickly enough. The main part of the wave energy is concentrated in the layer. Other examples are the *rectangular dielectric waveguide* (Fig. 6.3c) and the same on the conducting substrate (Fig. 6.3d; the *mirror dielectric waveguide*). The electromagnetic field of the waves on both coordinates in the figure plane, beyond the waveguide, decreases according to the exponential law.

Wave-guiding structures of the type shown in Fig. 6.3a–d are *open structures*. If, for instance, the ideally conducting electric (or magnetic – see later) walls are placed in the plane waveguide (Fig. 6.3a), as is shown by dashed lines, then this structure can be conditionally referred to as a structure of the semi-open type: the energy can propagate to the left and to the right in the figure plane. More complicated guiding structures of this class are possible, for instance, the double- or multi-layer *mirror waveguide* (Fig. 6.3e), the *mirror waveguide with an active layer* (Fig. 6.3f), on the basis of which the base elements of the type of an *amplifier, oscillator, detectors,* etc., can be constructed. These structures are convenient for constructing the

Fig. 6.3 Some examples of transmission line cross-sections with dielectrics. (**a**) The dielectric layer. (**b**) The dielectric layer on the conducting substrate. (**c**) The rectangular dielectric waveguide. (**d**) The rectangular dielectric waveguide on the conduction substrate. (**e**) The mirror waveguide. (**f**) The mirror waveguide with the active layer. (**g**) The dielectric waveguide of the circular cross-section. (**h**) The Goubau line. (**i**–**l**) Optical fibers. (Taken from Nefyodov and Kliuev [73], p. 65)

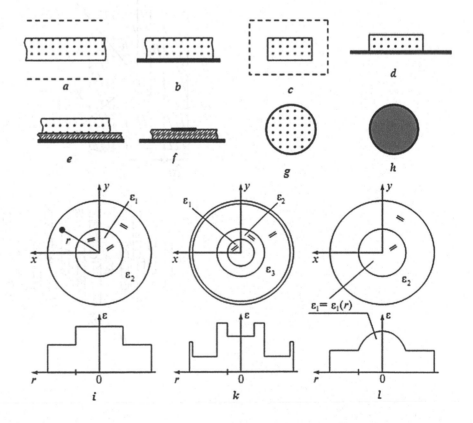

128

Fig. 6.4 Examples of transmission lines delay line structures $p/\lambda \ll 1$. An array from (**a**) circular and (**b**) rectangular conductors. (**c**) Two-dimensional array from rectangular waveguides. (**d**) The multi-wire line. (**e–n**) Arrays of various profiles. The common arrow indicates the direction of signal transmission (on z). (**o**) The array from anisotropic bands. (**p**) Two-dimensional periodic array from ideal cylindrical conductors. (**q**) Two-dimensional double-element array. (**r**) Double-element array from anisotropic conducting bands. (Taken from Nefyodov and Kliuev [73], p. 47)

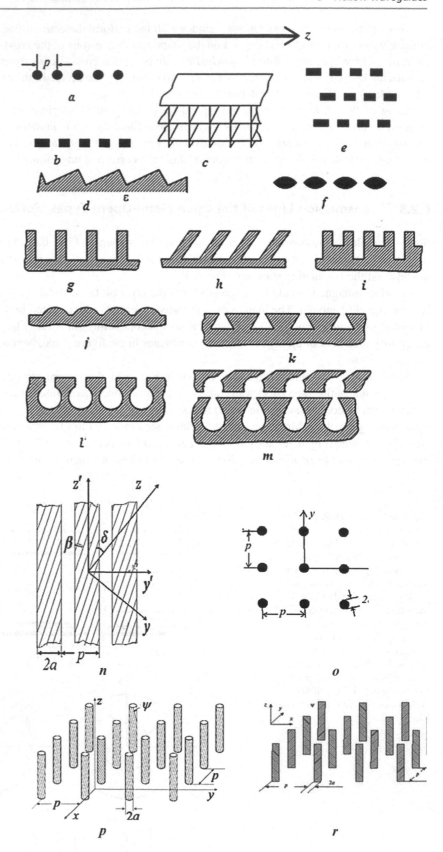

transmission lines between basic elements (at the creation of functional units) and the basic element itself for 3D-IC of the millimeter-range.

The dielectric waveguide of the circular cross-section (Fig. 6.3g) and also the single-wire of surface waves – the conducting rod covered by the dielectric layer (Fig. 6.3h; the *Goubau line*) – have wide distribution. The dielectric waveguide of the circular cross-section served as the basis for the creation of the wide class of transmission lines – *light guides*. Some examples of cross-sections of the light guides are shown in Fig. 6.3i–l. Profiles of refraction indices *n* of the fiber optical light guides are shown here. In the USSR, research into the fiber-optical light guides was begun at the initiative of academicians *V.A. Kotelnikov* and *A.M. Prokhorov*. Essential contributions to the development of native light guiders was made by Professors *D. I. Mirovitskiy* and *A.V. Sokolov*.

So-called delay lines (Fig. 6.4a–n) form the second group of transmission lines with surface waves. Usually, the delay line represents the part of the corrugated surface; thus, profiles can be manifold: "sine" corrugation (Fig. 6.4a), the comb structure (Fig. 6.4b), the helical waveguide (Fig. 6.4c), and others. The small size of the structure period *p* compared with the wavelength λ: $p/\lambda \ll 1$ is a typical property of the delay-line structures (see, for example, Wood, Ramo and Whinnery, Snyder and Love, Unger, Marcuse, and Buduris and Shevenie [103–107, 109]).

Such structures are usually referred to as *frequently periodic*. The periodic structure can be placed in a rectangular plane or other waveguide. Possible screens forming in these cases the plane or rectangular waveguide are shown by the dotted lines in Fig. 6.4a–c. In IC and 3D-IC devices of microwave and millimeter-waves, the periodic systems of the type shown in Fig. 6.4f are used when the band conductor is placed on the dielectric substrate mounted on the conducting base. Similar structures are used as emitting systems in the various *antenna-waveguide systems* of ultra-fast systems (phased antenna arrays, active and adaptive phased antennas arrays, and many others. In electronic microwave devices, for instance, structures of a cylindrical tube with periodic cut-outs (Fig. 6.4f) are used. The "similar" structures are applied in acceleration engineering.

In practice, so-called periodic resonance structures are widely used, when integer (or almost integer) numbers of half-wavelengths are packed up on the period $p : p = n\lambda/2$, $n = 1, 2, 3\ldots$ (*resonance condition*); see later Eqs. (6.7 and 6.8)). Some of these transmission lines will be considered sufficient in our lecture course.

It should be certainly noted that periodic structures in physics and engineering, and in *biophysics, bio-energy-informatics*, from old, attract the attention of scientists and engineers. In particular, many publications are devoted to the study of engineering applications of periodic systems.

6.2.4 Quasi-Optical Transmission Lines and Some Elements of the Quasi-Optical Path

At the beginning of the 1960s, in connection with penetrating into the millimeter-wavelength range and into the optical range, a new discipline appeared – *quasi-optics*. In a certain sense, it took an intermediate position between microwave electrodynamics and *geometric optics*. The well-known optics devices: mirrors, lenses, prisms, etc., formed the basis of quasi-optical engineering.

However, their analysis and also, in some cases, the construction itself needed an additional specification by electrodynamics methods. Examples of this class of devices are shown in Fig. 6.5c–e.

Seemingly, one of the first transmission lines of the quasi-optical type was the *lens line* (Fig. 6.5a), consisting of a number of similar dielectric lenses located on the same (but large enough compared with the size of the lens itself *a*) distance *L*. Thus, the usual *quasi-optical conditions* are $a \gg \lambda$, $L \ll a$, or, if to introduce into consideration the wave number of the free space $k = 2\pi/\lambda = \omega/c$, where λ, ω, *c* are the wavelength, the frequency, and the light speed respectively, the quasi-optical conditions take the form: $ka \gg 1$, $kL \gg 1$, $L \gg a$. Naturally, the same conditions are saved for other quasi-optical transmission lines.

The *diaphragm line* (see Fig. 6.6b) representing the periodically located (for instance, ideally conducting) shields with holes (see Sect. 9.3) is an interesting example.

The dielectric prism in breaking of the wide ($ka \gg 1$) waveguide (Fig. 6.5a, c) is the one of examples of quasi-optical devices, say, with the *mirror* in the waveguide breaking on an adequate angle ϑ (including approaching or even equal to 180°; Fig. 6.5b–d). Thus, the prism is used to overcome the relatively small (by the angle ϑ) transmission line rotation. In particular, in an example of prism operation in breaking of the wide waveguide, the main quasi-optics *principle of mutual compensation* ([52]; see Sect. 9.10) was discovered in due course. Slightly changed (compared with the prism constructed on the geometric–optical principal) shape of the prism is shown in Fig. 6.5c. This scheme takes into account that the lens is still not within the optical range; here, we can use, for example, the principle of *physical optics*.

The rotation mirror both in the shape (Fig. 6.5d) and in its opposition (Fig. 6.5e) can undergo similar variations. The *couplers* or *channel separators* (Fig. 6.5f, g) including adjustable ones (Fig. 6.5h) form a wide class of devices in our

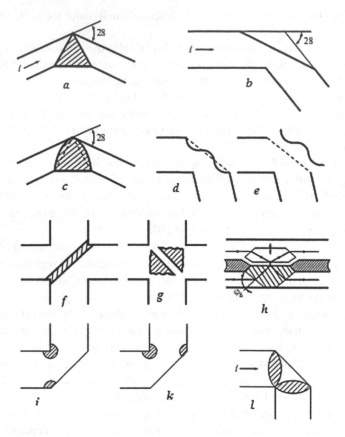

Fig. 6.5 Some quasi-optical devices. (**a**) The dielectric prism and (**b**) the mirror in breaking of the plane wide waveguide. (**c**) The compensated dielectric prism, and (**d, e**) the mirror in breaking of the plane wide waveguide. (**f, g**) Couplers. (**h**) Adjustable channel separator. (**i–l**) The combined mirrors in breaking of the plane wide waveguide. (Taken from Nefyodov and Kliuev [73]. p. 67)

frequency ranges. Thus, this adjustment can be both mechanical at the expense of displacement of the prism part, as shown in Fig. 6.5h, and electrical owing to the variation of the refraction index of the separate magnetic–dielectric prism. This principle is especially widely used in the optical range.

Finally, the combined devices are widely used. Thus, Fig. 6.5i–k shows variants of the mirror and dielectric cylinders (Fig. 6.5i, j), two lenses (Fig. 6.5k) in waveguide breaking, and many others.

In classic optics (partially known from secondary school and university physics) the *Porro* prism, the *penta-prism* and the *diamond prism*, which use a phenomenon of full internal reflection and are applied as mirrors, are widely used (at least as "zero" approximation). The *Dove prism* inverses an "image" that does not change; that of the propagation direction of the light beam. Interesting transformations with the almost parallel (paraxial) beam can be executed using the *Rochon* and *Wollaston prisms*. They give the possibility of obtaining from the one beam of two beams polarized in mutually-opposite planes and propagating in different directions. The Wollaston prism gives greater beam resolution compared with the Rochon prism. The last one keeps the propagation direction of one of the beams unchanged. If the wedge angles in the Rochon prism are small (so that beam separation will not practically happen), then we shall come to the wide class of *compensators*, from which, probably, the *Babine compensator* is the most often used. The main purpose of compensators is the determination of small admixtures of the polarized light in the mixture of natural and polarized light.

The significantly larger degree of separation of ordinary and extraordinary beams can be obtained in the *Nicol, Foucault, Glan–Thomson prisms*, and in their variations, such as, for instance, the *Cornu prism*, etc.

A series of examples for dielectric prism application in the quasi-optical transmission lines is shown in Fig. 6.7. Among them: two examples of the rotating prism in the surface wave line (Fig. 6.7a), the lens line (Fig. 6.7b), and also the prism as the *directional coupler* in the lens line (Fig. 6.7c). Other examples of dielectric prism utilization in quasi-optical transmission lines are presented in Sect. 9.3.

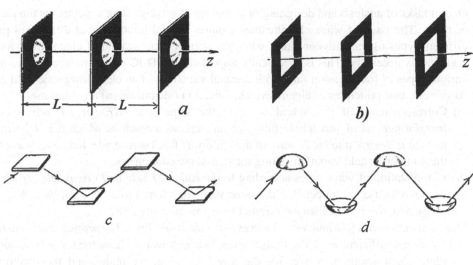

Fig. 6.6 Simplest quasi-optical transmission lines. (**a**) The lens (dielectric lens in absorbing infinite shields). (**b**) The diaphragm (holes, for example, of rectangular shape in absorbing infinite shields). The mirror quasi-optical lines with (**c**) plane and (**d**) focusing mirrors

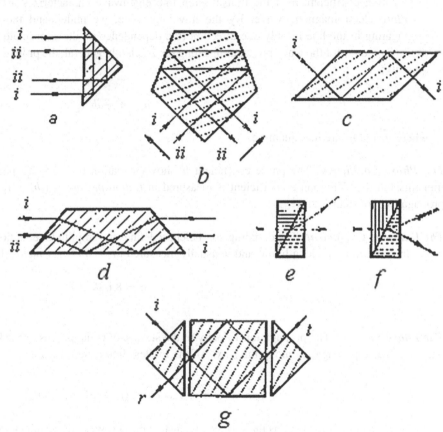

Fig. 6.7 Schemes of some dielectric prisms. (**a**) The Porro prism. (**b**) The pentaprism. (**c**) The diamond prism. (**d**) – The Dove prism. (**e**) The Rochon prism. (**f**) The Wollaston prism. (**g**) The interference Fabry–Perot modulator with prisms of full internal reflection

6.2.5 Main Characteristics of Transmission Lines

At selection of the transmission line type for application in the same device, the following electrical characteristics and parameters are of critical importance.

1. *The wave type, the structure of electromagnetic field and the critical frequency.* These characteristics are defined from a solution of the appropriate boundary task about eigenwaves of the transmission line. As a rule, transmission lines are used in the *main wave* mode, which has the least critical frequency f_{cr}. Nevertheless, in some cases, the preference is given to waves of *higher types* with a critical frequency exceeding the main wave frequency. Different examples of higher types of waves in the transmission lines are contained, for instance, in Nefyodov et al., Harmuth, and Kasterin [21, 22, 78].

One of tasks of analysis and designing of complicated wave-guiding structures is the problem of *eigenwave classification* of types. The issue of wave classification acquires special importance at diffraction problem solution at the joining of different types of transmission lines, when the correct identification of eigenwaves for the joining of transmission line segments is necessary. This is specifically important for 3D-IC of microwave and millimeter-wave ranges, in which the various types of transmission lines with unequal structures of an electromagnetic field are joined.

At present, two principles of eigenwave classification (identification) are often used.

(a) Correspondence of the hybrid wave to the same type (*HE*- or *EH*-waves) is defined by means of *limited transformations* of waveguide filling parameters, as a result of which the *HE-wave* is transformed to the *H-wave*, and the *EH-wave* into the *E-wave* of the uniformly filled waveguide. Indices at wave type designations usually indicate the number of field variations along the transverse coordinates.

(b) Classification of wave types according to the *ratio of phase and group velocity directions*. Waves whose phase and group velocities are directed to the same side, are referred to as *direct waves*. Waves with opposite directions of the phase and group velocities are referred to as *inverse waves* [96].

2. *Main characteristics (parameters) of waves in transmission lines. The propagation constant (propagation coefficient).* The propagation coefficient h of the transmission line eigenwave characterizes a variation of amplitude and phase of the *traveling* electromagnetic wave. By the traveling wave, we understand the electromagnetic wave of definite type propagating in the line in only one direction (the dependence on the coordinate z and time t is defined by the multiplier $\exp\{i(hz - \omega t)\}$ for the wave propagating along the z-axis). In general, the propagation coefficient of the wave is a complex quantity:

$$h = h' + ih'', \tag{6.1}$$

where h' and h'' are real quantities.

The Phase Coefficient   The phase coefficient h' shows variation of the wave phase at passing of the length unit of the transmission line. The phase coefficient is measured in *radian per meter* ($[h] = $ rad/m) and is equal to the real part of the propagation constant $h \equiv h'$.

The Damping Coefficient   The damping coefficient α defines the amplitude reduction of the electromagnetic wave at passing of the transmission line length unit and is usually measured in *decibels per meter* ($[\alpha] = $ dB/m). The damping coefficient

$$\alpha = 8.68 h''. \tag{6.2}$$

The Phase Velocity   The propagation velocity of the surface of equal phases of the harmonic wave is its phase velocity. The harmonic wave propagating along the z-axis of the lossless line is described as

$$\vec{E}\left(\vec{\rho}, z, t\right) = \vec{e}\left(\vec{\rho}\right)\exp\{i(hz - \omega t)\}. \tag{6.3}$$

The wave front of this wave as the surface of constant phases, which does not change in motion, should satisfy the equation $hz - \omega t = $ const. Using the usual definition of the velocity, we obtain the phase velocity of the plane wave

$$v_{\mathrm{ph}} = dz/dt = \omega/h. \tag{6.4}$$

The Wavelength   The wavelength λ is the distance passed by the surface of equal phases over the oscillation period T. As $T = 2\pi/\omega$, then

$$\lambda = v_{\mathrm{ph}} T = 2\pi/h. \tag{6.5}$$

The Dispersion Characteristic The function of the *phase velocity* vs frequency is referred to as a dispersion, and the dispersion characteristics represent the specific form of dependence given by the formula or the plot.

Transmission lines with *T-waves* have no dispersion and the phase velocity in them at any frequency is equal to the propagation velocity of the plane electromagnetic wave in the medium, which fills the transmission line $v_{ph} = c/\sqrt{\varepsilon\mu}$, where c is the light speed in a vacuum, ε, μ are permittivity and permeability of the medium respectively.

Transmission lines operating on the other wave types have dispersion. In general, the dispersion characteristic for the transmission line is absent in the explicit form (except for *H*- and *E*-waves); it is defined numerically on a personal computer from a solution of the dispersion equation for the transmission line.

Group Velocity *Group velocity* is the velocity of propagation of an *envelope* of the complicated signal (the group of waves). The concept of group velocity is introduced in the case of a dispersed transmission line and for complicated signals consisting of oscillations of several frequencies, when the concept of the phase velocity is inapplicable. The *group velocity* v_{gr} is equals [54, 125]

$$v_{gr} = d\omega/dh. \tag{6.6}$$

The characteristic and wave impedances of the transmission lines are the ratio of transverse components of $\vec{E}$ and $\vec{H}$ vectors. This parameter is defined by the wave type used in the transmission line. Introduction of the characteristic impedance Z of the transmission line has a sufficiently artificial character and it can be introduced, at least, in the three ways described in Sect. 6.13.

The wave impedance is introduced in a similar manner to the wave impedance of the long line as the ratio of a voltage and a current.

In spite of the conditional nature of introduction of the *characteristic impedance* concept for the wave-guiding structures, it is *convenient* because it allows the introduction of a series of simplifications and the theory of linear structures based on properties of the *scattering matrix* to be used.

Maximal Throughput Power It is restricted by the electric breakdown or by overheating of conductors and isolators of the transmission line. Power-carrying capacity in the transmission line is usually accepted to be equal to 25–30% of the *critical power*, which causes a breakdown or overheating in the mode of the purely traveling wave.

6.2.6 Classification of Electrodynamics Structures

Electrodynamics structures can be *infinite*, *semi-infinite*, and *finite* (for details see Gridin et al., Ufimtsev, Grinberg, and Nefyodov and Fialkovsky [16, 71, 76, 98]) in their construction with regard to dependence of their properties on the longitudinal coordinate (z). The transmission lines considered are typical examples of *infinite structures*. Another example of such structures is the infinite length plane, the unrestricted magnetic–dielectric layer, etc.

Some typical examples of the *semi-infinite structures* are shown in Figs. 2.3a–c, on the right. On the left, the appropriate *finite structures* are presented. Therefore, the semi-infinite plane (Fig. 2.3a) corresponds to a band; the wave *i* falls on both structures. The scattering fields are formed from their edges caused by diffraction of the incident wave on their structures. Semi-infinite structures are referred to as the *key structures*.

Many of them permit the *strict solution* or the sufficiently "good" *approximate solution* by any method. Having at our disposal the solution for the key structure (the *key problem*), we can construct the approximate solution for the *finite structure*. Sometimes, therefore, we can obtain an estimation of the *solution error*.

The system of two (or several) parallel bands (Fig. 2.3b) corresponds to the system of two (or, generally speaking, arbitrarily many) parallel semi-infinite planes (layers). One of its eigenwaves (*i*) comes to the open end of such a plane waveguide. As a result of diffraction on the open end of the plane waveguide, a part of energy of the incident wave is emitted into the free space, whereas a part is reflected, and if the mode in the transmission line is single-wave, from the open end, the wave (*i*) of the same type as the incident wave is reflected. In the finite structure (Fig. 2.3b, left) the superposition of incident and reflected waves from both ends of the structure occurs. Under definite conditions (see later), as a result of this superposition, the resonance oscillations are formed, and therefore, this structure is sometimes referred to as the *band open resonator.*[2] An example of micro-strip transmission line formation from the semi-infinite structure is a plane waveguide with a dielectric layer as shown in Fig. 2.3c, d (left). Thus, if the incident wave normally falls in the open end (the incidence angle $\alpha = 0$), then it is the same band open resonator (compare with Fig. 2.3b; on the left) with dielectric filling. If the incidence angle $\alpha \neq 0$ (Fig. 2.3d, left), then under definite conditions (the refraction angle $\beta = 0$), about which we shall speak later, the finite structure (Fig. 2.3c, d; on the left; see also Sect. 7.4, Fig. 6.4) can be the transmission line, which is usually referred to as the *asymmetric strip* (or *micro-strip*) line. If the shield (Fig. 2.3b, left; the band open resonator, vertical dotted lines) is introduced into the construction of the asymmetric strip line, then it will be the shielded (of course, partially, because energy can propagate in the transverse direction) asymmetric stripline. The presence of such a shield protects processes in the asymmetric stripline and in basing elements built on asymmetric strip lines, against the external impact, prevents the energy emission from the line, etc. For 3D-IC of sufficient long-wave range, such shields serve as the inter-floor overlap [14, 18, 110].

In the more general case (Fig. 2.3), we can form the finite structure from any infinitely-long (along the z-coordinate) transmission line by cutting some of the segment with the help, for instance, of ideally conducting planes (Fig. 2.3b) by vertical dotted lines.

Thus, superposition of incident (i) and reflected ($-i$) waves under the condition when at the distance $2l$, the integer number of half-wavelength is packed: $l = q\lambda/2$, $q = 1, 2\ldots$ or the same,

$$2kl = \pi q, q = 1, 2, \ldots \tag{6.7}$$

gives the so-called picture of *standing waves* and the presented condition is referred to as the *condition of a resonance*. If reflection from the planes restricted the transmission line is incomplete as, for instance, in structures in Fig. 2.3b, the mentioned resonance condition should be added, and it has the form:

$$2kl = \pi q + 2\pi p, \tag{6.8}$$

where p is the small in modulus complex quantity ($|p| \ll 1$), which is connected to the energy emission of the incident wave from the structure's open end into the free space, in addition to its conversion into other wave types (see Sect. 7.3). Thus, for open finite structures (of the type shown on the left in Fig. 2.3b, c), it is important that the resonance condition is fulfilled on the vertical coordinate, i.e., on the distance $2l$ the almost integer number of half-wavelength would be packed (because of complex adding p). Sometimes, it is useful to use a concept of the *complex resonance frequency* of the open resonator $\omega_{res} = \omega' - i\omega''$.

Of course, as a matter of fact, there is no complex frequency in nature at all, but its formal introduction is convenient for an analysis. So, for example, the Q-factor of the resonator is defined as $Q = \omega'/2\omega''$, which is clear and suitable [94, 106, 107]. We note that in open resonators of the optical range, usually $q \gg 1$ [111], and for millimeter and sub-millimeter ranges, the q quantity may be finite ($q = 1, 2\ldots,$).

6.2.7   Conclusions

We have become acquainted with the fundamentals of the element base of engineering electrodynamics; namely, with a concept of the transmission line (a waveguide), a resonator, which is arranged from a segment of this line with length-constant geometrical and physical parameters. In the future, we have to learn that a resonator can be constructed on the irregular transmission line segment or the transmission line with non-uniform physical parameters, etc. We described very briefly the main electrodynamics characteristics of transmission lines. Now, we have to assimilate the main laws of classical macroscopic electrodynamics with foundations of an analysis of different specific transmission lines and basing elements constructed based on it. Thus, the main task is understanding and assimilating the physical side of things, because if and only if such an understanding is achieved, we can begin an analysis and synthesis (creation of basing elements and functional units based in advance on the given characteristics) of any engineering electrodynamics devices.

---

[2] *Prokhorov A.M.* About the molecular amplifier and oscillator in sub-millimeter range (in Russian) // JETF, 1958, vol. 34, № 6, p.1658–1659.).

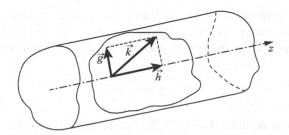

Fig. 6.8 Diagram of wave vector expansion of the free space $\vec{k}$ in the same arbitrary waveguide onto the longitudinal $\vec{h}$ and transverse $\vec{g}$ components. (Taken from Nefyodov and Kliuev [73], p. 150)

6.3 Eigenwaves of the Waveguide, the Variable Separation Method

6.3.1 Task Formulation, Eigenwaves

Solutions to uniform *Helmholtz equations* for field quantities (3.1), (3.2) or (the same) for scalar and vector potentials (3.7) satisfying the "natural" boundary conditions are referred to as *eigenwaves* of the waveguide (Fig. 6.8; as any other transmission lines). In the absence of the source $\left(\vec{j}_{\text{out}} = 0, \rho_{\text{out}} = 0\right)$, equations for the field quantities coincide, although, of course, the six equations for field quantities remain, whereas for potentials, we have only four equations (plus a condition of their calibration; *the Lorentz condition* [12, 43]). However, as we mentioned at the end of Sect. 4.2, in the cylindrical wave, we may always use the longitudinal field components $E_z \vec{e}_z$, $H_z \vec{e}_z$ in essence – *Hertz vectors* – and make with two equations only (for instance, a type for Eqs. (3.9)).

The *variable separation method* (the *Fourier method*) is an effective method of *wave equation* solution of the *Helmholtz equation* in addition to other equations of mathematical physics. We already became acquainted earlier (Sect. 5.9; (5.70)) with the first step of this approach, when we represented fields $\vec{E}\left(\vec{r}, t\right)$, $\vec{H}\left(\vec{r}, t\right)$, extracting from these the time dependence exp $\{-i\omega t\}$ as the multiplier:

$$\vec{E}\left(\vec{r}', t\right) = \vec{E}\left(\vec{r}'\right)e^{-i\omega t}, \quad \vec{H}\left(\vec{r}', t\right) = \vec{H}\left(\vec{r}'\right)e^{-i\omega t}. \tag{6.9}$$

Assumption that in the linear system all field components depend on time t in a similar manner was the foundation for such *variable separation* $\vec{r} \equiv \vec{r}\ (x, y, z)$ (in rectangular coordinate system) or $\vec{r} \equiv \vec{r}\ (r, \varphi, z)$ in the cylindrical coordinate system and t. Now we have to do one more step in this direction; namely, to make use of the fact that in the cylindrical wave all field components depend on the longitudinal coordinate z similarly; namely, as exp$\{ihz\}$ and then

$$\vec{E}(x, y, z) = \vec{E}(x, y)e^{ihz}, \quad \vec{H}(x, y, z) = \vec{H}(x, y)e^{ihz}. \tag{6.10}$$

Thus, (if the time function is exp$\{-i\omega t\}$) the wave (6.10) propagates toward the positive z-axis. The h quantity that appears in (6.10) is the *longitudinal wave number* (with dimension cm^{-1}) defining the *propagation constant* $\gamma = -ih$ and replacing (for the waveguide) the wave number of the free space $k = 2\pi/\lambda = \omega/c$ in the plane wave (see Chap. 5).

In general, h is the complex quantity $h = h' + ih''$; thus, the phase velocity v_{ph} and the wavelength in the waveguide Λ are determined as:

$$v_{\text{ph}} = \omega/h' = ck/h', \quad \Lambda = 2\pi/h' = \lambda k\sqrt{\varepsilon\mu}/h'. \tag{6.11}$$

The imaginary part of the propagation constant h'' defines the wave damping in the transmission line: exp$\{-h''z\}$. It is caused by either losses in the medium that fills the waveguide ($\varepsilon = \varepsilon' + i\varepsilon''$, $\mu = \mu' + i\mu''$), or at the expense of losses in the walls (because of their finite conductivity), or by emission from, say, longitudinal or transverse slots in one (or several) of the walls or by emission from some holes.

Waves in the guiding structure may be *fast*, $h < k$ (or $h' < k$, and then $v_{ph} > c$), and *slow*, $h > k$ ($h' > k$). The slow wave loses the wave character, does not carry an energy, and exponentially dampens: $\exp\{-h''z\}$.

The propagation constant h of the wave in the transmission line is connected to the wave number k as:

$$h = \sqrt{k^2 - g^2},\tag{6.12}$$

where the *transverse wave number* is designated as g. In the future, we shall see that g is defined and depends *only on geometry and dimensions of the cross-section* of the precisely considered waveguide.

Generally speaking, the wave number of the free space is the vector quantity $\vec{k}$ and, therefore, longitudinal and transverse wave numbers are also the vector quantities: $\vec{h}$, $\vec{g}$ (Fig. 6.8), so that

$$h = k\cos\alpha, \quad g = k\sin\alpha.\tag{6.13}$$

In (6.12) the only value of k is known to us ($k = \omega/c$; the operating frequency ω is given by the field source), values of g and h are subject to determination. In contrast to waveguides, in resonators, the k value, in essence, the *resonance frequency*, should be determined (see Chaps. 11 and 12).

If the homogeneous medium completely fills the waveguide cross-section, then k in Eq. (6.12) should be changed by $k\sqrt{\varepsilon\mu}$. Thus, two quantities are in Eq. (6.12) under the square root sign, one of which (k) depends only on the frequency (if parameters of filling ε, μ do not depend on it; no dispersion in the filling medium $\varepsilon \neq \varepsilon(\omega)$, $\mu \neq \mu(\omega)$), and the second (g) depends only on the geometry that defines the wave *dispersion* in the line $h = h(\omega)$. At very high frequencies, say, in the optical range, the material dispersion $\varepsilon = \varepsilon(\omega)$, $\mu = \mu(\omega)$ may be significant.

The longitudinal wave number h, as it follows from Eq. (6.12), for the hollow (nonfilled) waveguide is defined as

$$h = \sqrt{k^2 - g^2} \quad \text{or} \quad h/k = \sqrt{1 - (g/k)^2}.\tag{6.14}$$

If here are no losses in the waveguide walls and in the filling medium, then the propagation constant h is the real and positive number for fast waves $h > 0$, i.e., $k > g$. At $k = g$ the number h becomes zero ($h = 0$) and the wave stops propagating. The wavelength and the frequency of this type of wave, for which $h = 0$, are referred to as the *critical wavelength* λ_{cr} and the *critical frequency* ω_{cr} (or λ_{cr}) respectively. At frequencies $f < f_{cr}$ ($\lambda > \lambda_{cr}$), the wave does not propagate.

If there are losses at $h' = 0$, but then $h'' \neq 0$, i.e., the wave propagates with damping according to the exponential law $\exp\{-h''z\}$. At small losses, damping will be insignificant and the wave can "propagate" to the same distance. The idea of the creation of some miniature basing elements of the waveguide microwave engineering is based on this principle.

For all types of waveguide waves, the *critical wavelength* λ_{cr} exists. Hence, taking into consideration the *wavelength in the waveguide* as $\Lambda_{mn} = 2\pi/h_{mn}$ and the critical wavelength $\lambda_{cr} = 2\pi/h_{mn}^{(cr)} = 2\pi/g_{mn}$ (g_{mn} is the transverse wave number), we obtain

$$\Lambda_{nm} = \lambda\left(1 - \left(\lambda/\lambda_{mn}^{(cr)}\right)\right)^{-1/2}.\tag{6.15}$$

From Eq. (6.15) we see that always $\Lambda_{nm} = \lambda$. When λ approaches a critical value $\lambda \to \lambda_{mn}^{(cr)}$, the wavelength in the waveguide $\Lambda_{nm} \to \infty$, i.e., the wave seems to lose its wave character.

Now, taking into consideration Eq. (6.13), we write for the phase and group wave velocities in the waveguide the following equations that are suitable, generally speaking, for waveguides of any shapes of cross-sections (compare Fig. 6.8):

$$v_{ph} = \frac{c}{\sqrt{1 - \left(\lambda/\lambda_{mn}^{(cr)}\right)^2}}, \quad v_{gr} = \frac{c}{dh_{mn}/dk} = c\sqrt{1 - \left(\lambda/\lambda_{mn}^{(cr)}\right)^2}.\tag{6.16}$$

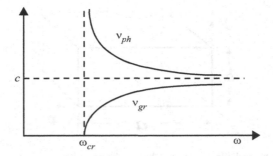

Fig. 6.9 Behavior of phase (v_{ph}) and group (v_{gr}) velocities of the wave in the waveguide as a function of harmonic wave frequency for absence of loss: $\varepsilon'' = \mu'' = 0$; c is the light speed (see later, formulas (6.16))

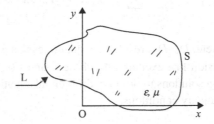

Fig. 6.10 Cross-section of the regular ($\partial/\partial z = 0$) cylindrical closed waveguide; L, S are a perimeter and area of the cross-section respectively

It is interesting to know that the product of these wave velocities in the waveguide is the constant quantity equal to

$$v_{\text{ph}} v_{\text{gr}} = c^2. \tag{6.17}$$

The function of velocities v_{ph}, v_{gr} versus frequency is shown in Fig. 6.9. We see that at the limit, at very high frequencies, they tend toward their own "natural" limit – the light speed c or to $c/\sqrt{\varepsilon\mu}$ - in the infinite space (or in the transmission line with a very large cross-section) with the refraction index $n = \sqrt{\varepsilon\mu}$.

In general, the variable separation method is applicable only when a contour of waveguide cross-section L coincides with the coordinate lines of the one of 12 known coordinate systems (rectangular, cylindrical, spherical, elliptical, etc.) [53, 112]. Then, in the rectangular system, we represent, for instance, the Hertz electrical vector in the form of a product of two functions

$$\Pi_z(x, y) = X(x)Y(y), \tag{6.18}$$

one of which depends on the x-coordinate only, and the second depends on y only.

Now, assuming that a solution in the form (6.18) must be substituted in the Helmholtz equation (in partial derivations), and taking into account the independence of x- and y-coordinates, to obtain two *ordinary dispersion equations* with regard to functions $X(x)$ and $Y(y)$. Finding out its solution, we should satisfy the given boundary conditions on the contour L (Fig. 6.10). The transverse number g plays the role of the *separation constant*.

6.3.2 The Total Field of the Guiding Structure

Mentioned in Eqs. (6.15) and (6.16), indices m, n are defined as a type of eigenwave in the transmission line. Strictly speaking, we have to write everywhere, instead of the propagation constant h, the quantity h_{nm}, emphasizing that there is an infinite totality of eigenwaves in the transmission line, and that the total field should be represented by their sum (independent eigenwaves)

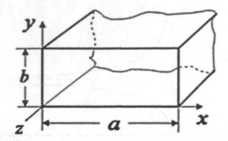

Fig. 6.11 The rectangular waveguide, geometrical dimensions, and the coordinate system

$$\Pi_{\text{total}}^{e} = \sum_{m=-\infty}^{m=\infty} \sum_{n=-\infty}^{n=\infty} D_m D_n X_m(x) Y_n(y) e^{ih_{mn}z}. \tag{6.19}$$

In the representation (6.19) of the total field, D_m, D_n are arbitrary quantities that are defined by the field sources $\left(\vec{j}_{\text{out}}, \rho_{\text{out}}\right)$ as a result of the task solution of transmission line excitation by these sources (see Chap. 12). The eigenfunctions $X_{mn}(x)$, $Y_{mn}(y)$ obtain a specific form after satisfying solutions to ordinary differential equations for $X_{mn}(x)$, $Y_{mn}(y)$, which are defined (for the given task geometry) by given boundary conditions.

6.4   The Rectangular Waveguide, Electric (E_{mn}) Waves

The geometry and dimensions of the rectangular waveguide cross-section are shown in Figs. 6.2a and 6.11.

The two-dimensional wave equation (the Helmholtz equation) for the Hertz electric vector $E_z \equiv \Pi^e(x, y)$ is:

$$\frac{\partial^2 \Pi^e}{\partial x^2} + \frac{\partial^2 \Pi^e}{\partial y^2} + g^2 \Pi^e(x, y) = 0, \quad g^2 = k^2 - h^2. \tag{6.20}$$

Substituting the assumed solution type $\Pi^e(x, y) = X(x)Y(y)$ in Eq. (6.12) and fulfilling the variable separation, we obtain the following equation

$$(1/X)d^2X/dx^2 + (1/Y)d^2Y/dy^2 + g^2 = 0. \tag{6.21}$$

As the quantity g^2 in Eq. (6.21) is the constant quantity, then the other terms of Eq. (6.21) should be constants as well; therefore, designating them as $-g_{x,y}^2$, we obtain from Eq. (6.21) the equation system:

$$\frac{d^2X}{dx^2} + g_x^2 X = 0, \quad \frac{d^2Y}{dy^2} + g_y^2 Y = 0, \quad g_x^2 + g_y^2 = g^2. \tag{6.22}$$

General solutions to ordinary differential Eq. (6.22) have the form:

$$\left\{ \begin{matrix} X \\ Y \end{matrix} \right\} = \left\{ \begin{matrix} A_1 \\ B_1 \end{matrix} \right\} \cos \left\{ \begin{matrix} g_x x \\ g_y y \end{matrix} \right\} + \left\{ \begin{matrix} A_2 \\ B_2 \end{matrix} \right\} \sin \left\{ \begin{matrix} g_x x \\ g_y y \end{matrix} \right\}. \tag{6.23}$$

Arbitrary constants $A_{1,2}$, $B_{1,2}$ and transverse wave numbers $g_{x,\,y}$ must be found out from the boundary condition (see Sect. 2.1; Eq. (2.11))

$$\Pi^e = 0 \tag{6.24}$$

on all waveguide walls. Satisfying this condition, we obtain $A_1 = B_1 = 0$ and $A_2 \sin(g_x a) = 0$, $B_2 \sin(g_y b) = 0$.. As $A_2 \neq 0$, $B_2 \neq 0$, the so-called *characteristic* of *dispersion equations* follows directly from here:

$$\sin{(g_x a)} = 0, \quad \sin{(g_y b)} = 0, \tag{6.25}$$

from which we should obtain the transverse eigennumbers of the waveguide. Usually, these characteristic equations are concerned with the class of *transcendent* equations and their solutions are found out by numerical methods only. The case of the rectangular (and also plane; Fig. 6.1d) waveguide constitutes a pleasant exception, and solutions to Eq. (6.25) can be obtained right away and in an analytic form. Evidently, fulfillment of the two last equalities in Eq. (6.25) is possible only for

$$g_x = m\pi/a, \quad g_y = n\pi/b, \quad m, n = 1, 2, 3, \ldots \tag{6.26}$$

Thus, the total solution of the boundary problem for the electric wave of the rectangular waveguide E_{mn} has the form:

$$\Pi_{mn}^e(x, y, z) = D^e \sin{(m\pi x/a)} \sin{(n\pi y/b)} e^{ih_{mn}z},$$
$$m, n = 1, 2, 3, \ldots; g^2 = (m\pi/a)^2 + (n\pi/b)^2. \tag{6.27}$$

Functions $\sin(m\pi/a)$, $\sin{(n\pi/b)}$ are the *eigenfunctions* of the two-dimension Laplace operator for the $\Delta_\perp$ rectangle, and numbers $g_{x, y}$ are *eigennumbers* or *characteristic numbers*.

The solution (6.26, 6.27) of Eq. (6.20) permits us to write the total field of electric waves E_{mn} of the rectangular waveguide as (compare with Eq. (6.19)):

$$\Pi_{\text{total}}^e = \sum_{m=-\infty}^{m=\infty} \sum_{n=-\infty}^{n=\infty} D_{mn} \sin{\frac{m\pi}{a}} x \sin{\frac{n\pi}{b}} y \, e^{ih_{mn}z} \tag{6.28}$$

Here, amplitudes D_{mn} are arbitrary and are defined from the waveguide *excitation problem* by given outside sources (see Chap. 12). Numbers m, $n \neq 0$ and their values are designated as a number of field maxima on coordinates x and y, m and n respectively.

In essence, the Hertz vector is written by expression (6.28). Using Maxwell's equation system (1.12–1.15), we can write the components of the rectangular waveguide's electric waves in detail:

$$\vec{E}_{mn} = D_{mn}\left[E_z \, \vec{e}_z^{\,0} - i\big(h_{mn}/g_{mn}^2\big)(m\pi/a) E_x \, \vec{e}_x^{\,0} + (n\pi/b) E_y \, \vec{e}_y^{\,0} \right] \exp\{ih_{mn}z\}, \tag{6.29}$$

$$\vec{H}_{mn} = i D_{mn}\big(h_{mn}/g_{mn}^2 W_{mn}^E\big) \left[\big(h_{mn}/g_{mn}^2\big)(n\pi/b) H_x \, \vec{e}_x^{\,0} - (m\pi/a) H_y \, \vec{e}_y^{\,0} \right] \exp\{ih_{mn}z\}, \tag{6.30}$$

where the following designations are used:

$$E_Z = \sin{(m\pi x/a)} \sin{(n\pi y/b)}, \quad E_x = \cos{(m\pi x/a)} \sin{(n\pi y/b)},$$
$$H_x = \sin{(m\pi x/a)} \cos{(n\pi y/b)}, \quad H_y = \cos{(m\pi x/a)} \sin{(n\pi y/b)},$$
$$W_{mn}^E = h_{mn}/\omega\varepsilon = w\sqrt{1 - \big(f_{\text{cr}}^{mn}/f\big)^2}, \quad h_{mn} = k\sqrt{1 - \big(f_{\text{cr}}^{mn}/f\big)^2}, \tag{6.31}$$
$$f_{\text{cr}}^{mn} = \big(c/2\pi\sqrt{\varepsilon\mu}\big)\sqrt{(m\pi/a)^2 + (n\pi/b)^2}, \quad w = 120\pi\sqrt{\mu/\varepsilon}.$$

The waveguide system of waves (Eq. 6.28) has the significant property of *completeness*, i.e., any field in the waveguide can be presented as an expansion by eigenwaves in the form (6.28), of course, the obtained (from the excitation problem – see Chap. 12) amplitudes D_{mn}. Another important property of eigenwaves is their *orthogonality* to each other. In other words, the integral over the waveguide cross-section of the product of two waves with different indices m and n is equal to zero. Thus, in the regular waveguide, the *eigenwaves are independent of each other*. We discuss this in detail in Sect. 12.1.

The wave $E_{11} (m = n = 1)$ is the simplest (or the lowest) electric wave of the rectangular waveguide. The field pictures of this wave are shown in Fig. 6.12, and distribution diagrams in sections xOy and xOz for the E_{11} wave and diagrams for some

Fig. 6.12 (**a**, **b**) The structure of E_{11} wave of the rectangular waveguide, the structure of its electromagnetic field for E_z, (**c**, **d**) field distribution diagrams on axes c, d ($m = n = 1$) and some others wave types (for m, n)

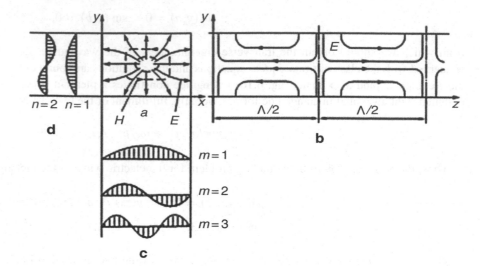

Fig. 6.13 Thee-dimensional pictures of electromagnetic field distribution in the rectangular waveguide cross-section for E_{11} wave: components (**a**) E_z, (**b**) E_x, (**c**) E_y, and (**d**) for E_{12} – the component E_z

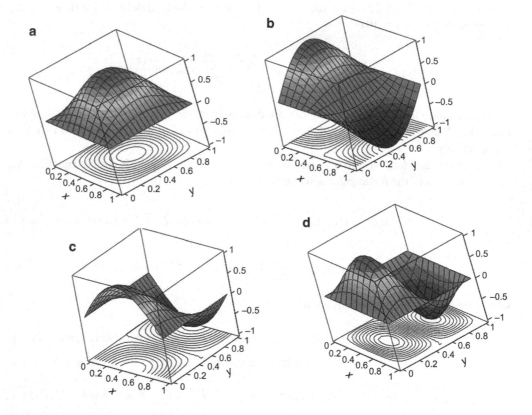

other waves are presented in Fig. 6.13. In Fig. 6.12a, b and following, the lines of the electric field E are shown by solid lines, whereas the lines of the magnetic field H are shown by dotted lines.

Construction of field pictures for higher (with regard to the E_{11} wave) waves is evident: the field picture of the E_{11} wave (Fig. 6.12a) should be represented mirror-like with respect to the x-axis and we have the field picture of the E_{m1} wave in the cross-section; and with respect to the x-axis – the picture of the E_{1n} wave, and so on. Really, the E_{11} wave has one maximum along the x-axis (Fig. 6.12c) and one maximum along the y -axis (Fig. 6.12d).

Three-dimensional pictures (Fig. 6.13) of electromagnetic field distribution of the Hertz vector component E_z in the cross-section of the rectangular waveguide for E_{mn} waves can be obtained using the MathCAD Program P.6.1, when we used designations $f(x, y) = E_z$, n is designated as a number of maxima along the x-axis, whereas $m-$ along the y-axis. In Program P.6.1 we have the possibility of looking at field dynamic pictures in the rectangular waveguide.

6.5   Rectangular Waveguide, Magnetic Waves

Magnetic H_{mn} waves of the rectangular waveguide satisfy the same Eq. (6.19) for the Hertz vector $E_z \equiv \prod^{(m)}(x, y)$, but the other boundary conditions on walls; namely, (see Sect. 2.1, (2.12)):

$$\partial \prod^{(m)} / \partial N = 0 \tag{6.32}$$

on walls; N is a normal to the wall (i.e., either on x, or on y).

Acting in a similar manner to the previous case (electric waves, Sect. 6.4), we obtain presenting representation for the Hertz vector:

$$\prod_{mn}^{(m)} = D_{mn}^{(m)} \cos \frac{m\pi}{a} x \cos \frac{n\pi}{b} y \exp\{ih_{mn}z\}, m, n = 0, 1, 2, \ldots \tag{6.33}$$

in which eigenvalues are:

$$g_{mn}^2 = (m\pi/a)^2 + (n\pi/b)^2. \tag{6.34}$$

Accordingly, the Hertz vector of the magnetic H_{mn} wave (6.33) is written and its eigennumbers are g_{mn} (6.34).

Again, using Maxwell's equation system, we can write components of the rectangular magnetic waves in detail:

$$\vec{H}_{mn} = D_{mn}\left[H_z \, \vec{e}_z^{\ 0} + i\left(h_{mn}/g_{mn}^2\right)(m\pi/a)H_x \, \vec{e}_x^{\ 0} + (n\pi/b)H_y \, \vec{e}_y^{\ 0} \right] \exp\{ih_{mn}z\}, \tag{6.35}$$

where

$$H_z = \cos(m\pi x/a) \cos(n\pi y/b), \quad H_x = \sin(m\pi x/a) \cos(n\pi y/b),$$

$$E_x = \cos(m\pi x/a) \sin(n\pi y/b), \quad E_y = \sin(m\pi x/a) \cos(n\pi y/b),$$

$$W_{mn}^H = \omega/h_{mn} = w\sqrt{1 - \left(f_{cr}^{mn}/f\right)^2}, \tag{6.36}$$

In contrast to electric waves E_{mn}, one of the indices m or n (but not both simultaneously) may take a zero value. These are H_{m0} and H_{0n} waves, for which the field does not depend on the y-coordinate (H_{m0}) or x-coordinate (H_{0n}).

The H_{10} wave (or H_{01}) is the lowest magnetic wave of the rectangular waveguide. Its field picture is presented in Fig. 6.14. It follows from Eq. (6.34) that at $g_{10} = \pi/a$ for $h_{10} = 0$ the critical wavelength is

$$\lambda_{cr}^{10} = \lambda_{cr} = 2a, \tag{6.37}$$

i.e., is equal to twice the length of the wide wall a (Fig. 6.11). Therefore, the H_{10} wave has the largest critical wavelength (the lowest frequency). The longer waves cannot pass through the waveguide with such a wide wall a at the λ wavelength. Thus, the size of the narrow wall b does not play any role. It defines the electrical strength of the waveguide. We should note that phenomena near critical frequency play an enormous role in waveguide electrodynamics and, say, in the theory and practice of *open resonators, waveguide joints, directional couplers*, and so on [14, 20, 21, 58].

In another limited case, when the operating wave $\lambda \ll \lambda_{mn}^{(cr)}$, in the empty waveguide $\Lambda_{nm} \approx \lambda$, and therefore, only slightly exceeds the wavelength in the free space λ. In other words, in this case, the waveguide dimensions are so large that waves in it propagate as in the free space.

Let us return to magnetic waves in the rectangular waveguide. Thus, the H_{10} wave is the main operating wave. As we see from Eq. (6.35), it has the maximal wavelength and its field does not depend on the y-coordinate. The second property, by the way, is typical of all waves of the H_{m0} type (and, naturally, of H_{0n}): there is no dependence on x (and accordingly on x for H_{0n}).

The electric field of the H_{10} wave is concentrated in the middle plane of the cross-section at $x = a/2$. This feature of its structure (constitution) makes convenient an approach to its excitation by the rod (for instance, by the internal coaxial conductor; see Sect. 12.2), which is located in the middle of the waveguide wide wall. We discuss the current distribution later (in Fig. 6.28c; in comparison with current distribution on walls of the circular waveguide). Calculations are performed

Fig. 6.14 The main operation wave of the rectangular waveguide H_{10}. Pictures of its fields for E_y in (**a**) transverse and (**b**) longitudinal planes, and (**c**, **d**) the distribution of intensities for them. The *solid lines* are designated as the electrical field lines, the *dotted lines* the magnetic field lines. Three-dimensional pictures of electromagnetic waves (**e**) H_{10} and (**f**) H_{11} in the transverse cross-section of the rectangular waveguide, (**f**, **g**) distribution of intensities on x and, and also (**h**, **i**) electric and magnetic field lines of H_{10} and H_{11} waves in the transverse waveguide cross-section

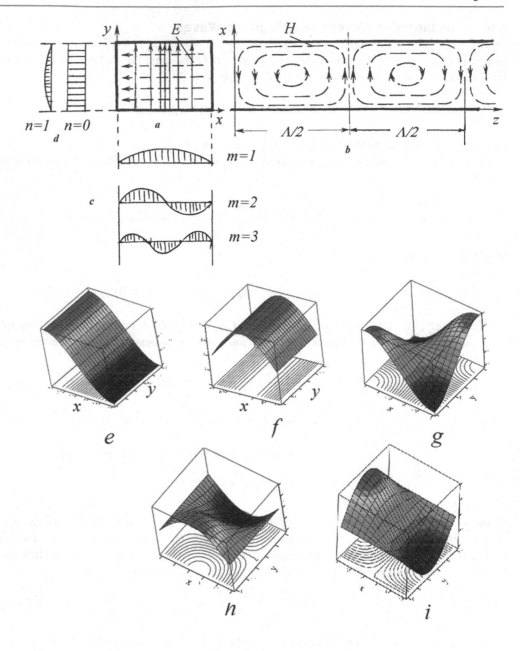

according to the Program P.6.2, in which there is the possibility of looking at dynamic pictures of field variations in the waveguide cross-section.

The wave H_{11} is another type of magnetic wave in the rectangular waveguide. Its field is shown in Fig. 6.14a–c. The set of H_{10}, H_{01}, and H_{11} waves allows a picture to be obtained of all the magnetic H_{mn} waves in the rectangular waveguide by means of a mirror-like representation of pictures of these main fields in the ideal walls (see Program P.6.2).

We have already drawn the attention of the reader to the importance of the ability to construct pictures of fields and currents in different electrodynamics structures.

In practice, especially when large powers are transmitted through the rectangular waveguide (equally, in some other cases; see, for example, Sects. 6.9 and 6.10), the hollow structures (without magnetic–dielectric filling) are used. Thus, the energy losses in the regular transmission line are possible only at the expense of the non-ideality of the wall material, i.e., in the presence of the finite (but small) conductivity in it. Determination of linear (per the line length unit) losses is usually performed using the small perturbation method or variation methods. We simply give here the final result for the main H_{10} wave of the rectangular waveguide [104]:

$$h'' = \left[(a/b) + 2(\lambda/2a)^2 \left[\sqrt{240\sigma a^3} \right] \sqrt{(\lambda/2a) - (\lambda/2a)^3} \right]^{-1}, \quad [Nep/m], \qquad (6.38)$$

in which h'' is an imaginary part of the longitudinal wave number of the H_{10} wave, σ is the conductance of the wall material (see Table 2.1 in Nefyodov [26]); we would like to remind the reader that h'' $[Nep/m] = h''$ $[dB/m]/8.86..$ The MathCAD program P.6.3 can be used for the calculation $h''(a/b, \lambda/2a)$ (see also Nefyodov and Kliuev [73]). There, we can find the three-dimensional picture for h''.

Thus, the function of H_{10} wave losses versus a ratio of waveguide wall dimensions has a sufficiently smooth character in its middle part and sharp growth at the edges for $a/b \to 0$ and 1. The approximate formula for h'' loses sense at $\lambda/2a \to 1$, i.e., at a frequency approaching the critical value: $f \to f_{cr}$. The relatively small damping increase with frequency growth can be explained by the "weak" dependence on the surface resistance upon a frequency proportional to $\sqrt{f}$.

It is interesting to note: the size of the narrow wall at its ideal conductivity has practically no effect on the main characteristics of the H_{10} wave, namely, to the phase and group velocities (Eq. (6.16)), in addition to the wave impedance: $W = 120\pi/\sqrt{1 - (\lambda/2a)^2}$. However, the situation changes for non-ideal walls, when damping at $b/a \ll 1$ is practically inversely proportional to b, which directly follows from Eq. (6.33).

Let us speak some words about a choice of side ratio a/b of the rectangular waveguide. Usually, the side ratio is $a/b \in (2, 0...2.4)$, which is rather far from the *golden ratio* $\Phi - 1.618$ and much closer to its square: $\Phi^2 - 2, 168$. The main point is that in "standard" cases, the side ratio *does not correspond to the beauty criterion*. Equally, this is concerned with the radii ratio of the coaxial waveguides (see Sect. 6.9).

An example of the *golden rectangle* is shown in Fig. 6.16. The side radio in it is $a/b = 1, 618$ It is remarkable that from the "large" golden rectangle $a \times b$ we can obtain the same gold rectangles of lesser cross-sections: *AFHD*, *AFEK*, *GFEL*, and so on. The golden spiral is inscribed in a "natural" manner into the golden rectangle. The center of this golden spiral is located on the intersection of diagonals of rectangles (dotted lines in Fig. 6.16). The golden spiral is the specific case of a well-known logarithmic spiral, which is widely used in the practice of microwave antennas. It can be described in the polar coordinate system by the function $r = \rho \exp\{k\alpha\}$, in which r is the length of the radius vector of the spiral, and α is the azimuth angle. The structure of the gold spiral corresponds to $k = -(\pi/2) \ln \varphi \approx 0.3$ (Fig. 6.15).

Thus, seemingly, there is a natural way to a choice of side ratio of rectangular waveguides of various ranges, corresponding to the general beauty criterion, which, by the way, meets the new ideas in radio physics related to self-similar (fractal) structures. In particular, they are one of directions in the construction of modern meta-structures, meta-materials with the specific properties. Structures of this class, obviously, can be built if we define the shape in Fig. 6.16 as the cell of some two-dimensional structure, for instance, the one- or two-dimensional periodic structure, which, of course, should provide the greater broadbandness of such a system without "breaks" in the amplitude frequency response and the appropriate narrowing of the antenna pattern.

6.6   The Rectangular Waveguide with Electric and Magnetic Walls, the Planar Waveguide Model

In the millimeter wavelength range, transmission lines of the simple planar waveguide type (Fig. 6.1d), the planar waveguide with ridges (Fig. 6.1e), the planar waveguide with the impedance insert (Fig. 6.17a) (an insert is shown by dotted lines), the same waveguide with impedance jumps near the edges (Fig. 6.17b), the same waveguide with the dielectric plate or rod (Fig. 6.17c), etc., have found wide application. In the upper part of Fig. 6.17c, the field distribution along the waveguide with the dielectric rod is shown.

In essence, the planar waveguide shown in Fig. 6.1d represents the single-dimensional structure ($\partial/\partial y \neq 0$), and its magnetic waves are defined by the Hertz vector as

$$\prod_{mn}^{(m)} = D_{mn}^{(m)} \cos \frac{n\pi}{b} y e^{ih_{mn}z}, \quad n = 0, 1, 2, \ldots. \qquad (6.39)$$

The presence of the transverse *TEM* wave in the planar waveguide for $n = 0$ and $h_{00} \equiv k$ is a significant feature. This feature can be equally attributed to H_{00} or to E_{00}. The electromagnetic field of this wave is located in the transverse plane (x, y) and has E_x, $H_y(E_Z = H_Z \equiv 0)$ components. We can use this circumstance in practice both at the stage of discussion and task

Fig. 6.15 The H_{11} wave of the rectangular waveguide. Its field pictures in (**a**) the transverse and (**b**) the longitudinal planes, and (**c, d**) intensity distribution of some magnetic wave components of different numbers. Three-dimensional pictures of electric and magnetic fields of (**e–g**) H_{10} and (**h–j**) H_{22} waves

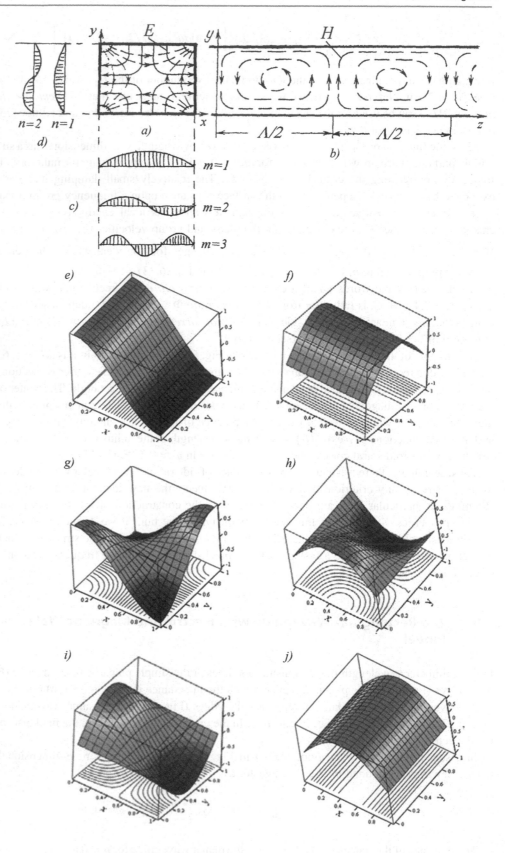

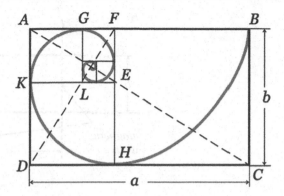

Fig. 6.16 The golden rectangle and the golden spiral inscribed in it

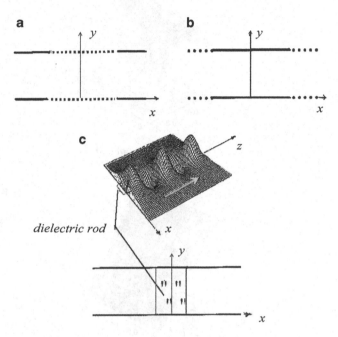

Fig. 6.17 Different variants of the planar waveguide. (**a**) With guiding impedance walls. (**b**) With impedance walls at the edges. (**c**) The planar waveguide with guiding dielectric rod; in the upper part of (**c**) the electric field distribution in this structure is given

statement, and at further device designing using the planar waveguide as a model of the rectangular waveguide with electric (for instance, horizontal) and vertical fictitious (virtual) magnetic walls. In Fig. 6.17a, the magnetic walls are shown by vertical dotted lines.

For a model with different walls (Fig. 6.18a), the magnetic eigenwaves can be written through the Hertz vector as:

$$\Pi_{mn}^{(i)} \equiv H_z^{mn} = D_{mn}^{(m)} \sin\frac{m\pi}{a}x \cos\frac{n\pi}{b}y e^{ih_{mn}z}, m = 1, 2, \ldots, n = 0, 1, 2, \ldots \tag{6.40}$$

Similarly, electric eigenwaves are defined as:

$$\Pi_{mn}^{(\dot{y})} \equiv E_z^{mn} = D_{mn}^{(e)} \cos\frac{m\pi x}{a} \sin e\frac{m\pi y}{b} e^{ih_{mn}z}, \quad m = 0, 1, 2, \ldots, n = 1, 2, \ldots \tag{6.41}$$

The transverse and longitudinal wave numbers for this model are

Fig. 6.18 (**a**) Model of the rectangular waveguide with horizontal magnetic and vertical electric walls. (**b–e**) Field pictures of the magnetic waves according to Eq. (6.29)

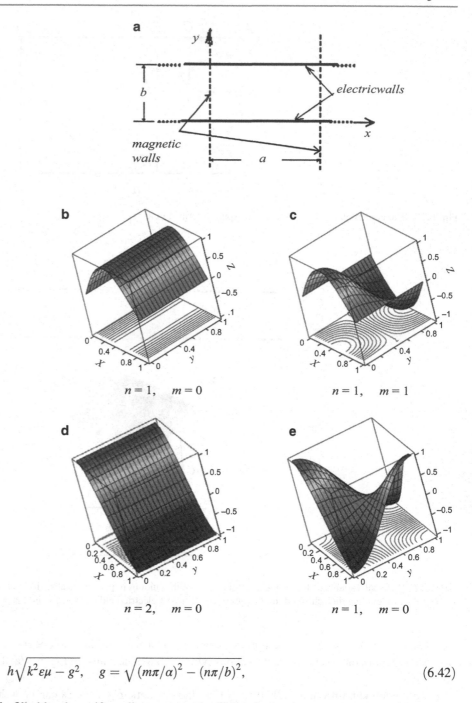

$$h\sqrt{k^2\varepsilon\mu - g^2}, \quad g = \sqrt{(m\pi/a)^2 - (n\pi/b)^2}, \tag{6.42}$$

if the model cross-section is completely filled by the uniform linear magnetic–dielectric (ε, μ).

In Appendix B (http://extras.springer.com/2019/978-3-319-90847-2), Programs P.6.4 and P.6.5 for observation of pictures of electric and magnetic waves of the plane waveguide can be found. The significant difference between the model with vertical magnetic walls and the conventional rectangular waveguide (with all four electric walls) consists in the fact that the *TEM* wave is the eigenwave of such a waveguide and it should be included in the complete system of eigenfunctions (such as the expansion (6.28)). We should note that the *TEM* wave will be the *only wave* at $a < \lambda/2$, $b < \lambda/2$. Such a model is of paramount importance for an understanding of physics and for the performing of estimated "engineering" calculations in the strip-slot transmission lines (Chap. 7). Their dimensions (at the necessity of the single-frequency mode) should be selected in such a manner. In some cases, and, in particular, in 3D-IC of microwave and millimeter-wave ranges, in *microstrip antennas*, etc., the basing element (or some part of it) operation on the one from higher types of waves turns out to be more promising (see Sect. 7.3).

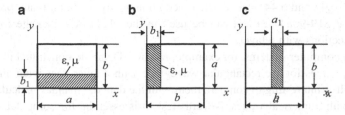

Fig. 6.19 Cross-sections of the rectangular waveguide with piecewise uniform cross-section filling by the dielectric. (Taken from Nefyodov and Kliuev [73], p. 165)

The vertical electric walls, which are usually used in models, makes this model fundamentally *closed*, and the *TEM* wave cannot exist in it. Wall separation of a large distance (at least, compared with the wavelength) $(2a \rightarrow \infty)$, changes a situation a little: the system remains completely closed. The model with magnetic walls (Fig. 6.17) is more attractive because accommodation of the current-carrying band in it (transfer from a nonfilled line to a symmetric stripline in Fig. 6.1f or to completely filled by magnetic–dielectric as in Fig. 6.1g) does not noticeably change the picture of the wave field. The closed asymmetric stripline (Fig. 6.1h), in which, strictly speaking, there is no "pure" *TEM* wave, is situated in a more complicated position. Nevertheless, in this case, the model with magnetic walls is frequently useful, for example, at low enough frequencies (decimeter waves and a long part of the centimeter range). In other cases, such a model can be used as zero approximation with, for instance, *variation* or *iteration* methods.

6.7   Rectangular Waveguide with the Dielectric Layer

The regular rectangular waveguides, like the waveguides of any other cross-section, often contain the single- and/or multi-layer magnetic–electric insertions (Fig. 6.19).

First investigations in the USSR into properties analysis of the waveguide with partial filling were carried out by professor *Y.V. Egorov* [57]. In contrast to waveguides with uniform filling, in which the field is usually presented in the form of expansions over full systems of electric E_{mn} (transverse-magnetic TM_{mn}) and magnetic H_{mn} (transverse-electric TH_{mn}) waves of type (6.28 and 6.32), it is more convenient to represent the total field of waveguides with piecewise-uniform filling of the cross-section by the magnetic–electric as expansions over the full system of eigenfunctions consisting of longitudinal-electric LE_{mn} and longitudinal-magnetic LM_{mn}.

Evidently, owing to the completeness of these systems of eigenfunctions ($E_{mn} + H_{mn}$ and $LE_{mn} + LM_{mn}$), the final results (the total electromagnetic field) will be the same; however, the introduction of the longitudinal waves LE_{mn} and LM_{mn} for analysis of structures with piecewise filling turns out to be good and suitable, because, each eigenwave of such a waveguide is thus the simple wave (not hybrid), as it would be when using wave systems of the *E*- and *H*-types. There are a number of examples (see, for instance, an analysis of microstrip antennas [127], dielectric waveguides (Chap. 9), in addition to Nefyodov [52], and many others). Thus, representation of the total electromagnetic field of the waveguide with piecewise-uniform filling of the cross-section turns out to be more appropriate (more adequate) for the physical side of the problem, which is determining in the educational process and in practical activity.

Representing fields in the empty and filled parts of a cross-section as the full system of eigenfunctions of LE_{mn} and LM_{mn}, and satisfying the continuity conditions of the field tangent components on the boundaries of dielectric–air (or dielectric–dielectric), we come to the following dispersion equations for determination of longitudinal wave numbers h of LE_{mn} and LM_{mn} waves:

$$(k_1 \varepsilon_2 / k_2 \varepsilon_1) \tan(k_1 b_1) + \tan(k_2 b_2) = 0 \quad \text{for} \quad LM - \text{waves,} \tag{6.43}$$

$$(k_2 \mu_1 / k_1 \mu_2) \tan(k_1 b_1) + \tan(k_2 b_2) = 0 \quad \text{for} \quad LE - \text{waves.} \tag{6.44}$$

Here $k_j / k = \sqrt{\varepsilon_j \mu_j - (m\pi/ka)^2 - (h/k)^2}$, $\quad m = 0, 1, 2, \ldots$ For LM – waves $m = 1, 2, \ldots$, for LE waves $m = 0, 1, 2, \ldots$; the second index $n = 0, 1, 2, \ldots$.

Obtaining the roots of Eqs. (6.43 and 6.44) can be performed numerically based on many programs (Appendix B – http://extras.springer.com/2019/978-3-319-90847-2) that can be used in the MathCAD package for the necessary geometry of rectangular waveguide cross-sections with a layer.

The possibilities of modern computer systems, for example, MathCAD, are demonstrated in Programs P.6.6, P.6.7, P.6.8, and P.6.9 for LE_{mn} and LM_{mn} waves for the rectangular waveguide with a layer near wide and narrow walls calculated in Nefyodov and Kliuev [1, 73]. It is interesting to compare them with the data of "manual" calculations by Nefyodov [52]. For the most part, they coincide with numerical values. Nevertheless, it is essential for computer determination of the roots of Eqs. (6.43 and 6.44) for these equations to be "modernized" after exclusion of singular points, which at "manual" calculation can be overcome rather simply. Here, we can again discuss the necessity for students of clear representations about process physics in the structures studied (and, of course, in many others).

Programs P.6.6, P.6.7, P.6.8, and P.6.9 use the "zero" approximation of the exact value of the propagation constant of the wave to increase the calculation speed for the waveguide completely filled by the uniform dielectric (magnetic): $h/k = \sqrt{\varepsilon\mu - (m\pi/ka)^2 - (n\pi/kb)^2}$. In programs P.6.6, P.6.7, P.6.8, and P.6.9 (Appendix B – http://extras.springer.com/2019/978-3-319-90847-2), the following designations are used: $e1 = \varepsilon_1$, $e2 = \varepsilon_2$, $m1 = \mu_1$, $m2 = \mu_2$, xx is the matrix of eigenvalues of waves in the completely filled rectangular waveguide. The presence of the i multiplier at element xx means that such a wave is below-cutoff (nonpropagating) (Fig. 6.20).

There are different approximate analytical representations of the roots of these equations (see, for instance, Nefyodov [52]). Below the reader can see the program in the MathCAD package

Fig. 6.20 Behavior of the left part of Eq. (6.26) depending on the argument and some of the results (for normalized propagation constant h/k) of the numerical solution of dispersion Eq. (6.43) for various positions of the dielectric plate near the edges (Eq. 6.44), according to Programs P.6.4, P.6.5, P.6.6, and P.6.7 for the rectangular waveguide. (Taken from Nefyodov and Kliuev [73], p. 167)

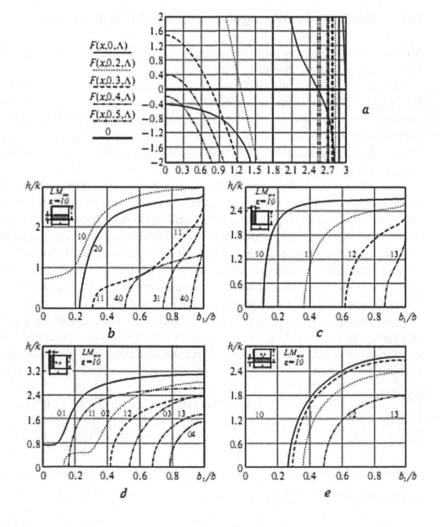

$$ka := 4.52 \quad kb := 1.96$$

$$Al := 0, 0.001..1 \quad x := 0, 0.001..4 \quad \Lambda := 0.3$$

$$\eta(er, Al, \Lambda) := \sqrt{1 + (er - 1) \cdot \left[Al + \left(\frac{1}{\pi}\right) \cdot \sin(\pi \cdot Al)\right] - \Lambda^2}$$

for an approximate search of the roots of the LM_{10} wave, which transforms to the H_{10} wave of the rectangular waveguide to be fulfilled by the dielectric. In this program we use the designations: $A1 = a_1/a \quad \Lambda = \pi/ka = \lambda_0/2a$.

From the presented calculation results of programs P.6.6, P.6.7, P.6.8, and P.6.9, the interesting physical knowledge follows about wave behavior in waveguides partially filled by the uniform layers of the dielectric. First of all, we note the clear physical contents and convenience of the use of longitudinal LE_{mn} and LM_{mn} waves, which are the *eigenwaves* of exactly these waveguides. Interesting physical consequences follow from Programs P.6.6, P.6.7, P.6.8, and P.6.9. In particular, these results clearly demonstrate the *dielectric effect* related to retraction of an electromagnetic field into the dielectric layer, for instance, at frequency growth; thus, the wave deceleration tends toward the value of $\sqrt{\varepsilon}$. The field in the nonfilled waveguide part decreases from the layer surface according to the exponential law, i.e., the wave has the character of a *surface wave*. This effect has manifold technical applications (see later Chaps. 7, 8 and 9, and also Nefyodov et al. [20, 21]). Dielectric structures of different classes find wide application in resonator or nano-devices, for instance, with utilization of graphene.

The examined models of the rectangular waveguides with a layer of dielectric permit the evident transformation to other interesting waveguiding structures: the planar waveguide with the dielectric layer (for $ka \to \infty$), which is composed on the basis of 3D-IC in the millimeter-wave range, the planar layer in the free space – the two-dimensional model of the dielectric waveguide and others.

A more general case is known; namely, the rectangular waveguide with the rectangular rod (Fig. 6.1c). The full analysis of such a complicated structure is extremely bulky. However, utilization, for example, of the variation method allows the approximate expressions for deceleration of the lowest wave type to be obtained [52]:

$$h/k = \sqrt{1 + (b_1/b)(\varepsilon - 1)[(a_1/a) + (1/\pi)(\sin(\pi a_1/a)) - (\lambda/4a)]^2}. \tag{6.45}$$

From here, the formula that was used in the above-mentioned program for calculation of the main wave deceleration is directly followed, when a layer is located in the middle of the rectangular waveguide cross-section.

6.8 Expansion of Waveguide Waves on Plane Waves

This method (also called the *Brillouin approach*) has the extraordinary clarity, and it is widely used in electrodynamics of transmission lines and basing elements, sometimes allowing essential simplification of complicated tasks or a deep understanding of its physical content. Let us take as an example the H_{10} wave of the rectangular waveguide. Its *Hertz vector* is defined in Eq. (6.20) for $n = 0, m = 1$, so that

$$\Pi^m = D \cos(\pi x/a) e^{ihz}. \tag{6.46}$$

Having changed in Eq. (6.38) $\cos(gx)$, $g = \pi/a$ using *Euler's formula* and using the designations in Eq. (6.13), we obtain

$$\Pi^m = D\left\{e^{ik(x \sin \alpha + z \cos \alpha)} + e^{ik(-x \sin \alpha + z \cos \alpha)}\right\}. \tag{6.47}$$

The form of H_{10} wave presentation as in Eq. (6.47) is the required expansion of the waveguide representation (6.46) in the form of two plane uniform waves in Eq. (6.47), which propagate under the angle to z – axis of the waveguide (Fig. 6.21). Sometimes, these waves are referred to as *partial*.

From the expressions for E- and H-waves (besides, obviously, H_{m0} and H_{0n} waves) of any indices, we clearly see that each of these waves can be represented as a sum, not of two, but of four plane waves. Thus, the *Brillouin approach* (sometimes, *Brillouin's* concept) is suitable for all waves of the rectangular waveguide. This situation is typical of other waveguides,

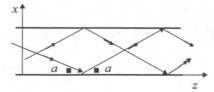

Fig. 6.21 Expansion of waveguide waves or plane waves

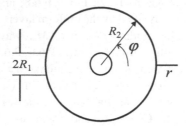

Fig. 6.22 Cross-section of the coaxial cylindrical (axis-symmetrical) waveguide: geometry, coordinates, and dimensions

although in much more complicated form. Thus, for a circular waveguide, we have an infinite number of plane waves (see Sect. 6.10).

In wide waveguides ($ka \gg 1$, $kb \gg 1$) a large number of waves can simultaneously propagate, which is approximately equal to $N \approx 2\pi ab/\lambda^2$. All these waves, according to (6.24), have different speeds, and therefore, different times of arrival at the receiving end of the transmission line. This is the great problem at transmission of the wideband radio signal through the backbone communication link. It concerns to a lesser degree transmission lines used not for the radio signal, but for transmission of microwave and millimeter-wave power, when problems of electrical durability, etc., are considered first.

6.9   Coaxial Cylindrical Waveguide, the Main Wave, Higher Types of Waves

6.9.1   Definition, Geometry

Geometry of the simplest variant of the axis-symmetrical *coaxial waveguide* (sometimes "the coaxial") is shown in Fig. 6.1b, c, and also in Fig. 6.22. However, the wide circle of electrodynamics structures with a multiply connected cross-section is covered by this term. So, for example, the transmission lines of a more complicated type, say, presented in Fig. 6.1e–k, and others, can be attributed to this class of structures. We particularly draw the reader's attention to the coaxial waveguide, and we have several reasons for this. *First,* the coaxial waveguide is widely used in practice. Thus, these applications have many adaptations. These are traditional transmission lines at low and very low frequency (including direct current, DC) on the *T* wave (or *TEM*), in essence, based on the *transverse electromagnetic wave*. The coaxial waveguide happens to be an effective transmission line at microwave and millimeter-wave ranges, and with some alterations (modifications) in the optical range.

Second, it has special significance at operation in the *higher types of waves*, which is barely discussed, not only in students' educational literature, but even in scientific books (excluding the authors of this lecture course and their Russian colleagues [22]). But we (and our colleagues) could determine (exactly by studying the properties of higher wave types of coaxial waveguides) the remarkable opportunities of so-called *near surface waves* and based on them construct a wide class of *resonant structures with unique characteristics.*

Third, the coaxial waveguide with the rectangular conductor represents the model of the symmetric strip transmission line, if we consider the side walls as "magnetic" (see Sect. 6.6).

Fourth, the *circular waveguide* (Fig. 6.2b) or the more complicated *elliptic waveguide* (Fig. 6.2c; Sect. 6.11) are specific cases of coaxial waveguides.

6.9.2 Coaxial Waveguide, Boundary Problems, Higher Types of Waves

Let the coaxial waveguide be formed by two ideally conducting circular cylinders with the radii R_1 and R_2, such that $\Delta = R_1/R_2 < 1$ (Fig. 6.1b). The electromagnetic field for $r \in (R_1, R_2)$ and $\phi \in [0.2\ \pi]$ satisfies the homogeneous *Helmholtz equation*, written in the cylindrical coordinate system, which is natural for the task:

$$\frac{1}{r}\frac{\partial}{\partial r}\left(r\frac{\partial \Psi_i}{\partial r}\right) + \frac{1}{r^2}\frac{\partial^2 \Psi_i}{\partial \varphi^2} + \chi^2 \Psi_i(r, \varphi) = 0, \tag{6.48}$$

This equation corresponds to ideal boundary conditions of the first (Eq. 6.24) or the second (Eq. 6.32) kinds on the surfaces of circular cylinders at $r = R_1, R_2$ and the *periodicity condition* on the φ coordinate: $\Psi_i(r, \varphi) = \Psi_i(r, \varphi + 2\pi n)$, $n = 0, 1, 2,$ ...

In Eq. (6.48), the dependence on the longitudinal coordinate z is preliminarily extracted: $\exp\{ih_{mn}^{E,H}z\}$. Applying the variable separation method and representing the required solution in the form of the product $\Psi_i(r, \phi) = R(r)\Phi(\phi)$, we obtain after substitution $\Psi_i(\varphi)$ on Eq. (6.48):

$$\frac{r^2}{R}\frac{d^2R}{dr^2} + \frac{r}{R}\frac{dR}{dr} + r^2\chi^2 = -\frac{1}{\Phi}\frac{d^2\Phi}{d\varphi^2}. \tag{6.49}$$

Thus, variables in Eq. (6.49) happen to be separated, and as there are functions of different arguments on the left and on the right in Eq. (6.49), then they should be independent and equal to some arbitrary constant ν^2. In this case, we proceed to two ordinary differential equations:

$$\frac{d^2R}{dr^2} + \frac{1}{r}\frac{dR}{dr} + \left(\chi^2 - \frac{\nu^2}{r^2}\right)R = 0, \tag{6.50}$$

$$d^2\Phi/d\varphi^2 + \nu^2\Phi = 0 \tag{6.51}$$

The equation for the radial function R is the *Bessel equation* and its solution can be written in two possible forms:

$$\begin{aligned} R(r) &= AJ_\nu(\chi r) + BY_\nu(\chi r), \\ R(r) &= \tilde{A}H_\nu^{(1)}(\chi r) + \tilde{B}H_\nu^{(2)}(\chi r). \end{aligned} \tag{6.52}$$

In solutions (6.52), we designate as $J_\nu(\xi)$, $Y_\nu(\xi)$ the *Bessel functions* of the first and second kinds, and through $H_\nu^{(1,2)}(\xi)$ the *Hankel functions*. The behavior of cylindrical functions $J_\nu(\xi)$, $Y_\nu(\xi)$, $H_\nu^{(1,2)}(\xi)$ was discussed by us earlier (see Chap. 4).

Equation (6.51) is well-known to us (see Chap. 4) and its solution can be represented in the one of the following forms:

$$\Phi(\varphi) = C\cos(\nu\varphi) + D\sin(\nu\varphi) \quad \text{or} \quad \Phi(\varphi) = \tilde{C}e^{i\nu\varphi} + \tilde{D}e^{-i\nu\varphi}. \tag{6.53}$$

In the first case in Eq. (6.53), we deal with so-called standing (on the azimuth φ coordinate) waves, whereas in the second case with traveling (on φ) clockwise and counter-clockwise waves. These representations are used in the open dielectric resonators, in microwave electronic devices, in accelerators of charged particles, in plasma devices, etc.

As the point $r = 0$ in solutions of Eq. (6.52) is excluded (when for $r \to 0$ the Bessel function $Y_\nu(r) \to -\infty$) by the geometry itself of the cross-section of the coaxial waveguide, we must retain the *Bessel functions* (both the *first* and the *second* kinds) in the total solution.

6.9.3 Electric and Magnetic Waves, Characteristic (Dispersion) Equations

Satisfying the *Helmholtz equation* to the boundary condition of the *first kind* (the *Dirichlet condition*; Eq. (6.24)), we come to the *characteristic equation* for *electric waves*

$$J_m(\eta)Y_m(\eta\Delta) - J_m(\eta\Delta)Y_m(\eta) = 0, \quad \Delta = R_1/R_2, \tag{6.54}$$

whose roots η_{ms} are normalized by the eigennumbers of the *electric waves* of the coaxial waveguide.

The roots of Eq. (6.54) are the transverse propagation constants η for electric waves $E_{0n}, E_{1n}, E_{2n}, n = 1, 2, 3$ of the coaxial waveguide can be obtained with the help of Program P.6.11. Of course, with its help η for waves of other numbers indicating the other search intervals may be determined.

In our lecture course, we used the simplest method for the determination of eigennumber values, i.e., roots of Eqs. (6.54 and 6.55), etc., the procedure root($F(x), x, a, b$) allowing in the indicated interval $x \in (a, b)$ the one root value satisfying the equation $F(x) = 0$ to be satisfied. In the MathCAD medium, there is a procedure called *polyroot* that allows the further roots of equation $F(x) = 0$ to be obtained according to the initial value of the a interval. From our point of view, this procedure is more bulky and not always convenient. An indication of the root search interval is, in essence, determination of the *zero approximation* to the root value. In physics and electrodynamics as a whole, in the overwhelming majority of cases, the exact zero approximation has decisive significance (at the level of understanding of physical content). In essence, this is the foundation of the comprehensive *method of small perturbations*, which gives excellent results to many problems, and especially, when studying linear processes, to which our lecture course is devoted.

Similarly, satisfying in the *Helmholtz equation* the boundary condition of the *second kind* (the *Neumann condition*; Eq. 6.29), we obtain the equation

$$J'_m(\widetilde{\eta})Y'_m(\widetilde{\eta}\Delta) - J'_m(\widetilde{\eta}\Delta)Y'_m(\widetilde{\eta}) = 0, \tag{6.55}$$

whose roots $\widetilde{\eta}_{ms}$ are the eigennumbers of the *magnetic waves* system. The sign ' is designated as the first derivative.

Roots of Eq. (6.55) are transverse eigennumbers of magnetic waves and can be found using Program P.6.10.

Longitudinal wave numbers of electric γ_{ms} and magnetic $\widetilde{\gamma}_{ms}$ waves are related to the wave number k by the usual relationships:

$$\gamma_{ms}^2 = k^2 - (\eta/R_2)^2, \quad \widetilde{\gamma}_{ms}^2 = k^2 - (\widetilde{\eta}/R_2)^2. \tag{6.56}$$

The system of eigenfunctions of electric and magnetic waves is *full*.

At $\Delta \to 0$, systems of electric and magnetic waves of the coaxial waveguide are transformed into systems of eigenwaves of the circular waveguide (see Sect. 6.10).

6.9.4 Field of Electric Waves

The components of the electric wave fields ($E_r, E_\varphi, H_r, H_\varphi$) can be obtained from *Maxwell's equations* with the help of differentiation operations of the longitudinal component of $E_Z(g_{ms} = \eta_{ms}/R_2)$:

$$E_Z(r, \varphi) = (g_{ms})^2 \left[J_m(g_{ms}r) - \frac{J_m(g_{ms}R_1)}{Y_m(g_{ms}R_1)} Y_m(g_{ms}r) \right] \cdot \left\{ \begin{array}{c} \cos \\ \sin \end{array} \right\} (m\varphi), \tag{6.57}$$

$$H_\varphi(r, \varphi) = ikg_{ms} \left[J'_m(g_{ms}r) - \frac{J_m(g_{ms}R_1)}{Y_m(g_{ms}R_1)} Y'_m(g_{ms}R_1) \right] \cdot \left\{ \begin{array}{c} \cos \\ \sin \end{array} \right\} (m\varphi), \tag{6.58}$$

6.9.5 Field of Magnetic Waves

Similarly, the "main" (determining) components of the magnetic waves can be determined as

$$H_Z(r, \varphi) = (\tilde{g}_{ms})^2 \left[J_m(\tilde{g}_{ms}r) - \frac{J_m'(\tilde{g}_{ms}R_1)}{Y_m'(\tilde{g}_{ms}R_1)} Y_m(\tilde{g}_{ms}r) \right] \cdot \left\{ \begin{matrix} \cos \\ \sin \end{matrix} \right\} (m\varphi),$$

$$E_\varphi(r, \varphi) = -ik\tilde{g}_{ms} \left[J_m'(\tilde{g}_{ms}r) - \frac{J_m'(\tilde{g}_{ms}R_1)}{Y_m'(\tilde{g}R_1)} Y_m'(\tilde{g}_{ms}r) \right] \cdot \left\{ \begin{matrix} \cos \\ \sin \end{matrix} \right\} (m\varphi).$$

On the basis of these solutions, we can write the full system of eigenfunctions of the coaxial waveguide as we did for the rectangular waveguide (Eq. 6.28). However, this infinite system of eigenfunctions should be added, in essence, by the "static" (for the field in transverse cross-section) solution, whose potential function is

$$\Psi_{00} \ln r / \sqrt{2\pi \ln (R_2/R_1)} = F \ln r \quad F = [2\pi \ln (R_2/R_1)]^{-1/2}.$$

The electromagnetic field pictures of some higher types of waves of the coaxial waveguides are shown in Fig. 7.12 in Nefyodov and Stratton [26, 29].

Let us see now in more detail the properties of the main and higher types of waves of the coaxial waveguide. This is interesting and instructive.

6.9.6 Main Wave, Properties

The *TEM* (transverse) wave (Fig. 6.23) is the simplest electric wave of the coaxial waveguide. Its components are obtained according to $\Phi_{00}(r)$ from Maxwell's equations in the following form:

$$E_r = UF\frac{1}{r}, \quad H_\varphi = IF\frac{1}{r}, \quad E_\varphi = E_z = H_r = H_z \equiv 0.$$

On the basis of these relationships, we can calculate the wave impedance of the coaxial waveguide on the main *TEM* wave (see Sect. 6.13).

The longitudinal wave number of the coaxial waveguide of the main *TEM* wave $h \equiv k$ (if the waveguide is not filled) and $h \equiv k\sqrt{\varepsilon\mu}$ (at fulfilling by the medium with the refraction index $n \equiv \sqrt{\varepsilon\mu}$). Thus, the phase and group velocities of the main wave of the coaxial waveguide coincide and are equal to the speed of light propagation in the filling medium. Therefore, the *dispersion*, i.e., dependence $k = k(\omega)$ for the *TEM* wave is usually absent; nevertheless, the dispersion is possible owing to the filling material $\varepsilon = \varepsilon(\omega)$, $\mu = \mu(\omega)$. By the way, it is noticeably manifested either at very high frequencies, for instance, within the light range, or in the materials with anisotropy, which are used in different *nonreciprocal devices, circulators, phase-shifters,* etc. [17, 20, 21, 36, 37, 128].

Some of the data of the main *TEM* wave of the coaxial waveguide [22] are interesting. So, at a constant external radius and voltage, the maximal strength of the electric field is minimal at $R_2/R_1 = 2.72$, which is close to the square value of the golden proportion $\Phi^2 = 2.618$. At constant a and $|E_{\max}|$ values, the largest power is obtained at $R_2/R_1 = 1.65$ ($\Phi = 1.618$). Finally, at a constant external radius and wavelength, the damping is minimal at $R_2/R_1 = 3.6$.

Fig. 6.23 Field picture of the main wave in the transverse cross-section of the coaxial waveguide (*E* lines are solid)

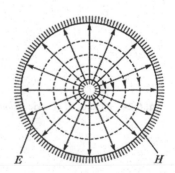

6.9.7 Numerical Investigation of Characteristic Equations

In general, determination of eigenvalues g_{ms} and $\tilde{g}_{ms}$ from Eqs. (6.48) and (6.49) should be carried out using numerical methods according to Programs P.6.10 and P.6.11. Detailed investigation of the behavior of eigennumbers of the coaxial waveguide was discussed in scientific [22] and educational [17–19, 28] literature and to save the course volume, we give the main results of some numerical calculations of eigennumbers g_{ms} and $\tilde{g}_{ms}$.

6.9.8 Asymmetric Magnetic Waves H_{ms}

The typical feature of asymmetric ($m \neq 0$) magnetic waves H_{ms} at $s > 2$ is "notches" in functions $\tilde{g}_{ms} = \tilde{g}_{ms}(\Delta)$. Thus, different ratios of radii Δ correspond to the one transverse number $\tilde{g}_{ms}$. a similar effect was observed for symmetric magnetic waves of the coaxial waveguide, when an impedance of the internal conductor has an *inductive character* [27]. The presence of such a peculiar degeneration leads to interesting practical applications and may be useful in the design of various elements, for instance, fillers of wave types, resonators, wave-meters, and other microwave devices (Sect. 6.10).

From the point of view of numerical results concerning eigennumbers, it is interesting to track the influence of the internal conductor diameter R_1 on eigennumbers g_{ms} and $\tilde{g}_{ms}$, and, as a result, on properties of electric (Fig. 6.24, *upper part*) and magnetic waves (Fig. 6.24, *lower part*). This influence is not the same for electric and magnetic waves.

First of all, we note the *weak influence* of internal conductor dimensions on the behavior of magnetic waves H_{m1}, which at $\Delta \to 0$ obviously take values corresponding to eigennumbers of the cylindrical circular waveguide (Sect. 6.10). The *reduction* (!) with Δ growth, instead of an increase, as is typical for all electric waves (Fig. 6.25b) and most magnetic waves (the Nefyodov-Rossiiski effect) is the second typical peculiarity of eigennumbers $\tilde{g}_{ms}$. These features are discussed in detail, as such *abnormal* behavior of H_{m1} waves allows the suggestion and creation of a wide class of devices with "*nonfocusing*" *mirrors* (both having and not having axis-symmetrical geometry).

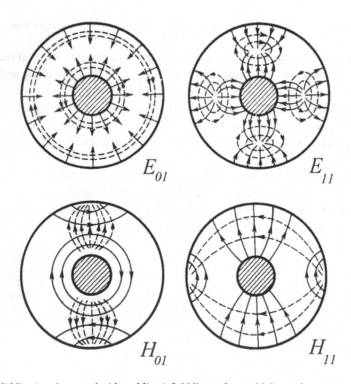

Fig. 6.24 Pictures of electric (*solid lines*) and magnetic (*dotted lines*) field lines of some higher order waves of the coaxial waveguide

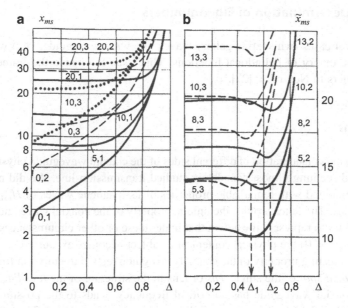

Fig. 6.25 Dependent on eigennumbers of (**a**) electric and (**b**) magnetic waves of the coaxial waveguide versus Δ. The number near the curves corresponds to values of m and s indices. *Solid lines* designate the odd second index (s), *dotted lines*, the even index. Vice versa for magnetic waves

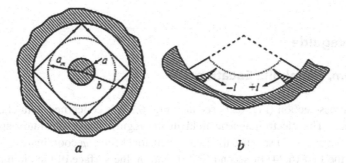

Fig. 6.26 (**a**) The transverse cross-section of the coaxial waveguide with the internal caustic of a_m radius (marked by the *dotted line*) and the geometric-optical beams. The portion of the cross-section with $b - a_m$ width is the zone of the *whispering-gallery waves*, (**b**) is the resonator on the whispering-gallery waves. (Experiments by P. E. Krasnooshkin). (Taken from Nefyodov and Kliuev [73], p. 183)

6.9.9   Whispering-Gallery Waves

The internal rod of the coaxial waveguide most of all "perturbs" the symmetrical electric waves E_{0s} (at small Δ), which can be tracked according to Program P.6.10, assuming $m = 0$). Its influence on the asymmetrical waves $E_{ms}(m \gg 1, s \approx 1)$ is manifested to a lesser degree; thus, the more m, the more the region $\Delta \in (0, \Delta)$, in which this influence turns out to be insignificant. The last circumstance is connected with a character of field distribution of E-wave at $m \gg 1$, $s \geq 2$. In this case, the field happens to be "pressed" to the surface of the external conductor and is described in terms of *whispering-gallery waves*. Until the internal conductor radius R_1 is less than the radius R_m of the zone boundary of the whispering-gallery waves (the *caustic* zone), the internal conductor does not factually affect the field (Fig. 6.26a).

The similar picture of waves traveling along he boundary $r = R_1$ takes place and for magnetic waves H_{ms}, $m \gg 1$, $s \geq 2$. At $r \in (R_m, R_2)$, the electromagnetic field has an oscillating character. Radii on internal caustics $R_m = mb/\eta_{ms}$ and $\tilde{R}_m = mb/\tilde{\eta}_{ms}$ are defined by m, s indices. If the azimuth index m is higher, and the radial index is lower, the field is pressed more severely to the boundary $r = R_2$.

The open resonator of the whispering-gallery wave (Fig. 6.26b) was first studied in experiments of 1947 by *P.E. Krasnooshkin* long before utilization of the idea itself of an open resonator in quantum electronics.

6.9.10 Note About the Determination of Eigennumbers

In the whole series of practical cases (whispering-gallery waves, the wide waveguide, etc.) we can manage with different asymptotic formulas from the theory of cylindrical functions. This issue is outside the framework of our course and we recommend that the reader refers to Nefyodov [27].

6.9.11 Some Conclusions

In enough material devoted to the consideration of different sides of the coaxial waveguide physics, we saw how manifold are properties of this "oldest" and seemingly most completely studied transmission line. We did not discuss these problems in vain, because we became acquainted with remarkable properties of its magnetic waves of H_{m1} type and, as a consequence, with *near-boundary oscillations*. The latter provide the unique property of the coaxial waveguides – *the uniform spectrum of natural frequencies*. It is difficult to represent the notion that under these or other circumstances (for instance, infinitely wide frequency bandwidth including $f = 0$) the physical content of a subject becomes exhausted.

We note one more most interesting property of the coaxial waveguide related to emission from its open end. In future, we shall see that the size of the *focal spot* of the single *lens* or lens in the structure of the *lens line* (Sect. 9.3) has an order of λ. However, the fact that the coaxial waveguide has no critical frequency, leads to the possibility of obtaining a spot from emission from the free end of the coaxial waveguide with dimensions much less than λ. This effect has great significance for application in IC and 3D-IC engineering, in medicine in millimeter-wave therapy, in radiation of biologically-active points, and many others. This problem was discovered and studied, probably, for the first time in the researches of Prof. A.V. Franzesson and colleagues.[3]

6.10 The Circular Waveguide

6.10.1 Boundary Problem, Helmholtz Equation

A waveguide with a circular cross-section (Fig. 6.2b, see also Fig. 6.27) is concerned with classical waveguiding structures and is widely used in practice. The electromagnetic field in the regular (with separate dependence on the longitudinal coordinate z and time t: $\exp\{\pm ihz - i\omega t\}$) satisfies the two-dimensional (in r, φ coordinates) *Helmholtz equation* (6.48), the ideal boundary conditions of the first (6.24) or second (6.29) kind on the surface of the circular cylinder for $r = R$, and the periodicity condition on the φ coordinate: $\Psi_j(r, \varphi) = \Psi_j(r, \varphi + 2\pi n)$, $n = 0, 1, 2, \ldots$

All actions related to variable separation and obtaining ordinary differential equations for Eq. (6.48) in the form (6.49) and (6.50) for the radial $R(r)$ and azimuth $\Phi(\varphi)$ functions remain the same as for the coaxial waveguide solution. The feature consists in the fact that in the solution for the radial $R(r)$ function, it must be assumed that the coefficient B in (6.44) is zero because at $r \to 0$ the *Bessel function* of the second kind (it is often referred to as the *Neumann function*) $Y_\nu(r) \to -\infty$.

Thus, the solution of the Bessel equation for the circular waveguide is written as

$$\Psi_i(r, \phi) = J_n(\chi r)\Phi(\phi). \tag{6.59}$$

Fig. 6.27 The circular waveguide: the geometry of cross-section, dimensions, and the coordinate system

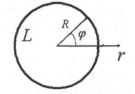

[3] *Zuev V.S., Franzesson A.V.* Sub-wave narrowing of the electromagnetic field. – Preprint № 31. FIAN, 1998.

Table 6.1 Roots of Eq. (6.60)

n	m			
	1	2	3	4
0	2.405	5.520	8.654	11.792
1	3.832	7.016	10.173	13.324
2	5.136	8.417	11.620	14.796
3	6.380	9.761	13.015	16.223

6.10.2 Electric Waves E_{mn}

From the boundary condition $\Psi = 0$ on $L(r = R)$ it follows that the differential equation for E_{mn} waves has the form (compare Eq. (6.51)):

$$J_n(\chi R) = 0, \tag{6.60}$$

i.e.,

$$\chi = \mu_{nm}/R, \tag{6.61}$$

where μ_{nm} is the *m-th* root of equation $J_n(\mu) = 0$. Determination of the roots of this equation can be carried out either numerically according to Program P.6.12 (Appendix B – http://extras.springer.com/2019/978-3-319-90847-2), or they can be taken from the appropriate tables of the Bessel functions.

Values of several roots of Eq. (6.60) for E_{nm} waves of the circular waveguide are presented in Table 6.1.

As a result, we can write the eigenfunctions of the boundary problems in the one of following forms:

$$\Psi_{inm}^{(1)} = N_{nm}^{(1)} J_n\left(\frac{\mu_{nm}}{R}r\right) \cos n\varphi, \quad \Psi_{inm}^{(1)} = N_{nm}^{(1)} J_n\left(\frac{\mu_{nm}}{R}r\right) e^{in\varphi},$$

$$\Psi_{inm}^{(1)} = N_{nm}^{(1)} J_n\left(\frac{\mu_{nm}}{R}r\right) \sin n\varphi, \quad \Psi_{inm}^{(1)} = N_{nm}^{(1)} J_n\left(\frac{\mu_{nm}}{R}r\right) e^{-in\varphi}, \tag{6.62}$$

where $\chi_{nm}^2 = (\mu_{mn}/R)^2$, $N_{nm}^{(1)}$ are uncertain constants.

The parameter $m = 0, 1, 2 \ldots$ in Eq. (6.62) means a number of field maxima on the radial coordinate r.

For electric waves, the condition $H_z = 0$ is satisfied; therefore, the current density has the longitudinal component i_z only, and thus the only longitudinal currents flow in the waveguide walls. Hence, the narrow longitudinal slots in walls can be used to field measure in waveguides, because they do not severely change the field structure. The transverse slots, on the contrary, are emitting and the wave field is highly distorted. The field pictures of the lowest waves E_{01} and E_{11} are shown in Fig. 6.28a, b. The volumetric distribution of components $E_z(r, \phi)$ can be calculated using Program P.6.12.

Program P.6.11 permits the distribution of the longitudinal component $E_z(r, \phi)$ for electric E_{mn} waves in the transverse section of the circular waveguide to be constructed.

6.10.3 Magnetic Waves H_{mn}

Solving for the same area (Fig. 6.2b) the *second boundary problem*, we again come to Eq. (6.60). However, the χ constant is defined now from the condition $\partial \Psi_i/\partial r = 0$ for $r = R$, i.e., the dispersion equation for H_{mn} waves have the form (compare Eq. (6.54)):

$$J_n'(\chi R) = 0. \tag{6.63}$$

From here, it follows that $\chi = \lambda_{nm}/R$, where λ_{nm} is the *m-th* root of equation $J_n'(\lambda) = 0$.. Determination of the roots of this equation is carried out either numerically using Program P.6.13, or they are chosen from tables of Bessel functions.

Values of several roots of Eq. (6.63) for H_{mn} waves of the circular waveguide are given in Table 6.2.

Fig. 6.28 Field pictures of (**a**) E_{01} and (**b**) E_{11} waves in the transverse section of the circular waveguide. In (**c**, and **d**) distributions of $E_s(r, \varphi)$ are shown, distributions of $E_z(r)$ of these waves on radius r and angle φ. (**e, f**) The files of E_{11} wave in the waveguide section. (Taken from Nefyodov and Kliuev [73], p. 189)

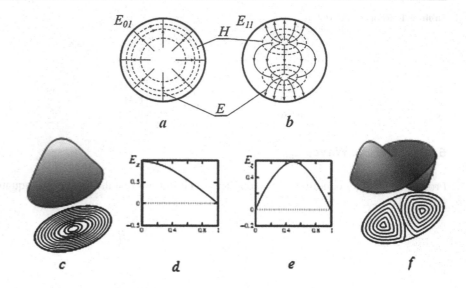

Table 6.2 Roots of Eq. (6.63)

n	m			
	1	2	3	4
0	3.832	6.016	10.173	13.324
1	1.841	5.331	8.536	11.706
2	3.054	6.706	9.969	13.170
3	4.201	8.015	11.346	14.586

Finally, we obtain the eigenfunction system for magnetic waves

$$\Psi_{inm}^{(2)} = N_{nm}^{(2)} J_n\left(\frac{\lambda_{nm}}{R}r\right)\cos n\varphi, \quad \Psi_{inm}^{(2)} = N_{nm}^{(2)} J_n\left(\frac{\lambda_{nm}}{R}r\right)e^{in\varphi},$$

$$\Psi_{inm}^{(2)} = N_{nm}^{(2)} J_n\left(\frac{\lambda_{nm}}{R}r\right)\sin n\varphi, \quad \Psi_{inm}^{(2)} = N_{nm}^{(2)} J_n\left(\frac{\lambda_{nm}}{R}r\right)e^{-in\varphi},$$

(6.64)

where $\chi_{nm}^2 = \lambda_{nm}/R$, $N_{nm}^{(2)}$ are uncertain constants.

The three-dimensional pictures of field distribution of H_{nm} waves of the circular waveguide in the cross-section are determined using Program P.6.13 and some of the data are shown in Fig. 6.29.

The volumetric distributions of the longitudinal component $H_z(r, \varphi)$ of H_{nm} waves in the cross-section of the circular waveguide can be found using Program P.6.13, in which k values should be taken from Table 6.2.

For magnetic waves $H_z \neq 0$, and therefore, the surface current density of H_{mn} waves, has both longitudinal and transverse components. The H_{01} wave occupies a special position because the *longitudinal currents are absent* for it (Fig. 6.29b). For this reason, the H_{01} wave has a unique property: with frequency growth, its damping factor does not increase, as for all other waves of the circular and rectangular waveguides, but it *decreases*. In due course, this was the basis for the design and development of the wideband backbone communication line on the H_{01} wave. In the USSR and abroad, for many years, serious investigations and research into construction design and technology manufacturing were carried out in this direction. However, owing to the arrival of *light guides* (see Sect. 9.2) that had essential advantages compared with the circular waveguides in many parameters (fewer linear losses, lower cost, etc.), investigations into the H_{01} wave were interrupted.

It is interesting to compare the field pictures of the H_{11} wave of the circular waveguide (Fig. 6.30b) and the H_{10} wave of the rectangular waveguide (Fig. 6.30c). They look very "like" each other. Nevertheless, if polarization of the $\vec{E}$ vector in the rectangular waveguide is stable, then in the circular waveguide it is not stable. Sometimes, it is inconvenient in practice, and to keep the $\vec{E}$ vector in the necessary position, we need to transfer to an operation on the elliptic waveguide (see also Sect. 6.11).

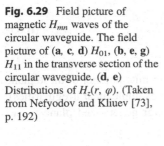

Fig. 6.29 Field picture of magnetic H_{mn} waves of the circular waveguide. The field picture of (**a, c, d**) H_{01}, (**b, e, g**) H_{11} in the transverse section of the circular waveguide. (**d, e**) Distributions of $H_z(r, \varphi)$. (Taken from Nefyodov and Kliuev [73], p. 192)

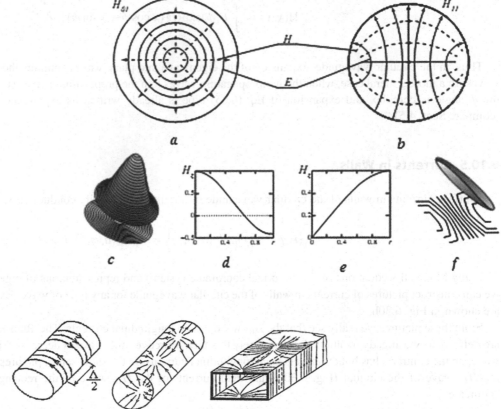

Fig. 6.30 Pictures of current distribution in waveguide walls for (**a**) H_{01} and (**b**) H_{11} waves of the circular waveguide, and (**c**) H_{10} wave of the rectangular waveguide

Waves of the circular waveguide with azimuth dependence of cos $m\varphi$ or sin $m\varphi$ are examples of *polarization degeneration*, because they transform to each other at $\vec{E}$ rotation by 90°. Degeneration of waves here consists in the fact that in the nonsimilar field picture (for instance, shifted by 90°), waves have the same speed, but at the receiving end of the transmission line, which is tuned, say, to the reception of the vertically polarized wave, the wave rotated by any angle (for instance, 90°), is received ineffectively or not received at all.

In a more general plan, *wave degeneration* in the transmission line relates to the equality of these transverse eigennumbers; therefore, the longitudinal eigennumbers, propagation speeds, etc., are similar at different field structures. Wave degeneration $H_{01} \leftrightarrow E_{11}$ is a remarkable example in the considered problem of radio signal transmission (and to a lesser extent of microwave and millimeter-wave power) on the wide ($kR \gg 1$) waveguide of the H_{01} wave. In due course, this became a complicated problem, and several variants of its solution were suggested. Apparently, the approach covering the internal tube surface by a spiral (almost circular) structure, which would have a slight influence on the propagation of the H_{01} wave (there are no longitudinal currents), but significantly change the E_{11} wave structure, and provides its strong absorption, is the most effective and most manufacturable.

6.10.4 About the Expansion of Circular Waveguide Waves into Plane Waves

Considering the similar issue for the rectangular waveguide, we noted that any wave can be expanded either onto two waves (waves of H_{m0} or H_{0n} types), or onto four plane (partial) waves. In the circular waveguide and in the waveguide of other forms of the cross-section, the Hertz vector is represented as

$$\Pi(x,y) = \int\limits_{-\pi}^{+\pi} F(\phi)\exp\{ig\,(x\cos\phi + y\sin\phi)\}d\phi. \tag{6.65}$$

Thus, in the circular waveguide, expansion of the eigenwave into plane waves contains the infinite number of these plane waves. In a more general case, when the nonpropagating waves join the propagating waves (for example, near irregularities), the ϕ angle is complex and expansion of Eq. (6.65) type is already written as expansion over generalized plane waves (compare Sect. 4.5).

6.10.5 Currents in Walls

The surface currents in walls of the circular waveguide of a radius with ideally conducting walls are determined as

$$\vec{j}_S(r,\varphi) = \left[\vec{r}_0, \vec{H}\,(a,\varphi,z)\right] = \vec{e}_\varphi H_Z(a,\varphi,z) + \vec{e}_Z H_\varphi. \tag{6.66}$$

Using Maxwell's equations (in a cylindrical coordinate system) and representations of eigenfunctions in the form (6.64), we can construct pictures of currents in walls of the circular waveguide for any type of wave. Examples for H_{01} and H_{11} waves are shown in Fig. 6.30b, c.

From these pictures we really see that the H_{01} wave has no longitudinal currents. The abnormally small linear ohmic losses are defined, as we already mentioned, by precisely this fact. Moreover, these losses decrease with frequency growth. The H_{11} wave, on the contrary, has both transverse and longitudinal currents. In the same sense, pictures of field current distribution of the H_{11} wave of the circular (Fig. 6.30b) and field currents of the H_{10} wave of the rectangular waveguides (Fig. 6.29c) "coincide."

Similar correspondence can be seen in a series of other transmission lines. So, for instance, there is some "similarity" in field pictures of waves in cylindrical and cross-shaped waveguides (Fig. 6.31b).

Let us give (for reference) values of conductor resistance for DC $- R_= = 1/\sigma t$, t is the conductor thickness, and for high-frequency $- R_\approx = \sqrt{\pi f/\sigma}$.

6.11 Elliptical Waveguides

We already mentioned that in practice the remarkable property of the elliptic waveguide is to keep (to fix) in a stable position of field polarization of the main operation wave. In other words, the transfer, say, from the circular waveguide to the elliptic one, decreases the *polarization degeneration of waves* of the circular waveguide, for instance, the H_{11} wave, for which the $\vec{E}$ vector is directed along the x-axis (Fig. 6.28) and other waves of the same type, but rotated by 90° with respect to the first.

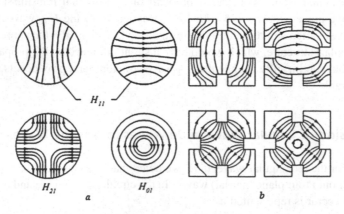

Fig. 6.31 Field pictures of some waves of (**a**) circular and (**a**) cross-shaped waveguides. (Taken from Nefyodov and Kliuev [73], p. 195)

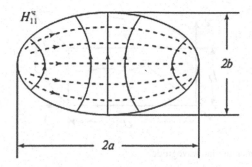

Fig. 6.32 Field picture of H_{11}^{even} and H_{11}^{odd} in the cross-section of the elliptic waveguide. (Taken from Nefyodov and Kliuev [73], p. 196)

The strict analysis of waves in the elliptic waveguide requires solution of Eq. (6.48) in the elliptical coordinate system. Solutions are expressed through the *Mathieu functions* and are not presented here (see, for instance, Nefyodov [27, 28]). The qualitative representation about the wave field structure in the elliptic waveguide can be obtained considering it as a gradual deformation of the circular waveguide. Thus, the H_{11} waves (Fig. 6.29c) of the circular waveguide are transformed into the H_{11}^{even} wave (Fig. 6.32) of the elliptic waveguide. Critical lengths of these waves depend on an eccentricity $e = \sqrt{1-(b/a)}$, where a and b are the large and small semi-axes of the ellipse (Fig. 6.32). At small ellipticity, critical wavelengths of H_{11}^{even} waves polarized in different axes differ slightly from each other. With e growth, the difference increases quite noticeably. The even H_{11}^{even} wave is the main wave of the elliptic waveguide. Its critical frequency can be calculated according to the approximate formula:

$$f_{\text{cr}}a = 8.7849\left(1 + 0.023e^2\right), \tag{6.67}$$

where f is a frequency, GHz; a is the large semi-axis (see Fig. 6.32). An error of f_{cr} determination by Eq. (6.67) does not exceed 1%.

In practice, the rectangular waveguides with side ratio $b/a = 0.5\ldots0.6$ are usually used (which is close to $\overline{\Phi} = 0.618$ [22]); thus, the maximal bandwidth of the single-wave mode is provided at relatively small damping. For example, at $b/a = 0.5$, the critical frequency of the first higher type (in this case, the H_{21}^e wave is this wave type) almost doubles the critical frequency of the main wave, and the damping factor of the main wave in the elliptical waveguide is less than in the rectangular waveguide with the same perimeter.

In antenna waveguide engineering, the flexible *corrugated* elliptical waveguides find application. They were produced in the USSR as segments with lengths of several hundred meters.

6.12 Examples of MathCAD Packet Utilization

In the present chapter, 15 examples of the utilization of MathCAD programs for electrodynamics tasks are examined: electric and magnetic fields of the rectangular waveguide, linear losses of H_{10} wave, electric and magnetic waves of the planar waveguide, surface waves with a dielectric layer near the narrow wall (four variants), electric and magnetic waves of the coaxial waveguide, electric and magnetic waves of the circular waveguide, and electric and magnetic waves of the circular horn. These examples of computer program utilization are considered sequentially in Appendix B (http://extras.springer.com/2019/978-3-319-90847-2).

6.12.1 Electric Waves in the Rectangular Wave Guide

The geometry and dimensions are presented in Fig. 6.33.

Program P.6.1 defines a and b parameters and wave indices, determines the equation for the electric field, and constructs the two- and three-dimensional vector field. Animation is possible.

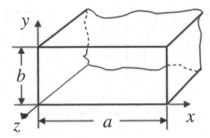

Fig. 6.33 Geometry and dimensions of the rectangular waveguide

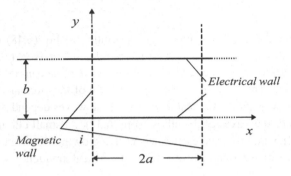

Fig. 6.34 Geometry and location of electric and magnetic walls

6.12.2 Magnetic Waves in the Rectangular Wave Guide

The same situation as in Sect. 6.12.1, but for magnetic waves.

6.12.3 Linear Losses of the H_{10} Wave in the Rectangular Waveguide

For the geometry of Fig. 6.33, Program P.6.3 defines the formula for linear losses and constructs the three-dimensional picture for various parameters.

6.12.4 Magnetic Waves in the Planar Waveguide

The geometry and dimensions are presented in Fig. 6.34.

Program P.6.4 defines indices and coordinate functions, functions of the cross-sections, and constructs the two- and three-dimensional pictures of field distribution in the cross-section of the waveguide and various components of the Hertz vector.

6.12.5 Electric Waves in the Planar Waveguide

A similar task but for electric waves. The essential difference from the model with vertical magnetic walls from the conventional rectangular waveguide (with all four electrical walls) consists in the fact that the *TEM*-wave is the eigenwave of this waveguide and must be included in the full system of eigenfunctions (like the expansion (6.28)). We note that the *TEM*-wave is the *only wave* at $a < \lambda/2$, $b < \lambda/2$. Such a model is of paramount significance for an understanding of physics and the fulfillment of estimation "engineering" calculations in strip-slot transmission lines (Chap. 7). Their dimensions (for the necessity of the single-wave mode) should be selected in this manner. In a series of cases, in particular, in microwave and millimeter-wave IC and 3D-IC, *microstrip antennas*, etc., the operation of the basing element (or its part) on the one of higher types of oscillations (waves) is more profitable (see Sect. 7.3).

6.12.6 Rectangular Waveguide with Dielectric Layer Near the Narrow Walls, LE$_{mn}$ Waves

Program P.6.6 defines the initial data, constructs the transcendent equation, determines the software for root determination, constructs the two-dimensional plot of propagation constants versus parameters.

6.12.7 Rectangular Waveguide with Dielectric Layer Near the Wide Walls, LE$_{mn}$ Waves

The same, but for the wide walls.

6.12.8 Rectangular Waveguide with Dielectric Layer Near the Narrow Walls, LM$_{mn}$ Waves

The same, but for the magnetic waves.

6.12.9 Rectangular Waveguide with Dielectric Layer Near the Wide Walls, LM$_{mn}$ Waves

The same, but for the magnetic waves. Programs P.6.6, P.6.7, P.6.8, and P.6.9 use the "zero" approximation to increase the calculation speed instead of the exact value of the wave propagation constant of a completely filled waveguide by the uniform dielectric (magnetic): $h/k = \sqrt{\varepsilon\mu - (m\pi/ka)^2 - (n\pi/kb)^2}$. In Programs P.6.6, P.6.7, P.6.8, and P.6.9 we use designations: $e1 = \varepsilon_1, e2 = \varepsilon_2, m1 = \mu_1, m2 = \mu_2, xx$ is a matrix of eigenvalues of m, n waves in the completely filled rectangular waveguide. The presence of i multiplier of xx element means that this wave is beyond cut-off (nonpropagating).

The interesting physical properties of wave behavior in the partially filled waveguides by uniform layers of dielectrics follow from the results of Programs P.6.6, P.6.7, P.6.8, and P.6.9. First of all, we have clear and understandable physical content and the convenience of longitudinal LE_{mn} and LM_{mn} wave utilization, which are eigenwaves of precisely these waveguides. From Programs P.6.6, P.6.7, P.6.8, and P.6.9, we see that the results clearly demonstrate the *dielectric effect* related to retraction of the electromagnetic field into the dielectric layer, for instance, for frequency increase; thus, wave deceleration acquires $\sqrt{\varepsilon}$. The field in the nonfilled part of the waveguide decreases according to exponential law, i.e., the wave has the character of the *surface wave*. This effect has numerous technical applications (see Chaps. 7, 8, and 9). Dielectric structures of various classes find wide application in resonators, nano-devices, for instance, with utilization of graphene.

6.12.10 Coaxial Waveguide, Magnetic Waves

The geometry and dimensions are presented in Fig. 6.35.

Program P.6.10 defines initial data (azimuth index and radii ratio), the dispersion equation for transverse eigennumbers, calculates the left-hand part of this equation for various parameters, and determines the root region on the specific calculation scheme. After this, the Program constructs the plot of the first three waves versus parameters and tables of roots.

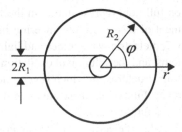

Fig. 6.35 Geometry and dimensions of the coaxial waveguide

6.12.11 Coaxial Waveguide, Electric Fields

The same, but for the electric waves. Roots of Eq. (6.54) are the transverse propagation constants η for electric waves E_{0n}, E_{1n}, E_{2n}, $n = 1, 2, 3$ of the coaxial waveguide can be obtained with the help of Program P.6.11. Of course, we can use it for the determination of η and waves of other numbers indicating other search intervals. Program P.6.11 allows construction of longitudinal component $E_z(r, \varphi)$ distribution for electric waves.

6.12.12 Circular Waveguide, Electric Fields

For the given geometry and initial parameters, Program P.6.12 defines the dispersion equation for determination of transverse eigennumbers for electric waves in the circular waveguide. The left-hand part of the equation is drawn as a function of η and with a special scheme, we define the roots. After that, the three-dimensional pictures of fields are constructed. Animation is possible. E_{mn} waves are in the transverse section of the circular waveguide.

6.12.13 Circular Waveguide, Magnetic Fields

The same, but for the magnetic waves.

6.12.14 Circular Horn, Magnetic Fields

For the given geometry and initial parameters, Program P.6.14 defines the dispersion equation for determination of transverse eigennumbers for magnetic waves in the circular horn. Then, we define the roots. After that, the three-dimensional pictures of fields are constructed. Longitudinal propagation constants in the circular horn are constructed for magnetic fields.

6.12.15 Circular Horn, Electric Fields

The same, but for the magnetic waves.

6.13 Summary

We have finished our examination of the problems of electromagnetic processes in guiding structures, which are important for electrodynamics as a whole. These structures are not only the base that is used for many modern radio engineering, radio physics, radar technology, radio astronomy, and many other systems and devices, blocks and modules of microwave and millimeter-wave systems. Moreover, these structures form the basis of the knowledge, which is extremely important for the specialist in any radio electronics field. Classification of the transmission line, which has already been performed (Sect. 6.2), shows the large number of various systems the specialist deals with. Besides, it is important that transmission lines are significant not only by themselves, which is, of course, appreciable, but by the fact that they constitute a base for development, for instance, closed (Chap. 10) and open (Chap. 11) resonance systems. Also important is that the methods of analysis, which are described in this chapter and based, say, on full field representation in the form of an expansion over the full system of eigenfunctions, are the basis for the development of theory and engineering in other areas of knowledge, for instance, for problem solution in strip-slot structures (Chap. 7, dielectric waveguides and lightguides (Chap. 9), the aforementioned closed (Chap. 10) and open (Chap. 11) resonance systems, and the solution of excitation tasks of waveguides and resonators. We paid special attention to the complicated waveguiding structures, in which several heterogeneous transmission lines immediately interact. We have already described such examples. First of all, these are three-dimensional ICs of microwave and millimeter-wave ranges, which are promising for present and future years.

Some transmission lines, for instance, the edge-dielectric, the waveguide with the dielectric plate, and the planar waveguide, directly serve as models with periodic structures (Chap. 8), dielectric waveguides, and lightguides (Chap. 9), etc.

By consideration of quasi-optical elements of the waveguide paths, we established the general operation principle of elements – the principle of mutual compensation (the Nefyodov principle) – to a great extent to widen the range of knowledge range on the action physics of the quasi-optical devices.

Applied programs in the MathCAD medium should essentially widen and reinforce the information studied in the materials of this chapter.

Checking Questions
1. What is the spectrum of eigenwaves of hollow closed transmission lines?
2. How does the Helmholtz equation differ from the wave equation?
3. Why do eigenwaves of waveguides, which are closed and even filled by the medium, have frequency dispersion?
4. When do the phase and group velocities in the closed waveguides coincide in value?
5. What is the difference between the concepts of the "electric" and "magnetic" wall?
6. Why is the transverse *TEM*-wave absent in the rectangular and circular waveguides? What should be changed in the waveguide construction for such a wave to exist?
7. How can we determine the wave impedance in the transmission line?
8. Under what conditions do whispering-gallery waves arise in the transmission line? What is the determining condition for the existence of whispering-gallery waves?
9. What is the sense of the near-surface wave concept? Which technical devices were created based on it?
10. In which cases is the frequency dispersion demonstrated in the asymmetrical stripline? How much the frequency dispersion is?
11. When is the elliptic waveguide used?
12. What sense and introduction procedure does the wave impedance of the transmission line consist in?
13. Using the known solutions of the static problems for transverse structures of regular transmission lines shown in Fig. 6.1a–c, draw the field lines of electric field strength for these structures.
14. Using Programs P.6.1 and P.6.2, construct the transverse electric and magnetic fields in cross-section of the rectangular waveguide for $a = 0.3$ and 0.8 and $b = 0.5$ and 0.9. Compare results with the case $a = b = 1$.
15. Using Programs P.6.4 and P.6.5, calculate the components of electric and magnetic fields for cases $m = 1$, $n = 2$ and $m = 2$, $n = 1$. Explain the results.

Strip-Slot and Ridge Dielectric Transmission Lines

The material of this chapter of our lecture course is of great scientific and practical interest. The progress of systems of ultra-fast information processing had required a transition to microwave integrated circuits, and understanding them is based on thephysics assimilation of strip-slot transmission lines and basing structures on them.

We start with the classification of strip-slot transmission lines (Sect. 7.1). The symmetric stripline (Sect. 7.2) is studied. We describe the field of the transverse electromagnetic (TEM) main wave and introduce the concepts of wave impedance, losses, the Q-factor, and the maximal operation frequency (Sect. 7.2.2). Further, the consideration of asymmetric strip transmission line follows (Sect. 7.3). We describe the field structure of the main wave (Sect. 7.3.2) and show the quasi-static approximation (Sect. 7.3.3); programs are given for numerical determination of the parameters of asymmetric striplines (Sect. 7.3.4). Section 7.3.3 is devoted to the system of coupled asymmetric striplines.

Taking into consideration the importance of these types of lines for practice, we present the fundamentals of the "strict" theory of the asymmetric stripline (Sect. 7.3.7) based on a solution to the crucial problem (Sect. 2.7). The solution is represented in the form of electric and magnetic eigenwaves, which are a superposition of two waves. This Section ends with the demonstration of some useful comparisons from the theory described and the data of the rectangular waveguide (Sects. 6.4, 6.5, 6.6, 6.7, 6.8, and 6.9). Such data rarely occur in scientific engineering and especially in the academic literature. Thus, in the present textbook, we show the nature of the different reactivity of the open end of the microstrip transmission line, the phenomenon of the whispering gallery waves in the disk microstrip resonator (Sect. 7.3.8).

Further, the reader's attention is switched to the wide class of slot transmission lines (Sect. 7.4 onward); the symmetric slot line (Sect. 7.4) is considered first. The operation principle and the waveguide model of the symmetric slot line are shown (Sect. 7.4.2). We describe two main analysis methods – the transverse resonance (the Kohh method) and the method of partial regions (the Kisun'ko method). For the main H−type wave the simple dispersion equation can be obtained for determination of propagation constants (Program P.7.1) from the integral equation, using the *Pistohlkors–Schwinger transformation*.

Owing to features of the slot lines, namely, sufficient narrow slots (compared with a wavelength), we use various approximate analysis methods of the conformal map method type (Sect. 7.4.1). In particular, for the exponential narrow slot, the delay is defined by the famous Feld formula. For an analysis at the actual slot sizes, we consider the waveguide model of the symmetric slot line (Sect. 7.4.2). The electromagnetic field of the slot line can be obtained from the model of the equivalent linear magnetic current (Sect. 7.4.3). On calculation of ε_{eff} and the wave impedance, the model may be used in the form of the transmission line with electric and magnetic walls (Sect. 7.4.3). Analytical representation for the main H-type wave deceleration is obtained from integral equations using the Pistohlkors–Schwinger transformation.

In connection with the wide design and creation of microwave and millimeter-wave 3-D IC, the asymmetric slot line has gained special significance (Sect. 7.5). Its eigenwaves are obtained for the integral equations using the standard Galerkin procedure (Sect. 7.5.2). This enables the line wave impedance and the effective permittivity to be obtained (Sect. 7.5.3, Program P.7.6). The approximate formula for wave impedance is also given. We describe the remarkable feature in the manufacturing of the very narrow slots. Lines of this class can be performed with practically any values of wave impedance.

The coplanar transmission lines have achieved wide distribution (Sect. 7.6). The quasi-T и H-type waves are the main wave classes of these lines. Being given the field in the slot, the capacity calculation can be performed with the help of the Ritz procedure and by obtaining the value of the wave impedance. The property analysis of the coplanar line can be provided via Program P.7.7.

© Springer International Publishing AG, part of Springer Nature 2019
E. I. Nefyodov, S. M. Smolskiy, *Electromagnetic Fields and Waves*, Textbooks in Telecommunication Engineering,
https://doi.org/10.1007/978-3-319-90847-2_7

At the end of this chapter, we are acquainted with the rather new class of electrodynamics structures – edge-dielectric lines (Sect. 7.7). They are applied within the millimeter wave range. The Helmholtz equation with appropriate boundary conditions and the additional condition on the edge (Sect. 2.5) can be reduced to the system of linear algebraic equations with the help of the Galerkin method. The solution is obtained by using the numerical approach. The analysis of numerical results is performed (Sect. 7.7.3).

7.1   Introduction: Classification of Strip-Slot Lines

The strip transmission lines were entered into the microwave engineering far enough, at the beginning of the 1950s [113, 115]. At first, there was some modification of the standard (at that time) and well-studied transmission lines (Fig. 6.1) of double-wire and coaxial transmission lines, the rectangular and plain waveguides, etc. As time went by, the natural variation happened, say, of the double-wide line of standard type (Fig. 7.1a), according to the mirror representation method, it was transformed to the single wire over the conducting plane (Fig. 7.1b). Then, a change of the wire of the circular cross-section into the more technological rectangular cross-section was used and its accommodation on the dielectric substrate. This was done with consideration of the evident constructive (construction rigidity) and electrodynamics (sharp reduction of transverse dimensions) (Fig. 7.1c). Measures for exclusion of the effects on transmission lines and on devices at its base were not less obvious, in addition from external sources and other radio engineering devices owing to emission from the open transmission line (Fig. 7.1c). Thus, different constructions of strip transmission lines appeared: symmetric (conditionally closed) striplines with dielectric filled the cross-section (Fig. 6.1g) and others.

We consider strip transmission lines in more detail than in our earlier discussion on a rectangular waveguide for the following reasons. First, because strip transmission lines and slot transmission lines are widely used structures, both in the form of transmission lines between basing elements in plain Integrated circuits (ICs) and three-dimensional integrated circuit (3D-IC) and as a basis for the creation of basing elements itself. The second consideration, in favor of more complete examination of symmetric striplines (Fig. 6.1f, g) and asymmetric striplines (Fig. 6.1i; Fig. 7.1e), consists in the fact that in the educational literature for university students, the promising and widespread symmetric striplines are described inadequately. Usually, the information included in the textbooks contains only knowledge about stripline parameters in static or, at best, in quasi-static approximation, which is insufficient for the microwave range and, even more, for the millimeter range. At the same time, the "standard" hollow metallic waveguides (rectangular, circular, and also coaxial but only on the main TEM wave) are considered and described many times in great detail in the usual textbooks.

Finally, information usually described in courses on electromagnetics does not permit a deep understanding of many physical peculiarities, by which devices base on strip transmission lines are so "rich" and therefore attractive for university study and further design. We had already mentioned that strip and slot lines are the basis for microwave ICs and 3D-ICs, and, consequently, more emphasis should be placed on it than is usually adopted. Below, we give various approximate formulas for characteristics of different types of strip and slot lines being guided first, by its physical contents and, second, by its adequacy for students regarding the fulfillment of annual and final projects in addition to performing preliminary investigations at the draft stage.

Many of the described results can be used as the "zero" approximation for further specification within frameworks of the perturbations method, the variation–iteration approaches, numerical, numerical–analytical methods, etc. For fulfillment of

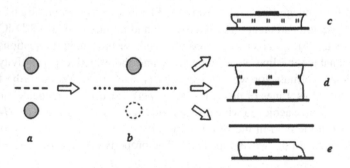

Fig. 7.1 Transfer from (**a**) the double-wire line to (**b**) its mirror representation and then to (**c**) the asymmetric stripline, the symmetric stripline with full filling by (**d**) the magnetic–dielectric and (**e**) the shielded asymmetric stripline. (Taken from Nefyodov and Kliuev [73], p. 211)

student annual and final projects, we can recommend computer packages MathCAD or the more professional MathLAB, and for further design – a version of high frequency structure simulator and many other packages [70, 129].

In the USSR, the publications by I.S. Kovaliov and M.I. Kontorovich constituted the first research into strip transmission lines.

7.2 Symmetric Stripline

7.2.1 Construction, Field Structure of the Main Wave

The historically first and, probably, the most widespread symmetric strip transmission line represents a thin metallic band of the finite width w located between two parallel metallic plates at the same distance from each of them (Fig. 7.1d). A gap between a band and grounded plates of the stripline is filled by a magnetic–dielectric ($\varepsilon > 1$, $\mu > 1$) with a high enough refraction index $n = \sqrt{\varepsilon \mu}$. This filling is necessary because of constructive considerations (attachment rigidity, stability to vibrations and so on) and has the purpose of reducing the dimension of microwave and millimeter-wave devices on symmetric striplines. Thus, the symmetric stripline is historically the first strip transmission line.

Now, we can represent the strip–slot structures filled by artificial dielectric media with different values of ε, μ, σ, χ. This provides huge new opportunities for the designer, both in reduction of weight and dimension parameters of equipment and for obtaining the principally new electrodynamics characteristics of transmission lines and basing elements (see, for example, Schwinger, and Rammo and Whinnery [90, 104]).

For a long time, it was considered that large ohmic and dielectric losses (a small linear Q-factor) are the main drawback of symmetric striplines. Nevertheless, losses in metals and dielectrics have essentially decreased over the last few years owing to new materials technologies, and therefore, the role of symmetric striplines in microwave ICs and 3D-ICs has again increased. On the other hand, the symmetric stripline is, in essence, the main type of transmission lines in 3D-ICs of microwave range (in its "shielded" variant; Fig. 7.2), because 3D-ICs of microwave and the millimeter-wave are "closer" to symmetric striplines with regard to construction. In that, horizontal shields play the role of inter-floor separators.

The transverse electromagnetic (TEM) wave is the main type of wave propagation along the strip transmission line, whose phase velocity is

$$v_{\mathrm{ph}} = c/\sqrt{\varepsilon}, \tag{7.1}$$

where c is the light speed in vacuum, ε is the relative permittivity of the uniform material, which completely fills the cross-section of the stripline.

7.2.2 Wave Impedance of the Symmetric Stripline

A wave impedance for the main wave type (TEM wave) is rather accurately defined by the conformal map method (see, for instance, Gupta and Singh, and Gunston [110, 115]), which leads to some equations containing the full elliptic integral of the first kind, which is not always convenient. In the fulfillment of annual and final projects, students can use the simpler and precise enough formula

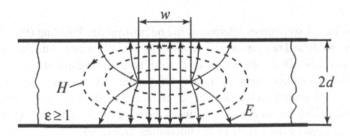

Fig. 7.2 The cross-section of the symmetric stripline (the structure Eq. (7.5) of the T-wave field). (Taken from Nefyodov and Kliuev [73], p. 212)

$$Z(x) = 94.172 \left[\sqrt{\varepsilon} (0.441 + x) \right]^{-1}, \qquad x = w/d, \tag{7.2}$$

where w is the band width, d is the dielectric thickness.

Equation (7.2) is valid for the case of the infinitely thin stripline conducting band.

The wave impedance of the symmetric stripline $Z(x)$ can be calculated using the Program P.7.1 from Appendix B (http://extras.springer.com/2019/978-3-319-90847-2). It is accurate and has great possibilities [70, 77, 113]. Usually, if not using the specific measures, the wave impedance of active devices is small, which leads to the utilization of symmetric (and asymmetric; see later) striplines with wide current-carrying bands. This may lead to the necessity of special measures, both in calculation models and in the implementation of necessary joints in practice [13–15, 107].

The Program P.7.1 has the possibility of taking into account the finite thickness t of the current-carrying band (compare Gvozdev and Nefyodov, and Neganov et al. [14, 17], where the numerous approximate analytic relationships are given), and also the total losses in the symmetric stripline. Total losses α in such a line can be divided into two components: losses in the conductor (metal) α_m and losses in the dielectric α_d. Because of its relative small size of the total losses, they are represented by a sum:

$$\alpha = \alpha_m + \alpha_d. \tag{7.3}$$

Losses in the dielectric depend on the tangent of the angle of the dielectric losses ($\tan \delta$) of the material used and are defined approximately as:

$$\alpha_d = \left(27.3 \sqrt{\varepsilon} / \lambda_0 \right) \tan \delta. \tag{7.4}$$

For practical calculation, the value of linear Q-factor of symmetric striplines has great significance for practical calculations. It is equal to:

$$Q = \left(27.3 \sqrt{\varepsilon} / (\alpha \lambda_0) \right) \left[1 - (\lambda_0 / 2\varepsilon)(d\varepsilon / d\lambda_0) \right]. \tag{7.5}$$

Calculation of losses in symmetric striplines with high enough accuracy can be performed by Program P.7.1.

The maximal operating frequency f_{max} of the symmetric stripline is defined by the possibility of excitation of transverse (waveguide) H-waves of the lowest order. For f_{max} expressed in gigahertz we know the approximate formula:

$$f_{max} = 15 \left[d \sqrt{\varepsilon} (w/d + \pi/2) \right]^{-1}, \tag{7.6}$$

where f_{max} in GHz, and d and w in cm. The value of f_{max} is defined by Eq. (7.6).

Starting from Eq. (7.6), it is necessary to choose the compromise dielectric thickness, at which the Q-factor of the symmetric stripline minimally decreases and the high cut-off frequency is saved for propagation of the H-wave.

To suppress the spurious H-wave, the constructive measures are used: the galvanic connection by the metallic rods of the metallic layers near the conductor or implementation of narrow longitudinal slots in the conductor of symmetric striplines.

7.3   Asymmetric Strip Transmission Line

7.3.1   Introduction

Asymmetric strip transmission line is sometimes referred as the microstripline. This line (Fig. 7.3) represents an example of the open structure, whose waveguide and resonance properties should be taken into consideration during calculations. Being the integral part of IC and especially 3D-IC, the asymmetric stripline is distinguished from other transmission lines by its simplicity and manufacturability. So-called planar technology allows production of complex basing elements and functional units on asymmetric striplines in the unified cycle and by large series, and to play the role of transmission lines between basing elements, between basing elements and functional units, between different functional units in microwave and millimeter-wave ranges. Structures with 2.5D technology have particularly interesting properties.

Fig. 7.3 Asymmetric stripline.
(**a**, **b**) Transverse and longitudinal sections. (**c**) Distribution of longitudinal currents.
(**d**) Structure of currents.
(**e**, **f**) Structure of the fields of surface waves in an asymmetric stripline. (Taken from Nefyodov and Kliuev [73], p. 217)

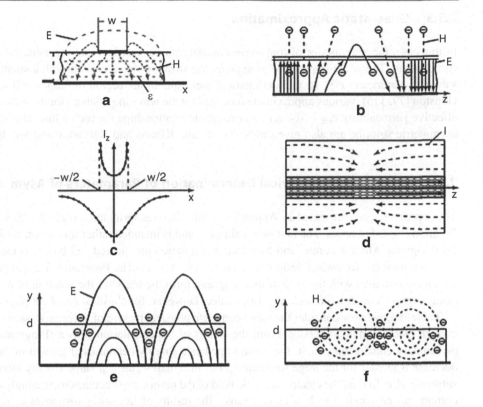

Taking into consideration the importance of asymmetric striplines for practice, on the one hand, and the deep physical contents of processes that happen in it, on the other, we shall more accurately (than had been done for other types of transmission lines) examine its operation principle and physics of processes. Unfortunately, we cannot consider this issue in detail as this is far beyond the framework of our lecture course. First, we present elementary information about asymmetric striplines.

7.3.2 Construction, Field Structure of the Main Wave

An asymmetric stripline is made by coating the metal layer with infinite width from the one side of a substrate and the conductor layer from another side (Fig. 7.3a). The quasi-TEM wave, whose structure is shown schematically in Fig. 7.3a, b, is the main (lowest) wave type propagating in the asymmetric stripline. Here, we should draw your attention to a "break" in the electric field line $\vec{E}$ at the transition from the substrate to the air, which obeys the Snell's law (see Sect. 4.3).

Longitudinal current distribution along the narrow conductor is described by exponential or logarithmic function; the same in the metal layer – by the bell-shape function of $I_z [1 + (x/d)^2]^{-1}$ type (Fig. 7.3 c). The total current distribution in asymmetric stripline conductors is presented in Fig. 7.3d.

The field structure of the surface waves is shown in Fig. 7.3e, f.

In spite of the external simplicity of construction, an asymmetric stripline differs significantly from the strip transmission line in its electrodynamic characteristics. The main difference consists in the fact that an asymmetric stripline represents the open (or almost open) structure and the development of its theory turns out to be connected to the whole series of the most complicated problems of diffraction, computational electrodynamics, and mathematics as a whole. At the same time, for the whole series of applications and asymmetric stripline utilization on 3D-IC of decimeter and centimeter ranges, various approximate relationships are rather useful. Therefore, at first, we consider the results of the quasi-static theory of asymmetric striplines, and then we proceed to electrodynamics approximation, in which we note only the principal issues of the theory.

7.3.3   Quasi-static Approximation

In this case, the magnitudes of transverse components of electrical and magnetic fields essentially exceed the longitudinal components. The wave impedance of asymmetric striplines is calculated with a small enough error (of $\pm 1\%$ order) for the substrate parameters $\varepsilon \leq 16$ and geometric dimensions in the region of $0.05 < w/d < 20$. In Nefyodov and Fialkosky, and Gunston [77, 115], various approximate formulas for the wave impedance (for wide and narrow conductors) are given, and for effective permeability $\varepsilon_{\text{eff}} = (c/v_{\text{ph}})^2$. Approximate relationships for such a line with suspended substrate and for the inverse asymmetric stripline are also given in Nefyodov and Kliuev, and Nefyodov and Smolskiy [1, 2].

7.3.4   Programs of Numerical Determination of Parameters of Asymmetric Striplines

These programs are presented in Appendix B (http://extras.springer.com/2019/978-3-319-90847-2): Program P.7.2-static и Program P.7.2-dynamics. The first one is simplest and is intended rather for demonstration of physical content of the task than for designing. More accurate, and therefore much more complicated and bulky, is the Program P.7.2. In the Program P.7.2 static we used the following designations: $u = w/h$, $v = \varepsilon_r$. The Program P.7.2 and parameters of the asymmetric stripline, which are obtained with the help of this program, form the basis for the solution of many tasks of microwave IC and 3D-IC designing, with acceptable accuracy for practice. However, for clarification of the physical content, these data are insufficient.

Therefore, we now consider the more complete physical picture of wave processes in asymmetric striplines based on strict enough electrodynamics theory. From the results of the calculation shown (Program P.7.2 static; c, e) we can make some preliminary conclusions. First, the wave impedance sharply decreases at growth of the band width w. In that, the speed of decrease is greater for the large substrate ε_r. For the small w, the ε_{eff} value sharply increases from its static values to ε_r of the substrate: d, e. In that, the electromagnetic field of the main wave becomes increasingly concentrated (retracts) in areas under a current-carried conductor. In a certain sense, the results of frequency responses of ε_{eff} that similar in a physical sense are presented: with substrate ε_r growth and frequency growth, the electromagnetic field is increasingly redistributed under the band.

These are the main qualitative results of numerical calculations of asymmetric stripline parameters. In students' annual and final projects, and for many engineering calculations the Program P.7.2 dynamics can be used, allowing accurate results to be obtained that are acceptable for many cases and for practice.

A deeper and fuller physical understanding of the operation principle of asymmetric striplines can be achieved using a stricter theory (see Sect. 7.3.6). This section is the one of the central in our lecture course and we must draw the reader's special attention to it.

7.3.5   Coupled Asymmetric Striplines

When designing and calculating the basing elements of ICs and 3D-ICs, it is frequently necessary to take into account a coupling between the strip transmission lines (Fig. 7.4). This coupling may be either "useful" (filters, directional couplers, etc.) or "harmful," providing the spurious coupling between non-interacting basing elements. Such estimations can be made in quasi-static approximation using known results (see, for instance, Nefyodov and Fialkosky [77]). In Appendix B (http://extras.springer.com/2019/978-3-319-90847-2), there is the Program P.7.4 allowing determination of the characteristics of two similar coupled asymmetric striplines [14, 16, 116]. An example of some calculation results concerning characteristics of coupled asymmetric striplines (ε_{eff} and Z) is presented there.

Results shown in Program P.7.4 for similar (in geometry) asymmetric striplines allow the following conclusions to be made. They are obtained for a large enough frequency range, from 1 to 100 GHz, in essence, from low frequency (almost static) to the millimeter-range. First, at rather low frequencies, characteristics of even and odd waves differ a little, but in the other case (in the millimeter-range), this difference for ε_{eff} achieves great values. Thus, at high frequencies, the field is pulling into the transmission line substrate and $\varepsilon_{\text{eff}} \rightarrow \varepsilon_{\text{sub}}$. It is interesting to note that at a low frequency (in the almost static case) the curves for $\varepsilon_{\text{eff}} = \varepsilon_{\text{eff}}(f)$ begin from the same values, and then behave in different ways for large and small ε values of a substrate. At large ε, the value of ε_{eff} increases with frequency growth, whereas at small ε, on the contrary, it decreases.

Dispersion of the wave impedance of coupled asymmetric striplines (Program P.7.4) for relatively small values of ε for even and odd waves leads to an increase in Z, whereas for large ε it leads to a decrease.

Fig. 7.4 Coupled asymmetric striplines. (**a**) Dispersion characteristics of even and odd waves of coupled asymmetric striplines: (**b**) effective permittivity and (**c**) wave impedance at $s = 0.2$ mm, $h = 0.638$ mm, $w = 1$ mm, and $t = 0.015$ mm

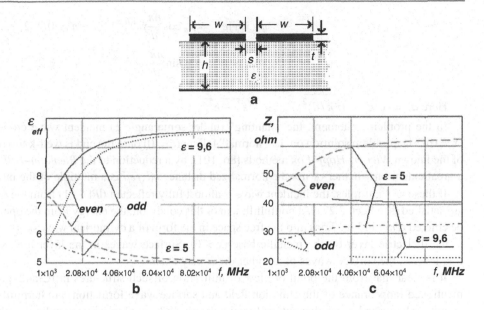

Fig. 7.5 The asymmetric stripline and its key structure. (**a, b**) Transverse and longitudinal sections. (**c**) Distribution of longitudinal currents. (**d**) The current structure. (Taken from Nefyodov and Kliuev [73], p. 224)

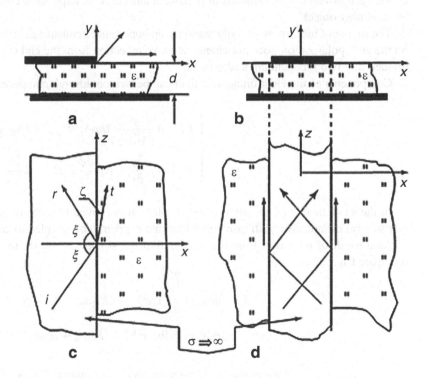

7.3.6 Fundamentals of the "Strict" Theory of the Asymmetric Stripline

We had already become acquainted with a concept of the key structure in electrodynamics problems (Fig. 2.3). The exact strict solution of the boundary problem for the key structure is a basis for construction of a full enough physical picture of phenomena in asymmetric striplines. Therefore, the word "strict" in this sub-section header is taken in quotes. The geometry, coordinate systems, dimensions of asymmetric striplines, and the corresponding key structure are shown in Fig. 7.5.

The key structure (semi-infinite toward to negative x) represents the open end ($x = 0$) of the planar waveguide completely filled by the magnetic and dielectric (ε, μ). The problem is formulated as: onto the open waveguide end ($x = 0$) from $x \rightarrow -\infty$ one of the planar waveguide waves surges (compare Sect. 7.7), which characterizes by y components of electrical φ^i and magnetic ϕ^i Hertz vector:

$$\varphi^i(x,y)e^{ihz} = a_n \cos\frac{n\pi}{d}y e^{i(\alpha_n x + hz)}, \quad n = 0, 1, 2, \dots$$

$$\phi^i(x,y)e^{ihz} = b_n \sin\frac{n\pi}{d}y e^{i(\alpha_n x + hz)}, \quad n = 1, 2, \dots$$

$$(7.7)$$

Here $\alpha_n = \sqrt{g^2 - (n\pi/d)^2}, \quad g = \sqrt{k^2 - h^2}.$

In the problem statement, the "slanting" (under some angle ξ) incident wave crowds (i) onto the open waveguide end ($x = 0$). The problem solution on the "normal" falling ($\xi = 0$) onto the end is well-known, is strict, and can be achieved by one of the known *Wiener–Hopf–Fok* methods [86, 101], by a reduction to the *Riemann–Hilbert*, etc. The final solution depends to a great extent on whether or not the normalized distance $kd\sqrt{\varepsilon\mu}$ is a multiple of the integer number of a half-wavelength.

If these are multiples, the incident wave is almost fully reflected ($|R| \to 1$) from the end, because the resonance conditions are satisfied (see Sect. 2.27). At nonfulfillment of this condition, the energy from the open end (besides the reflected wave) can be emitted (*emit* in Fig. 7.5a) into the free space in the form of a cylindrical wave or to excite surface waves (*surf* in Fig. 7.5a) in the dielectric layer on the metallic base ($x > 0$). Surface waves in the layer are excited for $\lambda_0 \geq 2a\sqrt{\varepsilon - 1}$. At higher frequencies, the surface waves of a higher order may exist.

It is clear that from the point of view of utilization of such structures in IC and, particularly, in 3D-IC, the phenomena mentioned (appearance of the emission field and surface wave formation) are harmful, as they can be reasons for spurious couplings between basing elements of functional units in IC or to promote couplings with other devices located on the same or even another object.

The saving of incident wave polarization is an important circumstance in the case of the normal incidence ($\xi = 0$) of a wave on the end: polarization does not change at wave reflection from the end ($x = 0$). The boundary (vector) task in this case is divided into two independent scalar tasks.

Components of the electromagnetic field can be written through components of Hertz vectors φ, ϕ from Eq. (7.7) as:

$$\begin{cases} E_x = \rho\dfrac{\partial^2\varphi}{\partial x \partial y} + kh\rho\phi, \quad H_x = -kh\varphi + \dfrac{\partial^2\phi}{\partial x \partial y}, \\[2mm] E_z = ih\rho\dfrac{\partial\varphi}{\partial y} + ik\rho\dfrac{\partial\phi}{\partial x}, \quad H_z - ik\dfrac{\partial\varphi}{\partial x} + ix\dfrac{\partial\phi}{\partial y}, \end{cases}$$

$$(7.8)$$

In the strict theory of the asymmetric stripline, it is proved that, in the general case, electric and magnetic polarization (separately) cannot satisfy all task conditions and happen to be coupled to each other, and the task as a whole is vectorial.

As a result of reflection from the open end of the structure ($x = 0$), the wave field is described by potential functions (compare Chap. 2):

$$\varphi(x,y) = [a_n e^{i\alpha_n x} - (R_{11}a_n + R_{12}b_n)e^{-i\alpha_n x}] \cos\frac{n\pi}{d}y,$$

$$\phi(x,y) = [b_n e^{i\alpha_n x} - (R_{21}a_n + R_{22}b_n)e^{-i\alpha_n x}] \sin\frac{n\pi}{d}y.$$

$$(7.9)$$

Here, $R_{ij}(i, j = 1, 2)$ are elements of the reflection factor matrix R. From Eq. (7.9) we see that when one polarization falls onto the end of the wave, the wave of another polarization arises. It reminds us (by something) of a situation of a wave falling onto the plane with anisotropic impedance (see Sect. 4.6), when the reflected field contains waves of both polarizations (see also Sect. 2.5). Strictly speaking, at the falling of the one wave onto the end, the whole spectrum of reflected waves arises (both polarizations; the discrete spectrum), waves of emission (continuous spectrum), waves in the structure "the dielectric list on the conducting surface" ($x > 0$; the discrete spectrum). However, in practice, the simpler constructions are sufficient, which, nevertheless, (commensurable, by the way, with technological errors of IC or 3D-IC manufacture) satisfy the needs of the design and manufacture processes with great accuracy.

Determination of the explicit form of the reflection factors R_{ij} is the most complicated problem of mathematical physics and theoretical electrodynamics. Strictly speaking, this is (to a definite degree) the unsolved problem of modern higher mathematics. Nevertheless, introducing some "natural" simplifications and finding out those waves that have small diffraction losses, we can obtain a solution that is close enough to reality, i.e., practically coinciding with experimental results.

Let us write this decision (at $n = 0$) in the following form:
E-waves:

$$\begin{cases} E_x(x,y,z) = \rho\, a_{nm} \begin{Bmatrix} \cos \\ i\sin \end{Bmatrix} (\alpha_{nm}^E x\;)\; \sin\left(\frac{n\pi}{d}y\right)\, \exp\{i h_{nm}^E\}, \\[2mm] H_z(x,y,z) = i a_{nm} \begin{Bmatrix} \cos \\ i\sin \end{Bmatrix} (\alpha_{nm}^E x)\; \cos\left(\frac{n\pi}{d}y\right)\, \exp\{i h_{nm}^E\}, \\[2mm] E_y = E_z = H_x = H_y \equiv 0. \end{cases} \tag{7.10}$$

H-waves:

$$\begin{cases} E_z(x,y,z) = \rho\, b_{nm} \begin{Bmatrix} \cos \\ i\sin \end{Bmatrix} (\alpha_{nm}^H x)\; \sin\left(\frac{n\pi}{d}y\right)\, \exp\{i h_{nm}^H\}, \\[2mm] H_x(x,y,z) = -i b_{nm} \begin{Bmatrix} \cos \\ i\sin \end{Bmatrix} (\alpha_{nm}^H)\; \cos\left(\frac{n\pi}{d}y\right)\, \exp\{i h_{nm}^H\}, \\[2mm] E_x = E_y = H_y = H_z \equiv 0. \end{cases} \tag{7.11}$$

Here, $n = 1, 2, \ldots$; upper signs in the first braces $m = 1, 3, \ldots$, and lower - $m = 2, 4, \ldots$; a_{nm}, b_{nm} are amplitudes. B (7.10, 7.11):

$$\alpha_{nm}^{E,H} = \sqrt{g/2d}\; s_{nm}^{E,H}, \tag{7.12}$$

$$h_{nm}^{E,H} = \sqrt{\varepsilon\,\mu\,(2\pi/\lambda)^2 - (n\pi/d)^2 - (\alpha_{nm}^{E,H})}, \tag{7.13}$$

$$s_{nm}^{E,H} = \frac{m\pi}{M + \beta' + i\beta_{E,H} + \nu_{E,H}}, \quad m,n = 1,2,\ldots\;,\quad M = a\sqrt{2g/d}. \tag{7.14}$$

Here, parameters β, ν are complex quantities, whose magnitudes are less than 1 and are tabulated in Nefyodov [24].

Formulas (7.10, 7.11), in spite of their simple form, are not obvious. It is known that each wave of E- and H-type represents a superposition of two waves traveling in different directions (remember the point of view of Brillouin; Sect. 7.8; Fig. 7.5, beams r and t):

$$h_{nm}^{E,H} \vec{z}^0 \pm \alpha_{nm}^{E,H} \vec{x}^0. \tag{7.15}$$

It is clear that for $h = 0$ the vectorial task has degenerated into the scalar one. However, Eqs. (7.10, 7.11, as 7.15), are true for $h_{nm} < \alpha_{nm}$ and even for $h_{nm} \approx \alpha_{nm}$. Although the similar result is known in the three-dimensional theory of open resonators with planar mirrors [77, 111], in previous cases, the magnetic dielectric layer was absent $\varepsilon = \mu = 1$ and we could neglect the polarization oscillation splitting (because small typical mirror dimensions compared with the distance between them).

At $n = 0$, the simple result is obtained for the arbitrary relationship of h_{0m} and α_{0m}:

$$\begin{cases} E_y(x,z) = \rho k \begin{Bmatrix} \cos \\ i\sin \end{Bmatrix} (\alpha_{0m} x) e^{i h_{0m} z}, \\[2mm] H_z(x,z) = a_n \begin{Bmatrix} i\sin \\ \cos \end{Bmatrix} (\alpha_{0m} z) e^{i h_{nm} z}, \\[2mm] H_x(x,z) = -h \begin{Bmatrix} \cos \\ i\sin \end{Bmatrix} (\alpha_{0m} x) e^{i h_{0m} z}, \\[2mm] E_x = E_z = H_y \equiv 0. \end{cases} \tag{7.16}$$

The presented Eqs. (7.10, 7.11, and 7.16) permit calculation of the wave impedance of the asymmetric split line, the linear losses in metal and in dielectric, in addition to line excitation by outside sources of higher types of waves (in cases when it is expedient to use the high wave order as the operating one).

7.3.7 Some Useful Comparisons

In conclusion regarding this, frankly speaking, not very simple material for initial studying and understanding, we consider some interesting details.

1. *Coupling of quasi-static approximation with the high-frequency one.* Quasi-static (Sect. 7.3.3) and high-frequency (Sects. 7.3.4 and 7.3.5) approximations considered in the previous sections for the main TEM wave in asymmetric striplines have their own application areas mentioned in the appropriate sections. At the same time, results of a mutual comparison of these different approaches are interesting and instructive. These results are presented graphically in Fig. 8.7 (reproduced from Nefyodov [26]), when the function of square delay is shown (ε_{eff}) for the quasi-TEM wave of asymmetric stripline versus $k_0 a$ (k_0 is the wave number of vacuums), i.e., in essence, this is the frequency function of the square of the delay coefficient. Long-wave asymptotic of the frequency function corresponds to this expression from Nefyodov [26]:

$$h = h_{\text{TEM}}\left[1 + (ka)^2 F_a + (kd)^2 F_d\right], \tag{7.17}$$

in which $F_{a,\,d}$ are complicated enough but restricted functions. According to Eq. (7.17), the longitudinal wave number is defined to an accuracy of $(kl)^4 \ln(kl)$, $l = \sqrt{a^2 + d^2}$.

   The high-frequency approximation, when the solution can be found by the method of successive approximations, is based on the exact solution for the semi-infinite key structure (see Fig. 7.5b, on the right).

   The long-wave asymptotic in accordance with Eq. (7.17) are true at $k\sqrt{a^2 + d^2} \leq 1$; at $k\sqrt{a^2 + d^2} > 1$ the error quickly grows. On the contrary, the result accuracy presented by high-frequency approximation is better for the higher frequency. Even for $2a \leq \lambda_s/2$, the result accuracy obtained according to the factorization method and the method of successive approximations (successive diffractions), may be high enough. Thus, in this region, the long-wave asymptotic is coupled to the high-frequency one. The obtained result of coupling of two different asymptotic approaches (quasi-static and high-frequency) is not evident in advance. Such a comparison turns out to be extremely fruitful because it extends our representations of a ratio of different asymptotic estimations, which is important for understanding physical sense.

2. *Comparison of the asymmetric stripline and the rectangular waveguide.* The electromagnetic field in the asymmetric stripline is mainly concentrated under the current-carrying conductor. When frequency grows, the field is increasingly concentrated in this area. The circumstance mentioned gives an occasion to compare the properties of such lines and the rectangular waveguide. Strictly speaking, the longitudinal wave number h in asymmetric striplines with the ideal conductor and ideal dielectric filling is the complex quantity: $h = h' + ih''$. The presence of the imaginary part of $h(h'')$ is concerned with diffraction emission of the wave from the line. Behavior of the normalized longitudinal wave number h'/k_0 and h''/k_0 (k_0 is the wave number in the air) depending on w and d for the main asymmetric stripline wave is shown in Fig. 7.6.

   Here, we see values of the longitudinal wave number of the rectangular waveguide with the cross-section $a \times b$. The presence of diffraction losses in the asymmetric stripline leads to wave existence in the post-critical region (as we see in Chap. 7, in the rectangular waveguide with ideal walls the propagation in this area is absent: $h' = 0$). The function of the wave impedance of asymmetric stripline versus kd is shown in Fig. 8.9 in Nefyodov, and Nefyodov and Fialkovsky [26, 58]. There, we can see the appropriate curves for the rectangular waveguide.

3. *More about a sign of edge effect reactivity in the key structure and in asymmetric striplines* (Fig. 7.7). When examining the key structure (for the asymmetric stripline) – the semi-infinite plain waveguide (Fig. 7.5a) – we noted that depending on the incident angle value ξ of the i wave on the edge ($x = 0$), various situations are possible. Up to now, we have been interested by the case when the refraction angle of the t wave was equal to zero (full wave reflection from the edge), when the wave in the real structure (Fig. 7.5b) propagates along the z axis. In this case, the incident angle is ξ_{cr} (Fig. 7.7a).

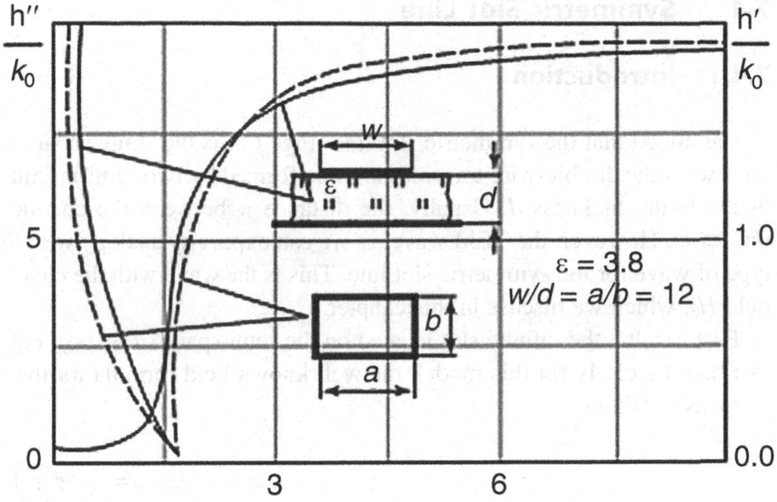

Fig. 7.6 Comparison of delay h''/k_0 and damping h'/k_0 similar in height to the asymmetric stripline and the rectangular waveguide; the left parts of curves are related to damping of the longitudinal wave number. (Taken from Nefyodov and Kliuev [73], p. 234)

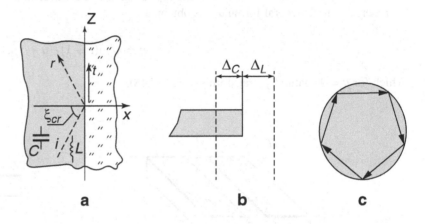

Fig. 7.7 To the note about the reactivity sign of the key structure edge. (**a**) The open end of the plain waveguide. (**b**) The open end of the asymmetric stripline. (**c**) Oscillation of the whispering gallery type in a disk open resonator

If the incident angle is less than ξ_{cr}, then the edge reaction is capacitive (C), and if it is more than ξ_{cr}, then the edge reaction is inductive (L). This leads to the fact that, for instance, a capacitive edge reaction must be compensated for by element extension ΔL (for example, the matching stub, etc.; Fig. 7.7b). When the grazing wave falls on the edge, $\xi > \xi_{cr}$, its reaction has is of an inductive nature and the stub, accordingly, should be shortened by (ΔL).

The whispering gallery mode, when the incident angle in the geometrical–optical picture of oscillations (Fig. 7.7c) is always in grazing mode with respect to the structure edge, is the usual operation mode of high-Q open resonators (see Chap. 12); micro-strip, dielectric, with metallic gratings, etc.) of the disk type. Thus, an amendment on the edge reaction can be noticeable. It should be noted that the oscillations of the whispering gallery type in acoustics have been known since ancient times. In electrodynamics of open systems, they were probably used for the first time by Russian professor P.E. Krasnushkin (see Fig. 6.20b).

Unfortunately, this fact does not attract the attention of readers in most books and student textbooks, although the significance of this effect is evident from the reasons mentioned here and because of the results of the whole series of devices. We tried here to fill this gap.

Thus, we considered some of the main issues of asymmetric stripline theory. It is clear that many problems remain "out-of-the frame", but the information already mentioned should, to our mind, convince the reader that only the detailed physical analysis of each electrodynamics structure can give a clear understanding of the essence of the problem and, therefore, to provide the correct engineering approach to designing of the chosen structure. In particular, it is concerned with the IC and 3D-IC of microwave and millimeter-wave ranges, which cannot be adjusted, aligned, etc., like meter and decimeter wavelength ranges.

7.4   Symmetric Slot Line

7.4.1   Introduction

It is assumed that the symmetric slot line (Fig. 7.8) is the simplest slot transmission line. In essence, the symmetric slot line represents the double-wire transmission line formed by two semi-infinite conductors located on the magnetic–dielectric layer of the finite thickness d. Usually, the distance w between the conductors, which forms the symmetric slot line, is small ($kw \ll 1$). However, the TEM wave, as we can expect by analogy with the double-wire transmission line, is not the operation type of wave for the symmetric slot line. This is the wave with the clearly expressed longitudinal component of the magnetic field H_z, which we discuss in this chapter.

Historically, the infinite slot located on the semi-space ($d \to \infty$); Fig. 7.8a, was the first physical model of the symmetric slot line. Precisely for this model, the well-known Feld formula for the main wave deceleration in the exponentially narrow slot was obtained:

$$\eta = \lambda \ /\lambda_{\text{slot}} = \sqrt{(\varepsilon + 1)/2}, \tag{7.18}$$

where η is the deceleration coefficient, and λ_{slot} designates the wavelength in the slot line.

Later, the more general formula was obtained:

$$\eta = \sqrt{\mu[(\varepsilon + 1)/(\mu + 1)]}, \tag{7.19}$$

which for $\mu = 1$, naturally, transforms to Eq. (7.18).

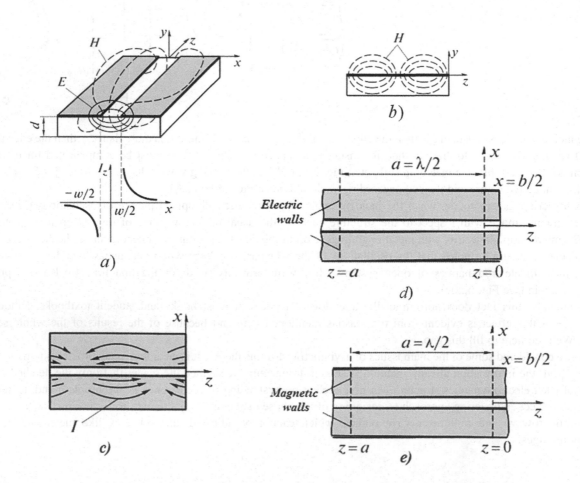

Fig. 7.8 The symmetric slot line. (**a, b**) Structures of the electromagnetic field. (**c, d**) Distribution of the longitudinal field in the x coordinate and the current picture. (**e, f**) Waveguide models of the symmetric slot line

In contrast to the asymmetric stripline (Sect. 7.3), in the symmetric slot line (as in other types of slot lines) there is the noticeable longitudinal component of the magnetic field H_z, which allows construction on the slot lines based on various nonreciprocal devices (for instance, phase shifters). There are some other differences. By now, this class of transmission line essentially extends, as does some of its representatives, for example, the asymmetric slot line, the generalized asymmetric double slot waveguide, and a series of related structures take on paramount significance in the construction of basing elements and functional units on 3D-IC.

7.4.2 Definition, Operation Principle, Waveguide Model of the Symmetric Slot Line

The symmetric slot line, like the strip transmission lines, asymmetric striplines and a series of other transmission lines, is the one of main basing transmission lines for the IC and 3D-IC of microwave and millimeter-wave ranges. The symmetric slot line represents the narrow slot cut in the conducting plane located on one of the sides of the plane-parallel magnetic–dielectric layer with thickness d forming the symmetric slot line substrate (Fig. 7.8a). The field structure of the main wave of the symmetric slot line is presented in Fig. 7.8a, b. Lines of the electric field (for $\varepsilon > 1$) are concentrated in the substrate, but in the slot, whereas the magnetic field lines have the form of ellipses transforming into curves of the saddle type (Fig. 7.8b), thus forming the main symmetric slot line wave, reminding the field construction of the wave of H_{10} type of the rectangular waveguide. The current distribution in the metallic layer (Fig. 7.8c) is exponential, and the current occupies the surface of the conducting metallic semi-planes of the relatively large area (Fig. 7.8d). Thereby, losses of the symmetric slot line (at finite metal conductivity) are significantly less than, for instance, for asymmetric striplines.

It is essential that the electromagnetic field of the main symmetric slot line wave, as we see already from Fig. 7.8a, contains all six components. Owing to slot width smallness (exponentially narrow slot, for which, by the way, relationships for deceleration Eqs. (7.18 and 7.19) were obtained), the electric field is mainly concentrated in the magnetic–dielectric region near the slot, and the transverse component exceeds it by approximately the order of the longitudinal component E_z. Components H_x, H_y, H_z of the magnetic field strength are approximately of the same order. The presence of the longitudinal H_z component allows classification of the main symmetric slot line wave as the H-type wave.

In the introduction, we already noted that there were no accurate and simultaneously available "engineering" methods for the analysis and development of the theory of the open symmetric slot line. There is also no clear physical understanding of the principle of the symmetric slot line operation, which is typical, for example, of the asymmetric stripline (according to Nefyodov and Fialkosky [77]). The clear understanding of physical processes in the symmetric stripline was obtained relatively recently (compared with the time of its intensive application in practice) and this is rather interesting, as we show here (see Sect. 7.3).

The above-mentioned is concerned with other slot lines, such as the asymmetric slot line, the asymmetric double slot waveguide, etc. Only in recent years, have we been able to essentially move forward in this direction and develop the theory of the symmetric slot line, in addition to offering algorithms for symmetric slot line analysis and synthesis satisfying the requirements of computer-aided design for IC and 3D-IC structures with different purposes.

The approximate relationships explaining the physics of symmetric slot line operation suitable for educational purposes (sometimes, in CAD) are presented below. Obviously, the new and extremely useful (to our mind) expressions, which, as a rule, students do not encounter in the educational literature, are presented.

In contrast to asymmetric striplines, the slot line analysis in quasi-static approximation is impossible as the capacitance between the conducting planes that form this line is infinite, and, therefore, other electrodynamic approaches are necessary.

7.4.3 Electromagnetic Field of Symmetric Slot Lines

Usually, the engineering practice of the approximate calculation of symmetric slot line parameters requires knowledge of the main components of the electromagnetic field. For $w/\lambda \ll 1$, the infinitely lengthy slot may be replaced by the equivalent linear magnetic current. Thus, the longitudinal component of the magnetic field outside the slot is [14]:

$$H_z(r) = A, H_0^1(kr), \quad k^2 = \gamma_z^2 + k_0^2, \tag{7.20}$$

where $H_n^1(\varsigma)$ is the Hankel function of the first kind of n-th order, $\gamma_z = i\,2\,\pi/\lambda$ is the longitudinal wave number, $k_0 = 2\pi/\lambda_0$ is the wave number of the wave in the free space, $\lambda = \lambda_0\,(\varepsilon_{\text{eff}})^{1/2}$ is a wavelength in the symmetric slot line, and ε_{eff} is the effective permittivity of the line.

Transverse components of the electromagnetic field in symmetric slot lines in cylindrical coordinates are defined from Maxwell equations according to Eq. (7.20):

$$H_r = -\left(h_r/k^2\right)\partial H_z/\partial r = A\left[1 - (\lambda/\lambda_0)^2\right]^{-1/2} H_1^{(1)}(kr),$$

$$E_\varphi = \frac{i\omega\mu}{k^2}\frac{\partial H_z}{\partial r} = -120\pi(\lambda/\lambda_0)\left[1 - (\lambda/\lambda_0)^2\right]^{-1/2} H_1^{(1)}(kr). \tag{7.21}$$

At calculations of ε_{eff} and the wave impedance of the symmetric slot line, we often use a model in the form of the transmission line with electric and magnetic walls (Fig. 7.8e, f); this allows (with a known degree of accuracy) representation of the symmetric slot line in the form of the rectangular waveguide. The waveguide models in Fig. 7.8e, f assume the wave propagation without losses in the longitudinal direction (z) of the symmetric slot line.

The transverse resonance (the Kohn method [113]) and the partial regions method [118] are the main methods for slot-line analysis. The first method leads to bulky enough expressions, whereas the second method leads to the system of linear algebraic equations, with which we must be very accurate so as not to encounter nonphysical solutions. Many publications are devoted to the partial regions procedure and its implementation in specific cases and, first of all, publications of G.V. Kisun'ko (cited in Balanis [118]). After Kisun'ko, a great contribution to the development of this method was made by G.I. Veselov, V.V. Nikolsky, S.B. Raevsky, Y.N. Feld, among many others. The partial regions method was repeatedly used for the determination of parameters of shielded symmetric slot lines taking into account the thickness t or current-carrying conductors, in addition to other types of strip-slot lines.

We note that the wave impedance in the symmetric slot line is determined to be ambiguously similar to its definition in the theory of closed waveguides and other guiding systems with non-TEM waves (see Sect. 6.13). At the derivation of Z expression, we usually use the "energy" definition of wave impedance (see Sect. 6.13). The field structure of the symmetric slot line is such that the wave impedance (in the case of the very narrow slot) can be obtained as a ratio of the maximal voltage in the slot and a current flowing in the longitudinal direction on metallic half-planes: $Z = V/I$.

The wave impedance of the symmetric slot line has the form (see, for instance, Gvozdev [14]:

$$Z = 296.1\sqrt{\varepsilon_{\text{eff}}}\left\{(1 - \varepsilon_{\text{eff}})\left[\ln\left(k_0 d\sqrt{\varepsilon_{\text{eff}} - 1/4}\right) + \ln\gamma\right]\right\}^{-1}, \tag{7.22}$$

where $\ln\gamma = 0.5772$ is the Euler constant.

The drawback of the partial regions' method is the relatively large calculation time on the computer, which increases the more complicated the structure of the transmission line. Therefore, this method is more convenient for investigations of physical properties and for studying the transmission lines of complicated cross-sections than for the designing of slot lines, and moreover, for including in CAD systems. Of course, a whole series of "tricks" sometimes permits the use of the partial regions method for necessary tasks and to achieve, therefore, acceptable accuracy and calculation time.

In the general case, calculation of the main characteristics of symmetric slot lines is carried out by Program P.7.5. For the main wave of the H-type, from integral equations, using the Pistohlkors–Schwinger transformation, we can obtain the simple dispersion equation for the structure shown in Fig. 7.8a [17]:

$$\left(\frac{h}{k_0}\right)^2 = \left[\frac{3a}{\pi b}\mu(\varepsilon_1 + \varepsilon_2)l + \mu\left(\varepsilon_1\frac{d_1}{b} + \varepsilon_2\frac{d_2}{b}\right) - \frac{3}{k_0^2 d_1 d_2}\right]\bigg/\left(1 + \frac{6al}{b}\right), \tag{7.23}$$

in which $l = -\ln\sin(\pi w/2a)$, $b = d_1 + d_2$.

At the narrow enough slot ($w \ll b$), the expression (7.23) leads to the well-known Feld result:

$$(h/k_0)^2 = \mu(\varepsilon_1 + \varepsilon_2)/2 \tag{7.24}$$

For slot dimensions equal to the waveguide width, this transforms into the known expression

$$(h/k_0)^2 = (1/2)\mu(\varepsilon_1 + \varepsilon_2) + (\lambda/2a)^2. \tag{7.25}$$

Some calculated curves of the wave impedance and the wavelength in the shielded symmetric slot line uniformly filled by a dielectric and located symmetrically in the shield section are presented in Gvozdev [14].

7.5 Asymmetric Slot Line

7.5.1 Definition, Features

Asymmetric slot lines are composed based on many basing elements of the IC and 3D-IC of microwave and millimeter-wave ranges. It is formed by metallic half-planes deposited on different sides of the plain parallel dielectric layers (substrates) (Fig. 7.9a, b).

Depending on the mutual location of the half-planes, two main variants of asymmetric slot lines are possible: with overlapping (Fig. 7.9a) and without it (Fig. 7.9b), in addition to zero overlapping, when the half-plane edges (ridges) are situated exactly one against the other.

It is essential that in the asymmetric slot lines, the constructive–technological feature is absent related to implementation of narrow conductors and slots, which is a complexity rather typical for basing elements based on the symmetric slot line in the planar ICs. This allows fulfillment of asymmetric slot lines with practically any value of wave impedances and provides matching with other types of transmission lines. Restrictions are imposed by the presence of energy emission (the open asymmetric slot line) from the side of large wave impedances (the large distance between half-planes), whereas from the small wave impedances due to possibility of wave excitation of the waveguide types and surface waves. In addition, the asymmetric slot line has a great wide-band property and the simplicity of constructive implementation of basing elements on its base in combination with asymmetric striplines and symmetric slot lines, which simplifies many respects, including the semiconductor device.

Nevertheless, until a relatively recent time, the asymmetric slot line was theoretically studied only in the so-called ridge model used in almost all planar devices of the millimeter-range. In this waveguide model, the significant influence on the processes of electromagnetic energy transmission is exerted by the walls of the "standard" waveguide, in the E-plane, in which the dielectric substrate is placed with metal layers deposited in it. From the point of view of 3D-IC technology and functional features of the circuit, we recommend conducting assembly of dielectric and metallic layers in the case of a wider range of dimensions ratio. Below, taking into consideration the fundamental importance of asymmetric slot lines for 3D-IC and the

Fig. 7.9 The asymmetric slot line: (**a**, **b**) are structures of the electromagnetic field; (**c**, **d**) are distributions of longitudinal currents corresponding to (**a**, **b**); (**e**) is the distribution of the total current

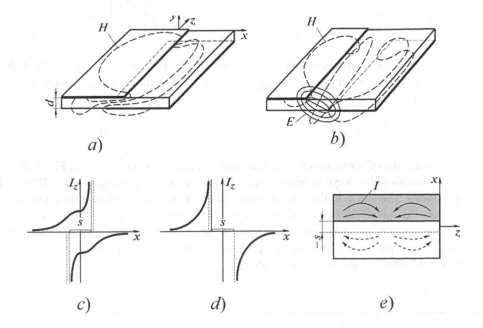

insufficiency of the development of its theory, the results of the asymmetric slot line theory and of experiments are described in more detail than was done for other types of strip transmission lines.

7.5.2 Integral Equations for Asymmetric Slot Lines, Eigenwaves, Physical Properties

Asymmetric slot lines represent the transmission line, in which the current-carrying conductors (in the simplest case these are half-planes) are located in dielectric layers at different layers. The case of the shielded 3D-IC model representing the rectangular waveguide filled with dielectric layers of similar thickness (which is typical for 3D-IC) is the simplest. Some parts of the boundaries of the dielectric layers are metallized: thin layers with ideal conductivity are deposited on them.

The general approach to obtaining the integral equations with respect to magnetic or electric currents consists of two stages: (1) expression on unknown coefficients in the Fourier transformation, and (2) expansion of fields in layers into series of Fourier transforms of currents on their boundaries and substitution of these coefficients into an equation of boundary conditions not used at the first stage. Fulfillment of this procedure leads to the following system of integral equations:

$$\int_0^a \left\{ \begin{array}{c} \|Y(x,x')\| \\ \|Z(x,x')\| \end{array} \right\} \left\{ \begin{array}{c} \vec{j}(x') \\ \vec{i}(x') \end{array} \right\} dx' = \left\{ \begin{array}{c} \vec{i}(x) \\ \vec{j}(x) \end{array} \right\}, \tag{7.26}$$

where $\|Y(x,x')\|$, $\|Z(x,x')\|$ are the Green tensor functions, $\vec{j}(x)$ is the magnetic current, $\vec{i}(x)$ is the electric current, and a is the size of the shield base.

The system of the integral Eq. (7.26) is general enough. From it, the system of integral equations of the fourth order for asymmetric slot lines is directly followed. In this case, the one cell $\|Y_n\|$ serves as the conductivity matrix with substitution in it of the sizes of the dielectric layers and their permittivity. For the algebraization of the integral equation system, we use the *Galerkin* method. The significant point of this approach is the problem of basing function selection. As we already mentioned, the surface currents in the asymmetric slot line (due to diffraction effects) should have the rather complicated form far from the classical power-series distribution of $(1 - x^2)^{-1/2}$ type, which is typical, for instance, for narrow symmetric slot lines (Sect. 7.2). For this reason, it is extremely difficult to select the convenient basis. However, from the theory of function approximation, we know that in a similar situation, the piecewise-defined polynomials, which easily represent the highly oscillating functions, give good results.

Then, using the standard Galerkin procedure for the integral equation, we obtain the system of linear algebraic equations with respect to unknown quantities $a_i^{(n)}, b_i^{(n)}$:

$$\|A\| \cdot \left\{ \begin{array}{c} a_i^{(n)} \\ b_i^{(n)} \end{array} \right\} = 0. \tag{7.27}$$

Here, $\|A\|$ is a matrix obtained as a result of the application to Eq. (7.27) of the Galerkin method. The compatibility condition for the system of linear algebraic Eq. (7.27) allows the dispersion equation for the asymmetric slot line to be obtained:

$$\det \|A\| = 0. \tag{7.28}$$

The results of some numerical calculations on Eq. (7.28) are shown in Fig. 7.10.

Functions of the wave impedance (Fig. 7.10a) and deceleration λ_0/λ (Fig. 7.10b) are presented versus the s parameter. This parameter is negative in the case of spaced metal ridges and is positive for asymmetric slot lines with overlapping layers.

At spacing of the metal ridges ($s < 0$), deceleration λ_0/λ aspires to deceleration of the H_{01} wave of the three-layer waveguide. On the other hand, an increase in the overlapping degree ($s > 0$) leads to the fact that the wave in the asymmetric slot lines acquires more and more properties of the H_{01} wave propagating in the rectangular waveguide, whose height is equal to the thickness of the dielectric substrate.

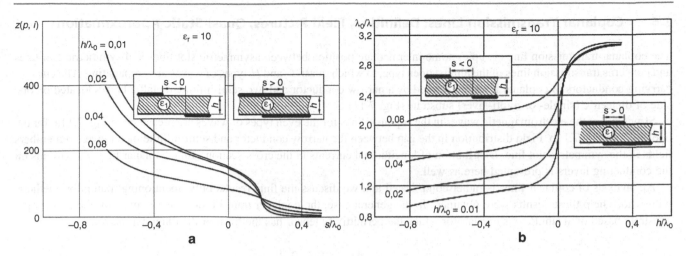

Fig. 7.10 (**a, b**) The wave impedance and deceleration of the asymmetric slot line (accurate calculation according to the high frequency structure simulator method)

7.5.3 Wave Impedance of Asymmetric Slot Lines and Effective Permittivity

At the fulfillment of students' annual and final projects and at the first stage of the designing of 3D-IC elements, it is desirable to have a model of asymmetric slot lines that is simple and clear in physical aspects. We can achieve this by the mathematical processing of experiments:

$$\varepsilon_{\text{eff}} = (1/4)\varepsilon\,[3 + \text{th}\{(S - q)/2(\varepsilon - 1)\}], \tag{7.29}$$

in which $q = S(2 - \varepsilon) + \varepsilon(\varepsilon - 1)/4$, $S = s/d$.

The MathCAD program for ε_{eff} analysis in asymmetric slot lines in quasi-static approximation is described in Appendix B (http://extras.springer.com/2019/978-3-319-90847-2) (Program P.7.6), where we introduce the following designations: Eeff (ε, S) is the value of ε_{eff} on Eq. (7.29) (first iteration); eeff(ε, S) is the value of ε_{eff} (second iteration). It is interesting that the value of the deceleration coefficient of the asymmetric slot line main wave, at not very high values of ε of the dielectric layer, is well enough described by the Feld approximate formula (7.18); and, as we mentioned, it is suitable for the exponentially narrow slot.

The wave impedance of the asymmetric slot line can be approximately calculated according to the following relationship obtained by the method of conformal mapping for the finite width of conductors [14, 47]:

$$Z = \left(120\pi/\sqrt{2\varepsilon}\right) K'(k')/K(k'), \tag{7.30}$$

where $k' = 0.515 + 0.5\,\text{th}\,(w/d - 0.75)$.

The asymmetric slot line has a series of interesting properties. First, in the case of conductor fold-over ($s > 0$), the wave impedance of the asymmetric slot line practically coincides with the wave impedance of the stripline. The complete quantitative coincidence is observed at an overlapping value of the half-wavelength order and more. In the case of conductor underlapping ($s < 0$), the asymmetric slot line looks like the symmetric slot line to some extent in its characteristics (Sect. 7.4).

In asymmetric slot lines, the constructive–technological features related to the obtaining of narrow conductors and slots are absent, which allows asymmetric slot line implementation with practically any value of wave impedance. We note that in practice a large number of varied asymmetric slot line variants is used.

7.6   Coplanar Transmission Lines: Definition, Field Pictures, Quasi-Static Approximation

The coplanar transmission line occupies some intermediate position between asymmetric slot lines and symmetric slot lines and concerns transmission lines of the almost open type, in which quasi-T- and H-types of waves are propagating. The current-carrying conductors of coplanar lines are formed by a narrow conductor and two semi-infinite metallic layers located on one side of dielectric (single- or multi-layer) substrate (Fig. 7.11).

Structures of the electromagnetic waves in the coplanar lines for the even types of waves are presented in Fig. 7.11a, for the odd types in Fig. 7.11b. Field distribution in the gap between the narrow conductor and semi-infinite metal layers remembers fields in the symmetric slot line. Distribution of longitudinal currents in the cross-section of the coplanar line and currents on the conducting layers is presented here as well.

An analysis of coplanar lines is rather complicated and we discuss the final results only, mentioning, naturally, methods with whose help these results were obtained. In the general case, the coplanar transmission line is made on the anisotropic substrate based on a single-axis crystal, for which the permittivity tensor has the form of the diagonal matrix:

$$\|\varepsilon\| = \varepsilon_0 \left\| \begin{matrix} \varepsilon_\perp & 0 & 0 \\ 0 & \varepsilon_\perp & 0 \\ 0 & 0 & \varepsilon_{\mathrm{II}} \end{matrix} \right\|.$$

Most frequently, the symmetric (with respect to the $x = 0$ plane) construction of the coplanar line is used. The field in the coplanar line represents a sum of even and odd solutions corresponding to location in the plane $x0z$ of the magnetic or electric walls.

The concept of the linear capacitance of the coplanar line comprises the basis of quasi-static approximation. For this capacitance we can write:

$$C = \frac{Q}{U} = \int_a^b \int_a^b \int_0^\infty E_x(x)\, G(\alpha; x|x')\, E_x(x')\, d\alpha dx dx' \left[\int_a^b E_x(x)\, dx \right]^{-1}, \tag{7.31}$$

Fig. 7.11 The coplanar transmission line and field structures of (**a**) even and (**b**) odd waves, and also the distribution of (**c, d**) longitudinal and (**e, f**) total currents

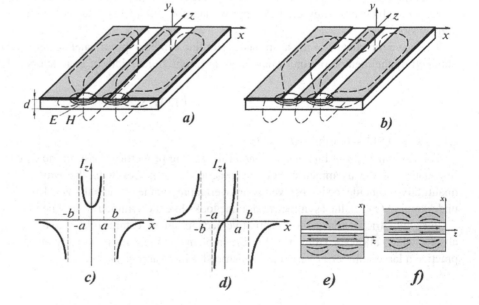

where Q is the linear charge, U is the voltage in the slot plane of the coplanar line, the Green function $G(\alpha; x|x') = 2\alpha^{-1}F(\alpha) \sin(\alpha x) \sin(\alpha x')$, in which

$$F(\alpha) = (2\varepsilon_0/\pi)\left\{1 + \left[\left(1 + \sqrt{\varepsilon_{\mathrm{II}}\varepsilon_{\perp}}\,\mathrm{th}\sqrt{\varepsilon_{\perp}/\varepsilon_{\mathrm{II}}}\,d\alpha\right)/\left(\mathrm{th}\sqrt{\varepsilon_{\perp}/\varepsilon_{\mathrm{II}}}/\sqrt{\varepsilon_{\mathrm{II}}\varepsilon_{\perp}}\right)\right]\right\}.$$

where indices $\perp$ and II designate the perpendicular and parallel components in the definition of the $\|\varepsilon\|$ matrix.

The voltage U is the integral of E_x (denominator in Eq. (7.31)).

Calculation of capacitance is provided by the variation method with the help of the Ritz procedure. The field in the slot is assumed as

$$E_x = [1 - 2(x - s)/w]^{-1/2} + \sum_{k=1}^{N} A_k T_k[2(x - s)/w]\left\{1 - [2(x - s)/w]^2\right\}^{-1/2}, \tag{7.32}$$

where $w = b - a$, $2s = a + b$, $T_k(\xi)$ are Chebyshev polynomials of the first kind, and A_j are varied parameters. The first term in Eq. (7.32) corresponds to field distribution in the coplanar line without the dielectric. The second term is introduced to take account of the dielectric. Satisfying the condition $\partial C/\partial A_j = 0$ for $j = 1, 2, \ldots, N$, we obtain more accurate field representation in the slot and the C value.

For the case of an isotropic dielectric of the substrate ($\varepsilon = \varepsilon_{\perp} = \varepsilon_{\mathrm{sub}}$):

$$C = 2(\varepsilon + 1)\varepsilon_0 K(k)/K'(k), \tag{7.33}$$

where $K'(k) = K(k')$, $k = a/b$, $k' = (1 - k^2)^{1/2}$.

For large values of a/b in Eq. (7.33), it is enough to take into account only one term ($N = 0$). Terms of higher order are necessary to take into consideration lower values of a/b, when the interaction degree increases between the narrow slot gaps in the coplanar line.

The wave impedance of the coplanar line in this approximation is:

$$Z \approx (60\pi/K(k))\left[(\varepsilon - 1)/2\right]^{-1/2}. \tag{7.34}$$

It is interesting to note that the second term in Eq. (7.33) defines, according to the Feld formula (7.18), deceleration of the main wave in the slot line on the infinitely thick substrate for $d \to \infty$ (Fig. 7.8a).

The program for analysis of the coplanar line properties is presented in Appendix B (http://extras.springer.com/2019/978-3-319-90847-2) (Program P.7.7) [1, 73]. It gives the reader the possibility of obtaining the general physical representations about properties of the "symmetric" coplanar line in quasi-static approximation.

The two-dimensional and three-dimensional pictures presented in Program P.7.7 allow the main conclusions about the physical properties of the coplanar lines to be drawn. Thus, the effective permittivity $\varepsilon_{\mathrm{eff}}$ rather sharply decreases with the reduction in the substrate permittivity ε. This means that the electromagnetic field of the main wave becomes increasingly displaced with ε reduction from the substrate in the air. When the distance between slots grows, the $\varepsilon_{\mathrm{eff}}$ increase occurs, i.e., the electromagnetic field "draws" into the substrate.

7.7 Ridge Dielectric Transmission Lines

7.7.1 General Notes

We already mentioned about ridge dielectric lines in Sect. 1.14 and examples of cross-sections of some its variants (Fig. 6.2k–m) were shown there. Here, we discuss in more detail about class of transmission line [1, 16]. The ridge dielectric lines represent a wide class of new transmission lines. In the planar ICs, the slot line on the suspended substrate has a wide distribution (Fig. 7.12a; fin-line). Sometimes, it is referred to as the waveguide–slot transmission line, thus underlining the essential influence of the side walls on its properties. However, in 3D-IC, for a whole series of reasons, it is more convenient to

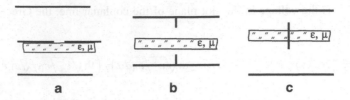

Fig. 7.12 Cross-sections of transmission lines based on the waveguiding magnetic–dielectric layer. (**a**) The slot line on the suspended substrate. (**b**) The symmetric ridge dielectric line. (**c**) The inverse ridge dielectric line

Fig. 7.13 The ridge dielectric line with ridges of different heights. (**a**) The floor fragment of a shielded 3D-IC of millimeter-wave range (the part with the ridge dielectric line with ridges of different heights). (**b**) The calculation model –half of the circuit section

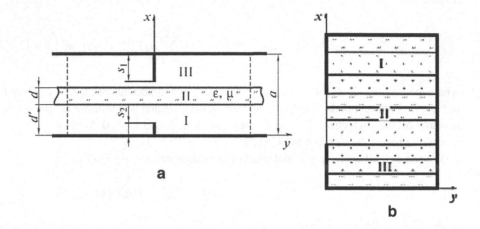

use the name "ridge dielectric line" in its initial canonic type (Fig. 6.2k; Fig. 7.12b). The diffraction-coupled ridge dielectric lines (Fig. 6.2l) are also used for more complicated combinations of ridge dielectric lines and the planar structures (Fig. 6.2m).

Here, we briefly consider the symmetric ridge dielectric line (in Fig. 7.12b its cross-section is shown), which is the typical (although the simplest) representative transmission lines of the millimeter-range, in which the dielectric layer is waveguiding and thin ridges promote the energy concentration of the operating wave on the restricted (on width) part of the layer. Below, we consider two variants of the ridge dielectric line. The system of four integral equations with respect to tangent field components was used for an analysis. Its algebraization is conducted according the Galerkin method [70, 112]. Full investigations of different classes of ridge dielectric lines were conducted by T.Y. Chernikova under supervision from E. Nefyodov.

In the sub-millimeter-wave range and in the optical range, the dielectric waveguides (light guides) of restricted width (Chap. 9) have a wide range of applications. Therefore, the element base of millimeter-wave ICs is based on their fundament. In the intermediate zone (between the short millimeter-waves and the sub-millimeter-waves), the combined metal–dielectric 3D transmission lines and resonance structures are mainly used. Implementation of basing elements for 3D-IC in the millimeter-wave range based on the rectangular dielectric waveguide and kindred transmission lines still meets with significant constructive–technological difficulties.

Transmission lines, in which the unified dielectric layer, common to each circuit "floor", which guides and supports the operating surface wave, comprising the basis of the transmission lines, is more promising for the creation of this range of 3D-ICs. The energy concentration of this wave on the restricted part of the layer is provided with the help of dielectric straps on the main layer, metallic ridges (bands), frequently periodic lattices, etc. The ridge dielectric structures made based on the ridge dielectric lines have definite constructive–technological possibilities. Ridge dielectric structures with ridges of different heights displaced relative to one another are the generalized structures.

7.7.2   Task Statement

Ridge dielectric lines without metal (vertical) shields represent the almost open transmission line. The upper metallic shields are shown by solid lines in Fig. 7.13a, whereas the side magnetic shields are shown by dotted lines. Development of a complete electrodynamics theory of such a ridge dielectric line is a complicated problem. In general cases, the electromagnetic field in the range–dielectric line has all six components, and the differential equation solution is usually obtained by numerical methods only. Taking into consideration that the field of the main wave in the ridge dielectric line has a priori the two-dimensional surface character, we can proceed to the close model introducing the fictitious electric and/or magnetic walls (in Fig. 7.13a they are shown by vertical dotted lines) removed enough from the region of field concentration.

The electromagnetic field in the ridge dielectric line should satisfy the Helmholtz equation

$$\left(\Delta + k^2 \varepsilon \mu\right) \vec{\Pi} \,(x, y, z) = 0, \qquad (7.35)$$

where $\vec{\Pi}\,(x, y, z) \equiv \vec{\Pi}\,(\vec{r})$ is the electrical or magnetic Hertz vector, ε, μ are piecewise constant functions of x. We assume that metallic surfaces have the ideal conductivity, and ridges have zero thickness. An account of finite conductivity and finite ridges thickness can be carried out, for example, using the perturbation method. Taking into account the cross-section symmetry of the ridge–dielectric line taking into consideration of the calculation model, we examine half of a transmission line cross-section, dividing it into three rectangular partial regions: I, II, III (Fig. 7.13b). Thus, in each partial region, the wall in the plane $y = 0$ can be both electric and magnetic depending on the ridge dielectric line variants. Each of partial regions (I, II, III) we shall consider to consist of three layers (maybe not the same thickness), which is important for the development of the basing elements theory for the 3D-ICs of millimeter-wave range based on the ridge dielectric lines.

The choice of x−component of electric (e) and magnetic (m) Hertz vector $\vec{\Pi}$: $\Pi_x^p(x, y) \exp\{ihz\}$ (where $p = e$, h; h is the propagation constant), is caused by a priori known or assumed information about eigenwaves of the laminated region. Obviously, in limited cases of ridge absence of their decreasingly small value, eigenwaves of the ridge–dielectric lines and the laminated region coincide.

Functions $\Pi_x^p(x, y)$ should satisfy boundary conditions on the Γ contour of the region cross-section (Fig. 7.13b): at $x = \text{const}$ it should be $\Pi^m = d\Pi^e/dx = 0$, at $y = \text{const}$ for the electric wall

$$\Pi^e = d\Pi^m/dy = 0. \qquad (7.36)$$

If the wall is magnetic, then

$$\Pi^m = d\Pi^e/dy = 0. \qquad (7.37)$$

Continuity conditions of field components on the media boundaries (for $x = \text{const}$) have the form (2.9). Conditions at the edge are given in Eq. (2.42).

We represent fields in partial regions I and III in the form of expansions

$$\Pi^{pq}(x, y, z) = \sum_{n=0}^{\infty} A_n^{pq} X_n^{pq}(x) Y_n^{pq}(y) e^{ihz}, \qquad q = \text{I, III}, \qquad (7.38)$$

and in the region II:

$$\Pi^{p\text{II}}(x, y, z) = \sum_{n=0}^{\infty} \left[B_n^{p\text{II}} X_{n1}^{p\text{II}}(x) + C_n^{p\text{II}} X_{n2}^{p\text{II}}(x) \right] Y_n^{p\text{II}}(y) e^{ihz}. \qquad (7.39)$$

Here, functions $Y_n^{pq}(y)$ $(q = \text{I}, \text{II}, \text{III})$ satisfy the one-dimensional Helmholtz equation and boundary conditions at $x = 0$, a and conditions of layer separation boundaries, which consist of the partial regions I and III. Functions $X_n^{p\text{II}}(x)$ are a sum of two terms, one of which corresponds to the metal boundary $x = x_1$.

Then, we compose the integral equation system with respect to tangent components of the electric field on two lines of separation of partial regions at $x = x_j$, $j = 1, 2$. The algebraization of the integral equation system is performed according to the Galerkin method. Thus, we selected the Chebyshev T_n, U_n polynomials of the first and second kind as the basing functions system (in this case, the field peculiarities on edges are considered automatically for $x = x_j, j = 1, 2, y \to 0$, when the edge is located in the uniform dielectric). These expansions have the form

$$f_{1,2} = Y_1 \sum_{}^{M} V_i^{1,2}\, T_{2i-1}(\bar{y}), \quad g_{1,2} = Y_2 \sum_{}^{L} W_i^{1,2} U_{2i-1}(\bar{y}) \tag{7.40}$$

where $Y_1 = \left(1 - \bar{y}^2\right)^{-1/2}, \quad \bar{y} = (b-y)/b, \quad Y_2 = \left(1 - \bar{y}^2\right)^{1/2}$.

The compatibility condition of the system of linear algebraic equations obtained gives the required differential equation:

$$\det \ \|D(h,k)\| = 0. \tag{7.41}$$

7.7.3 Numerical Results

Frequently, utilization of the Galerkin method based on the weighted Chebyshev polynomials is accompanied by extraction and analytical summation of slowly converging series, which are included in the matrix elements of the algebraic equations. However, at selected dimensions of the ridge dielectric lines, the calculation of propagation constants and critical frequencies can be carried out without such an improvement in series convergence. Thus, the effectiveness of the method changes insignificantly.

Dependences on deceleration $h/k = \sqrt{\varepsilon_{\text{eff}}}$ for two of the lowest and one of the highest waves in the ridge dielectric line, when the layer is situated between two shields because precisely this position is typical for 3D-IC of millimeter-wave range, are presented in Fig. 7.14.

The left parts of curves ($s_1/a = 0$) correspond to the case of the ridge dielectric line with a single ridge (Fig. 7.14b), in which the layer is located between the shields. At a similar ridge height, the results coincide with data for the symmetric ridge dielectric line.

All curves have the typical feature related to the fast growth of deceleration at the ridge approaching the dielectric surface, which is caused by a field concentration increase in the ridge region. At further ridge submersion into the dielectric after achievement of a maximal value h/k, the field re-distribution occurs in the ridge dielectric line, which leads to a gradual decrease in deceleration: the wave is forced out from the layer and becomes more "waveguide."

Fig. 7.14 The single-ridge ridge dielectric line at $\varepsilon = 10$, $ka = 2.0$, $kb = 2.3$, $s_2/a = 0.25$, $d/a = 0.3$, $d' = 0.35$. 1 The curve for the symmetric ridge dielectric line mounted in the middle ($x = a/2$) of the electric wall. 2 The curve for the case when instead of the electric wall, the magnetic wall is installed. 3 The curve for the waveguide wave ($h/k < 1$). (**a**) edge begins from the wall, (**b**) edge between walls

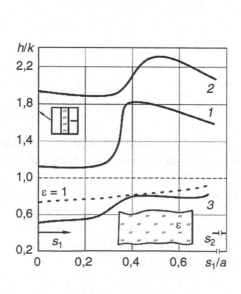

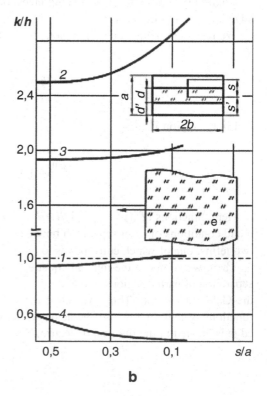

a

b

7.8 Examples of MathCAD Packet Utilization

7.8.1 Calculation of the Wave Impedance

The wave impedance of symmetric slot line $Z(x)$ can be calculated using the Program P.7.1 from Appendix B (http://extras. springer.com/2019/978-3-319-90847-2). It is relatively accurate and has large possibilities. Usually, if not to use the specific measures, the wave impedance of active devices is small, which leads to utilization of a symmetric transmission line (and a asymmetric transmission line; see Sect. 7.8.2) with wide current-carrying bands. This may lead to the necessity of special measures both in calculation models and in the implementation of necessary joints in practice.

The Program P.7.1 has the possibility of taking into account the finite thickness t of the current-carrying band, where the numerous approximate analytic relationships are given), and also the total losses in symmetric striplines. Total losses α in symmetric striplines can be divided into two components: losses in the conductor (metal) α_m and losses in the dielectric α_d. Because of the relatively small total losses, they are presented by a sum (see Sect. 7.2.2):

$$\alpha = \alpha_m + \alpha_d. \tag{7.3}$$

Losses in the dielectric depend on the tangent of the dielectric losses angle ($\tan\delta$) of used material and are defined approximately as:

$$\alpha_d = (27.3 \sqrt{\varepsilon}/\lambda_0) \tan \delta. \tag{7.4}$$

For practical calculation, the value of linear Q-factor of the symmetric stripline has great significance for practical calculations. It is equal to:

$$Q = (27.3 \sqrt{\varepsilon}/(\alpha\lambda_0)) \ [1 - (\lambda_0/2\varepsilon)(d\varepsilon/d\lambda_0)]. \tag{7.5}$$

Calculation of losses in symmetric striplines can be performed by Program P.7.1 with enough accuracy.

7.8.2 Numerical Determination of Parameters of Asymmetric Striplines

These MathCAD programs in Appendix B (http://extras.springer.com/2019/978-3-319-90847-2): Program P.7.2-static and Program P.7.2-dynamics. The first one is the simplest and intended rather for demonstration of the physical content of the tasks than for designing. More accurate and therefore more complicated is Program P.7.2. In the Program P.7.2 static the following designations are used: $u = w/h$, $v = \varepsilon_r$. Program P.7.2 and parameters of asymmetric striplines obtained with its help constitute the basis for the solution of many tasks of IC and 3D-IC design in microwave and millimeter-wave ranges with sufficient accuracy for practice. However, these data are inadequate for understanding the physical sense.

That is why, we shall further consider the more complete picture of wave processes in the asymmetric striplines based on adequate strict electrodynamics theory. From the presented results (Program P.7.2-static; c, e) we can already draw some preliminary conclusions. First of all, the wave impedance sharply decreases at growth of the band width w. Thus, the more ε_r of the substrate, the greater the falling speed. At small w, the ε_{eff} value sharply increases from its static values to the substrate ε_r: cases d, f). Thus, the electromagnetic field of the main wave increasingly concentrates in (draws into) regions under the current-carrying conductor. In a certain sense, the frequency functions of ε_{eff}, similar in the physical content, are presented: with substrate ε_r growth and with frequency growth, the electromagnetic field increasingly re-distributes under the band.

There are main qualitative data of numerical calculations of microstripline parameters. At fulfillment of students' annual and final projects in these directions, as well as at many engineering calculations, one may use the Program P.7.2 (dynamics) permitting to obtain accurate enough results acceptable in many cases and for practice.

The stricter theory (see Sect. 7.3.7) gives a deeper and fuller physical understanding of the principle of the microstripline operation. This is one of the central sections in our lecture course and the reader should pay special attention to it.

7.8.3   Characteristics of Two Similar Coupled Asymmetric Striplines

The presented results for similar (in geometry) coupled asymmetric striplines permit some conclusions to be drawn. They are obtained for a large enough frequency range: from 1 to 100 GHz, in essence, from low frequency quasi-static to millimeter-wave range. First, at relatively low frequencies, the characteristics of even and odd waves differ a little, but in another case, in millimeter-waves, such a difference for ε'_{eff} achieves large values. Thus, on high frequencies, the field is drawn into the transmission line substrate and $\varepsilon'_{eff} \rightarrow \varepsilon$ of the substrate. It is interesting to note that in low frequencies (in quasi-static cases), curves for $\varepsilon'_{eff} = \varepsilon'_{eff}(f)$ begin from the same values, and then behave in different ways for large and small substrate ε. For large ε the value of ε'_{eff} increases with frequency growth, whereas for small ε the value of ε'_{eff}, on the contrary, decreases.

The dispersion of the wave impedance of coupled asymmetric striplines (Program P.7.4) for relatively small values of ε for even and odd waves leads to Z growth, whereas large ε leads to its decrease.

7.9   Summary

It is evident according to the conducted classification of strip-slot and edge–dielectric transmission lines that we have to deal with such a large branch of radio physics and radio engineering. In the last few decades, the circle of problems of the theory, engineering, and electrodynamics modeling occupied the significant areas of engineering and theory. So, for example, over the decades, the knowledge of processes in the microstripline was restricted by the quasi-static approach. It was shown later that even the initial crucial structure has no strict solution. The almost strict theory described here permits moving ahead very far forward in a deep understanding of operational physics of the microstripline and also to show the connection of this solution with the quasi-static approach. We have demonstrated the significance of a deep understanding of physics. Unfortunately, numerical approaches, which give the necessary results as a whole, do not always facilitate a physical understanding of the processes in the line. Sometimes, they require large computing resources.

The asymmetric slot line (Sect. 7.7) turns out to be the one of much-needed line constructions for 3D IC at microwave and especially millimeter-wave ranges. To date, it has been investigated using mainly numerical methods. Unfortunately, nobody can yet suggest an analytical solution. This is a wide field of activity for young researchers and engineers.

In connection with the general development tendency of integration engineering, which, as a rule, is reduced to the continuous increase in the frequency range, the type and shape of transmission lines changes accordingly. Thus, transition to the millimeter-wave range requires utilization of dielectric waveguides and lightguides (Chap. 9), the edge–dielectric lines considered here (Sect. 7.7). Here, we can expect great investigations into the most complicated electrodynamic structures. Obviously, this requires the knowledge of the generalized fundamentals of electrodynamics (Chap. 4).

Checking Questions

1. Which requirements should satisfy the geometry of the symmetric slot line to have the possibility of using the Feld formula?
2. Why is the static approach unsuitable for analysis of the symmetric slot line?
3. Can you determine toward which side the main wave propagates using the field picture of the symmetric slot line (Fig. 7.8) and considering that at $y > 0$ the H_z the component is directed to the positive direction of the z-axis?
4. Why are the nonreciprocal devices built mainly based on the asymmetric slot lines instead of the symmetric striplines or asymmetric striplines?
5. How can we use the method of transverse resonance to analyze asymmetric slot lines?
6. Who developed the fundamentals of the partial regions method and what is the sense of this method?
7. Which advantages do asymmetric slot lines have compared with symmetric slot lines?
8. What is the main property of the ridge dielectric line?
9. What will the wave impedance be for the symmetric stripline (Fig. 7.2) for the main wave at the thickness of the current-carrying conductor $t = 0$ and $t = 0.025$ mm, $2d = 4$ mm, $w = 2.3$ mm, $\varepsilon = 2, 55$ at the frequency $f = 1.2$ GHz? The conductor is located between the conducting shields of the symmetric stripline.
10. How can we change the width of the current-carrying conductor in point 9 upper so that the wave impedance of the symmetric stripline would be 50 and 75Ω for the same conditions?
11. What will be the wave impedance of the asymmetric stripline (Fig. 7.3) for the main wave at the thickness of the current-carrying conductor $t = 0$ and $t = 0.025$ mm, $2d = 4$ mm, $w = 3.4$ mm, $\varepsilon = 2.55$ at frequencies $f = 2.2$ and 6.8 GHz?

12. Compare expressions for fundamental electric Eq. (7.19) and magnetic Eq. (7.23) waves of the rectangular waveguide with electric Eq. (7.33) and magnetic Eq. (7.34) waves of asymmetric striplines. Under which conditions do these waves transfer to each other?

13. Which differences are there between the fast and slow waves of higher order in asymmetric striplines?

14. What is the sense of the Vavilov–Cherenkov effect applied to the theory of strip transmission lines?

15. When was the Vavilov–Cherenkov effect discovered and what is its essence?

16. In which physical sense is the critical frequency concept in hollow waveguides and strip transmission lines? How does the influence of dielectric substrate loss manifest in the strip transmission lines and losses in the dielectric, which partially or completely fill the section of the hollow waveguide?

17. What is the role of the reflection factor (in Eq. (7.32)) in the key semi-infinite structure (Fig. 7.4a, b) for understanding process physics in the operation of asymmetric striplines?

18. What does the effect of polarization splitting of waves consist of in asymmetric striplines?

19. Which are the common properties in the wave behavior of asymmetric striplines and the rectangular waveguide?

20. By what is TEM wave dispersion defined in the asymmetric stripline?

Periodic Processes and Structures in Nature, Science, and Engineering

8

This chapter begins from the general review of periodic processes and structures in nature, science, and engineering (Sect. 8.1). Such processes and structures began their complete life from the discoveries of the Russian scientist D. Mendeleev. Periodic structures are widely used at present in active and adaptive phased antenna arrays, on the one hand, and in nano-devices base on meta-materials, on the other. Floquet's theorem and the principle of the phase synchronism are physical fundamentals of periodic structures (Sect. 8.2.1). In a series of knowledge branches, the periodic systems play the role of slowing systems with their own deceleration coefficient. In the whole case series, slowing structures are the constitutive parts of devices based on the already considered hollow waveguides (Chap. 6).

Structure periodicity permits the use of Floquet's theorem for its analysis and to represent the total field in the form of an infinite sum of spatial harmonics (or Hartree harmonics) (Sect. 8.2.2). Thus, we offer two types of model for consideration: infinite and finite models. The brief physical analysis of various operation modes of the slowing system is provided. A wavelength for the n-th harmonic and the phase and group velocity are determined (Sect. 8.2.3).

The equivalent circuit of the infinite periodic system is suggested (Sect. 8.3) and the characteristic impedance of the periodic structure is defined. Section 8.4 is devoted to an analysis of the periodic system of finite length.

As an example, the periodic waveguiding structures on the slotted lines in the E-plane are examined (Sect. 8.5). The boundary problem reduction to the integral equation with respect to the surface current density forms the basis of the analysis. Passing on to Pistohlkors–Schwinger variables, we obtain the Fredholm integral equation of the first kind with the singularity of a logarithmic type. The results obtained are determined by the boundary problem solution for eigenwaves in the generalized periodic structure (Sect. 8.5.1). As the second example, we examine the structure with a metallic partition in the center of the E-plane of the rectangular waveguide, which was studied previously in the pure type (Sects. 6.4 and 6.5).

A class of corrugated slot transmission lines (Sect. 8.6) was initially offered for use in rocket-space engineering. An analysis was based on the utilization of equivalent Weinstein–Sivov boundary conditions (Sect. 2.1.5). We examine two cases for corrugation with longitudinal and transverse grooves. Properties of the symmetric corrugated slot line are discussed (Sect. 8.6.2). It is shown that a magnitude and a location of the opacity band are especially interesting in the designing of controllable microwave and millimeter-wave devices.

In practice, the periodic structures from the wire lattices are widely distributed (Sect. 8.7). We consider cases of electric and magnetic polarization. Reflection and transmission coefficients are determined. For the case of magnetic polarization, we note conditions of complete wave transmission through the lattice (the Maliuzhinets effect, Sect. 8.7.2). We concluded that failure to take into consideration the transverse currents in the lattice conductors may lead to noticeable errors (the Lamb error).

8.1 Introduction

The world of periodic processes and structures is truly infinite. Beginning from the descriptive fundamentals of nature, the periodic Mendeleev element system, to modern active and adaptive phased antenna arrays, on the one side, and nano-devices based on meta-material, on the other. Evidently, the main attention we shall attract is to the application of periodic structures in radio electronic and radio physics applications. Apparently, the widest "utilization" such structures had achieved was in radio physics and microwave and millimeter-wave electronics. In recent years, great interest in studying periodic structure is expected in medicine (for instance, in millimeter-wave therapy), in biology, etc.

© Springer International Publishing AG, part of Springer Nature 2019
E. I. Nefyodov, S. M. Smolskiy, *Electromagnetic Fields and Waves*, Textbooks in Telecommunication Engineering,
https://doi.org/10.1007/978-3-319-90847-2_8

The most interesting examples of periodic structure application are matching coverings [119–121] in optics, the construction of large antennas for radio telescopes, and in the periodic or quasi-periodic movement of charged particles in crossed electric and magnetic fields. For example, we can clarify (antireflect) the solar elements because semiconductors have a large refraction index and appropriately large losses at reflection. On the telescope mirrors, the interference coverings are deposited to increase the reflection. We can achieve the transmission/reflection in any part of the spectrum by a combination of the deposited layers including implementation of filters with extremely narrow bandwidth. Onto the main and secondary mirrors of the telescopes, the coverings from the aluminum (Al) layers, the quartz (SiO_2), the titanium dioxide (TiO_2), and the quartz again are deposited. This combination gives the uniform reflection over the whole wavelength range. Such mirrors with clarification reflect 95% of light, whereas usual mirrors with an aluminum layer reflect only 86–88%.

The motion of charged particles in complicated fields is interesting both from the engineering and the general point of view of investigation of nature processes. A trajectory of the charged particle is like a reel in the line of magnetic induction. This phenomenon is used for magnetic thermo-isolation of the high-temperature plasma, i.e., the fully ionized gas at a temperature on the order of 10^6 K. A substance in such a state can be obtained in the scientific installation of the Tokamak type when studying controllable thermal–nuclear reactions.

Similar phenomena occur in the Earth's magnetic field, which is a defense for all alive against charges particles flows from the far space. The fast charged particles from space (mainly from the Sun) are "locked" by the Earth's magnetic field and form so-called radiation zones, in which particles, as in magnetic traps, move forward and backward along the spiral trajectories between the North and South magnetic poles during a period on the order of parts of seconds. Only in polar areas, some of the particles intrude into the higher layers of an atmosphere causing the auroras. Radiation zones of the Earth extend from distances of 500 km order to tens of the Earth's radius. It should be remembered that the magnetic South pole is located near the geographic North pole (in the north-west of Greenland). The nature of the Earth's magnetism has not been studied in detail until now.

The fast charged particles from the Sun (mainly electrons and protons) fall into the magnetic traps of radiation zones. Particles may leave zones in the polar region, causing auroras.

Obviously, we can provide a large number of periodic structures and processes. Let us return to the usual classical electrodynamics course.

8.2   Definition, Floquet Theorem

8.2.1   Deceleration Coefficient, Phase Synchronism

In Chaps. 5, 6, and 7 we studied electromagnetic processes, whose phase velocity is mainly equal to or more than the light speed in the medium, which fills the waveguiding system $\left(c/\sqrt{\varepsilon\mu}\right)$. However, in the whole engineering areas (microwave and millimeter-wave electronics, accelerators, antenna systems, etc.) the so-called *slowing structures*, in which the phase velocity, as a rule, is less than the light speed ($v_{ph} < c$) or much less than c, are used. One of the main characteristics of the slowing structure is the *deceleration coefficient*

$$k_{dec} \equiv \eta = v_{ph}/c = \lambda_{line}/\lambda_0 = k/h, \tag{8.1}$$

where $\lambda_{line} \equiv \Lambda = 2\pi/h$ is the wavelength in the transmission line and h is its propagation constant.

Periodic slowing structures are an integral part of *electronic devices* with a lengthy interaction of the electron flow with the slowing structure or (more precisely) with one of the waves of this slowing structure. The *traveling wave tubes*, the *backward wave tubes, magnetrons, orothrons,*[1] and many other microwave electron devices and quantum devices, in addition to other radio equipment, may serve as examples.

The fundament of the electron flow and a wave interaction is the equality of velocities – *the condition of the phase synchronism*:

$$v_{ph} = v_{el}. \tag{8.2}$$

The similar condition we saw on examination of the wave degeneration effect with the same phase velocities in waveguides, which transform to one another (exchange energy) on various irregularities of the transmission line.

---

[1] In Russian technical literature, an orothron is often referred to as the *oscillator of diffraction emission*.

Periodic systems are used in parametric amplifiers, in *delay lines*, the *linear accelerators*, in *antenna engineering*, etc. Examples of transmission lines with periodic structures are shown in Fig. 8.1.

The main properties of periodic guiding structures can be revealed from the *Floquet theorem*, according to which the field distribution in any infinite periodic transmission line is repeated periodically owing to the *periodicity of boundary conditions*. Hence, fields in adjacent cells of the periodic structure can differ only in amplitude and phase; thus, the field phase in each cell is defined by initial conditions on the open boundaries of this cell.

8.2.2 Floquet Theorem

Let the infinite periodic transmission line, shown in Fig. 8.2, have the spatial period L. Therefore, it is natural to represent the electromagnetic field components $\Psi_i\{\overline{1,\infty}\}$ propagating along the z axis as

$$\Psi_i(\vec{\rho}, z) = f_1(\vec{\rho}, z)\exp\{i(-\omega t + \gamma_0 z)\} \tag{8.3}$$

where the function $f_i(\vec{\rho}, z)$ is the periodic function with L period, and γ_0 is the wave propagation constant in the infinite periodic structure (averaged value of propagation constant of the electromagnetic wave).

Let us expand the function $f_i(\vec{\rho}, z)$ in a series on *Fourier harmonics*, which are usually referred to as *spatial harmonics* (or *Hartree harmonics*) at examination of periodic structures:

$$f_i(\vec{\rho}, z) = \sum_{-\infty}^{\infty} a_n \exp\{-i2\pi n z/L\}, \tag{8.4}$$

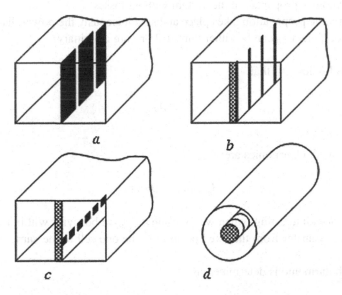

Fig. 8.1 Waveguide periodic structures. (**a**) With metallic diaphragm. (**b, c**) With partially metallized dielectrics with a one-sided picture. (**d**) The circular waveguide with the dielectric rod partially covered by metal

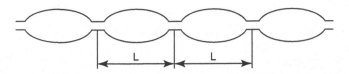

Fig. 8.2 The model of the infinite periodic structure

where

$$a_n = \frac{1}{L} \int\limits_{z}^{z+L} f_i(\vec{\rho}, z) \exp\{i(2\pi n z/L)\} dz; \quad n = \overline{0, \pm\infty}. \tag{8.5}$$

Taking into account the expansion (8.4), we write the expression for the field Ψ_i component as

$$\Psi_i = \sum_{-\infty}^{\infty} a_n \exp\{i(-\omega t + \gamma_n z)\}, \tag{8.6}$$

where $\gamma_n = \gamma_0 + 2\pi n/L$ and coefficients a_n are determined by (8.5).

Equation (8.6) is the mathematical formulation of the *Floquet theorem*. The zero harmonics a_0 is referred to as the main harmonic and characterizes the direct traveling wave. The propagation constant γ_0 corresponds to it, which is less than all the other γ_n in value.

8.2.3 Brief Physical Analysis

Let us consider the expression

$$i\gamma_0 = i\beta_0 + \alpha_0 \tag{8.7}$$

and stop on three main cases:

1. $\beta_0 = 0$. In this case, all spatial harmonics dampen from the cell to the cell. The main harmonic does not undergo the phase shift at transfer to the next cell; however, harmonics of higher orders have a phase shift.
2. $\alpha_0 = 0$. In this case, all harmonics propagate in the system without losses.
3. If $\alpha_0 \neq 0$, $\beta_0 \neq 0$, the harmonic propagation takes place and simultaneously the power dissipation occurs. If the periodic system has no losses, the coefficient γ_0 can be either pure real or pure imaginary.

Let the transmission line be lossless, so that

$$\gamma_0 = \beta_0, \tag{8.8}$$

where β_0 is the real number.

Phase constants β_n for n-th spatial harmonics are:

$$\beta_n = \beta_0 + 2\pi n/L. \tag{8.9}$$

We note that the case $\beta_n \neq \beta_0$ is interesting, when we examine the wave behavior within the limits of the cell. The phase shift for all spatial harmonics, at transfer from the fixed position in the one cell to the same position in the adjacent cell, is equal to $\beta_0 L$.

The wavelength for the n-th harmonic is determined as:

$$\lambda_n = 2\pi/|\beta_n| = 2\pi/(\beta_0 + 2\pi n/L) \tag{8.10}$$

and its *phase velocity* is:

$$v_{\text{TM}n} = \omega/\beta_n = \omega/(\beta_0 + 2\pi n/L). \tag{8.11}$$

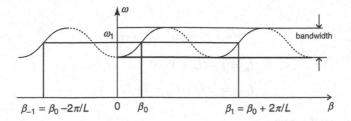

Fig. 8.3 Typical ω-β diagram for the periodic waveguiding structure

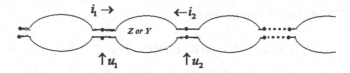

Fig. 8.4 The equivalent circuit of the infinite periodic waveguiding structure

Thus, λ_n and $v_{\text{TM}n}$ decrease as the harmonic number n grows. Besides, $v_{\text{TM}n}$ is negative if the harmonic number n is the negative number.

The group velocity is equal to

$$v_{\text{gr}n} = d\omega/d\beta_n = d\omega/d\beta_0, \tag{8.12}$$

i.e., at a given frequency, all harmonics have the same group velocity.

The typical $\omega - \beta$ diagram for periodic waveguiding structure is shown in Fig. 8.3. Some range of frequency ω variation corresponds to the bandwidth ($\beta_0 \neq 0$). The first interception point of the line $\omega = \omega_1$ with the right part of curve $\omega = f(\beta)$ corresponds to the fundamental harmonic with the phase constant equal to β_0.

8.3   Equivalent Circuit of the Infinite Periodic Structure

Let us consider the infinite periodic transmission line with known parameters of each cell, i.e., we assume that the appropriate electrodynamics problem is solved. The equivalent circuit of the infinite waveguiding structure is shown in Fig. 8.4, according to which the periodic system represents the series (cascade) connection of the infinite number of two ports with the same parameters $\widetilde{Z}$ (or $\widetilde{Y}$).

Relationships between true voltages and currents are determined in two ways: either with the help of resistance matrix $\widetilde{Z}$:

$$\begin{bmatrix} U_1 \\ U_2 \end{bmatrix} = \begin{bmatrix} Z_{11} & Z_{12} \\ Z_{21} & Z_{22} \end{bmatrix} \begin{bmatrix} I_1 \\ I_2 \end{bmatrix} \tag{8.13}$$

or the conductance matrix $\widetilde{Y}$:

$$\begin{bmatrix} I_1 \\ I_2 \end{bmatrix} = \begin{bmatrix} Y_{11} & Y_{12} \\ Y_{21} & Y_{22} \end{bmatrix} \begin{bmatrix} U_1 \\ U_2 \end{bmatrix}. \tag{8.14}$$

Utilization in (8.13 and 8.14) of the absolute values of voltages and currents simplifies the investigation of the periodic system, especially because we cannot always carry out the normalization.

It is convenient to change the current I_2 direction so that all currents are directed to the right. Then, the output current and the output voltage of the cell become the input current and the input voltage of the next cell. Then, for the n-th cell with the isotropic medium ($Z_{12} = Z_{21}, \quad Y_{12} = Y_{21}$)

$$U_n = Z_{11}I_n - Z_{12}I_{n+1},$$
$$U_{n+1} = Z_{12}I_n - Z_{22}I_{n+1},$$

(8.15)

$$I_n = Y_{11}U_n + Y_{12}U_{n+1},$$
$$I_{n+1} = -(Y_{12}U_n + Y_{22}U_{n+1}).$$

(8.16)

In accordance with the *Floquet theorem*

$$U_n = U_0 e^{-i\gamma_0 nL}, \qquad I_n = I_0 e^{-i\gamma_0 nL}.$$

(8.17)

Substituting expression (8.17) into Eq. (8.15), we obtain

$$U_n = U_0 e^{-i\gamma_0 nL} = I_0 \left(Z_{11}e^{-i\gamma_0 nL} - Z_{12}e^{-i\gamma_0(n+1)L}\right),$$
$$U_{n+1} = U_0 e^{-i\gamma_0(n+1)L} = I_0 \left(Z_{12}e^{-i\gamma_0 nL} - Z_{22} \ e^{-i\gamma_0(n+1)L}\right).$$

(8.18)

From equation system (8.18) it follows that

$$\cos \gamma_0 L = (Z_{11} + Z_{22})/2Z_{12}.$$

(8.19)

In a similar manner, from Eq. (8.16) we can obtain the equality

$$\cos \gamma_0 L = -Y_{11}Y_{22}/2Y_{12}.$$

(8.20)

Thus, Eqs. (8.19 and 8.20) define the γ_0 constant through elements of $\widetilde{Z}, \widetilde{Y}$ matrices.

If to define the characteristic impedance of the periodic line on terminals of any cell as a ratio of a voltage and a current under the condition that the line is infinite (there are no reflected waves), i.e.,

$$Z = U_n/I_n,$$

(8.21)

then it is not difficult to obtain the following formula:

$$Z = iZ_{12} \sin \gamma_0 L + (Z_{11} - Z_{22})/2.$$

(8.22)

Formulas (8.19, 8.20, and 8.22) are true for waves propagating along the z axis; for waves propagating in the opposite direction, the γ_0 constant must be changed to $-\gamma_0$.

8.4   Periodic Structures of the Finite Length

If the periodic structure of finite length is loaded, as a rule, on the same complex load Z_{load}, it is necessary to take into consideration the presence of waves reflected from Z_{load}. If the periodic waveguiding structure contains m cells and is loaded to the impedance Z_m (Fig. 8.5), then

Fig. 8.5 The periodic line of the finite length

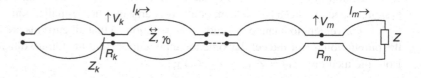

$$U_m/I_m = Z_m, \tag{8.23}$$

where

$$U_m = U_m^+ + U_m^-, \quad I_m = I_m^+ + I_m^-; \tag{8.24}$$

indices "+" and "−" indicate belonging, accordingly, to incident and reflected waves.

In the m-th section, the reflection coefficient R_m can be defined as

$$R_m = -I_m^-/I_m^+. \tag{8.25}$$

Then, the following useful relationships [58, 75] are obtained:

$$R_m = \frac{Z_m - iZ_{12}\sin\gamma_0 L - (Z_{11} - Z_{22})/2}{Z_m + iZ_{12}\sin\gamma_0 L - (Z_{11} - Z_{22})/2}, \tag{8.26}$$

$$Z_m = (Z_{11} - Z_{22})/2 + iZ_{12}\sin\gamma_0 L(1 + R_m)/(1 - R_m). \tag{8.27}$$

Having determined in the k-terminal plane the reflection coefficient R_m in the form

$$R_k = -I_k^-/I_k^+, \tag{8.28}$$

we can express the impedance Z_k through the impedance Z_m and parameters of the equivalent circuit as follows [58]:

$$Z_k = (Z_{11} - Z_{22})/2 + Z_{12}\sin\gamma_0 L \times$$
$$\times \left[\frac{(Z_m - (Z_{11} - Z_{22})/2)\ \cos\ [(m-k)\gamma_0 L] - Z_{12}\sin\gamma_0 L\ \sin\ [(m-k)\gamma_0 L]}{(Z_m - (Z_{11} - Z_{22})/2)\ \sin\ [(m-k)\gamma_0 L] + Z_{12}\sin\gamma_0 L\ \cos\ [(m-k)\gamma_0 L]}\right]. \tag{8.29}$$

The final load can be obtained through its conductance $Y_m = 1/Z_m$. In this case, it is suitable to use the conductance Y_k on the k-th terminal:

$$Y_k = (Y_{11} - Y_{22})/2 - Y_{12}\sin\gamma_0 L \times$$
$$\times \left[\frac{(Y_m - (Y_{11} - Y_{22})/2)\cos\ [(m-k)\gamma_0 L] + Y_{12}\sin\gamma_0 L\sin\ [(m-k)\gamma_0 L]}{(Y_m - (Y_{11} - Y_{22})/2)\sin\ [(m-k)\gamma_0 L] - Y_{12}\sin\gamma_0 L\cos\ [(m-k)\gamma_0 L]}\right]. \tag{8.30}$$

Under the condition

$$(m-k)\gamma_0 L = l\pi, \qquad l = \overline{1,\infty} \tag{8.31}$$

Eqs. (8.29 and 8.30) simplify and take the form

$$Z_k = Z_m, \quad Y_k = Y_m. \tag{8.32}$$

* * *

The expression (8.31) is true for segments of the periodic system equal to the *integer number of a half-wavelength*. In this case, the segment can be used as the transmission line in the *bandwidth*. The periodic system can be considered as a *filter*, whose parameters can be easily calculated; however, it will have neither a maximally plain characteristic, as in the *Butterworth filter*, nor an approximation of the staircase characteristics, as in the Chebyshev filter, nor (particularly) as the *Zolotariov filter* (with an *elliptical amplitude response*).

8.5   Periodic Waveguiding Structures of Slotted Lines in the *E*-Plane

8.5.1   Problem Statement: Boundary Problem Reduction to the Integral Equation

Recently, an interest in waveguiding periodic structures on the slotted lines located in the *E*-plane has sharply increased. In particular, based on waveguide slotted lines, the *pass-band filters* with low losses suitable for operation at a frequency range 18–170 GHz were constructed. Such structures can be created by means of metal bands or metallized dielectric substrates in the *E*-plane of the waveguide (Fig. 8.1). Thus, the metallic substrate may have both the one-sided and the two-sided picture. Thus, here we clearly see the "joining" of 3D-IC ideology with structures of the nano-type and meta-type, for instance, when there are several lattices of these types [14, 67].

Wave propagation in media with periodically changing parameters is accompanied by the appearance of new qualitative features, which are most noticeable when the wavelength becomes comparable with the line spatial period. First of all, the appearance of *pass-bands* and *opacity-bands* should be attributed to these features.

Let us examine the structure (Fig. 8.6a) from periodically located, ideal, infinitely thin conductors accommodated in the $x = 0$ plane between two regions $x < 0$ and $x > 0$, each of which can be any of those shown in Fig. 8.6b–d. The electromagnetic waves propagate along the *z*-axis, and field variations along the *y*axis are absent.

According to the *Floquet theorem*, functions of wave field components with respect to the *z*-coordinate should have the form:

$$\vec{E}(x,z) = \sum_{-\infty}^{\infty} \vec{e}_n(x)\, e^{i\gamma_n z}, \quad \vec{H}(x,z) = \sum_{-\infty}^{\infty} \vec{h}_n(x)\, e^{i\gamma_n z}, \tag{8.33}$$

where

$$\gamma_n = \gamma_0 + 2\pi n/L; \tag{8.34}$$

γ_0 is the unknown propagation constant, L is the spatial period, $\vec{e}_n(x)$, $\vec{h}_n(x)$ are some vector functions of the *x*-coordinate.

The task of wave analysis, which propagates in the generalized periodic structure (Fig. 8.6a), is reduced to a solution of the *integral equation* with respect to the j_z component of the surface current density on metallic bands. This equation will have a simpler form if transferred to new variables u, v (*variables of Pistohlkors–Schwinger*) with the help of expressions

$$u = s^{-1}\cos(\pi z/L), \quad v = s^{-1}\cos(\pi z'/L), \tag{8.35}$$

in which $s = \sin[\pi(w_2 - w_1)/2L]$.

Introducing the new unknown function $f(u) = j_z(z)\,dz/du$, we write the following integral equation ($u \in [-1, 1]$) (see also Neganov et al., Nefyodov and Fialkovsky, Ufimstev et al. [18, 19, 58, 71]):

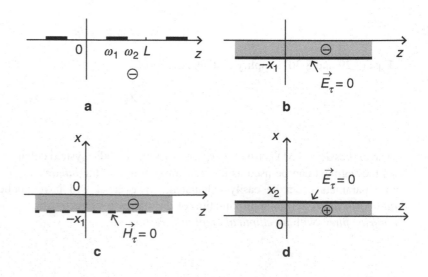

Fig. 8.6 Examples of some periodic structures. (**a**) The periodic structure from ideally conducting thin bands located on the media boundaries *1* and *2*. (**b**) The model or the periodic structure for the rectangular waveguide with a metallic diaphragm in the center of the *E*-plane (compare Fig. 8.3a). (**c**, **d**) Models of the periodic structure for the rectangular waveguide with the metallic diaphragm covered by the dielectric layer (compare Fig. 8.3b), relatively, at $x \in [-x_1, 0]$ and $x \in [0, x_2]$

$$(Z_0 - 2t_1 \ln 2s) \int_{-1}^{1} f(v)dv + \int_{-1}^{1} f(v)G_1(v,u)\,dv +$$

$$+ t_1 \int_{-1}^{1} f(v) \ln \left| \frac{v\sqrt{1 - s^2u^2} + u\sqrt{1 - s^2v^2}}{(u^2 - v^2)\left(u\sqrt{1 - s^2v^2} - v\sqrt{1 - s^2u^2}\right)} \right| dv = 0, \tag{8.36}$$

where

$$G_1(u,v) = \sum_{m=1}^{M} \Delta_m \{T_{2m}(sv)\,T_{2m}(su) + U_{2m-1}(sv)\,U_{2m-1}(su)\} \times$$

$$\times \sqrt{(1 - s^2v^2)(1 - s^2u^2)}, \quad \Delta_m = Z_m + Z_{-m} - \frac{2t_1}{m}, \quad t_1 = \lim_{m \to \infty} \{mZ_m\}, \tag{8.37}$$

Z_m is the impedance of the $x = 0$ plane for the m-th Fourier harmonic; the sign minus before m index means that the wave propagates in the negative z direction.

A kernel of the *integral* Eq. (8.36) has the *singularity of logarithmic type* and hence, the Eq. (8.36) is the *Fredholm integral equation of the first kind*. Summation on m in (8.37) is restricted by some number of M, because coefficients Δ_m at large enough values of m are small.

Let us present the kernel of the *integral* Eq. (8.36) in the form of asymptotic expansion over s parameter. As

$$G_1(v,u) = K_1(v,u) + K_2(v,u) + O\left(s^4\right), \tag{8.38}$$

where

$$\begin{cases} K_1(v,u) = R_1 - 2s^2 R_2 \left(u^2 + v^2\right), & R_1 = \sum_{1}^{M} \Delta_m, \\[3mm] K_2(v,u) = 4s^2 R_2 vu, & R_2 = \sum_{1}^{M} m^2 \Delta_m, \end{cases} \tag{8.39}$$

then (8.36) is divided into two independent equations. These *integral equations* have the following form:

$$(Z_0 - 2t_1 \ln 2s) \int_{-1}^{1} f(v)\,dv + \int_{-1}^{1} f(v)K_1(v,u)\,dv -$$

$$- 2t_1 \int_{-1}^{1} f(v) \ln |v - u|\,dv = 0, \tag{8.40}$$

for even wave types ($f(u)$ is the even function) and

$$\int_{-1}^{1} f(v)K_2(v,u)\,dv - s^2 t_1 u \int_{-1}^{1} vf(v)\,dv - 2t_1 \int_{-1}^{1} f(v) \ln |v - u|\,dv = 0 \tag{8.41}$$

for odd wave types ($f(u)$ is the odd function). We should note that at derivation of Eqs. (8.40, 8.41), the terms with the order of $O(s^4)$ were missed.

For *even wave types*

$$j_z = a_0 \sin(\pi z/L) e^{-i\gamma_0 z} \left(s^2 - \cos^2(\pi z/L)\right)^{-1/2} + O(s^2); \tag{8.42}$$

and the propagation constant γ_0 is determined from the dispersion equation

$$Z_0 - 2t_1 \ln s + R_1 + O(s^4) = 0. \tag{8.43}$$

Accordingly, for *odd wave types* the dispersion equation is:

$$4t_1 + s^2(4R_2 - t_1) + O(s^4) = 0, \tag{8.44}$$

$$j_z = a_1 \sin(2\pi z/L) e^{-i\gamma_0 z} \left(s^2 - \cos^2(\pi z/L)\right)^{-1/2} + O(s^2). \tag{8.45}$$

Equations (8.42)–(8.45) determine the solution of the boundary problem for *eigenwaves* in the generalized periodic structure. They have a large enough degree of generality: they are true for any waveguide with metallic bands, periodically located in its *E*-plane. We must know only expressions for Z_m coefficients. Thus, we can *estimate their accuracy*.

This point is presented by us as determining both for the students of the subject of our lecture course and for specialists/ designers. At the first stage of the approach to designing, it is desirable to know the parameter estimations of the system under investigation. This is not always simple or, more precisely, is always not simple. Therefore, such an approach in the engineer's hands is very convenient and productive.

In isotropic structures with periodically metallized dielectric, at small values of the *s*parameter ($s \ll 1$), odd wave types cannot propagate. Therefore, in the future, we shall consider even wave types.

8.5.2 Periodic Structure with a Metallic Diaphragm

As a first *example*, we consider a structure with a metallic diaphragm in the center of the *E*-plane of the rectangular waveguide (Fig. 8.1a). In this case, it is necessary to examine the task geometry (Fig. 8.4a), for which the lower region ($x < 0$) corresponds to Fig. 8.6b, whereas the upper region ($x > 0$) corresponds to Fig. 8.6d for $x_2 = x_1$.

For such a line

$$Z_n = \omega\mu_0/2r_n \cot r_n x_1, \quad t_1 = \omega\mu_0 L/4\pi, \quad r_n = \sqrt{k^2 - \gamma_n^2}. \tag{8.46}$$

At $L \leq x_1$, coefficients Δ_m ($m = 2, 3\ldots$) modulo are much lower than Δ_1; therefore, $R_1 \cong \Delta_1$ and instead of (8.43) we have for $\gamma_0 \ll \pi/L$:

$$\begin{aligned}
\frac{\tan(r_0 x_1)}{r_0} = -\Big\{ &\left[(\gamma_0 + 2\pi/L) - k^2\right]^{-1/2} + \\
+ &\left[(\gamma_0 - 2\pi/L) - k^2\right]^{-1/2} + (L/\pi)\,(\ln(1/s) - 1)\Big\}.
\end{aligned} \tag{8.47}$$

Roots of Eq. (8.47) exist in the condition

$$(2m - 1)\,(\pi/2) \leq r_0 x_1 \leq m\pi, \quad m = 1, 2, 3, \ldots \infty. \tag{8.48}$$

From Eq. (8.47), the approximate formula follows for the determination of the wave's lower cut-off frequency ($\gamma_0 = 0$):

$$k_{\text{low}} = x_1^{-1}\left[(\pi/4) + \sqrt{(\pi/4)^2 + A_1}\right], \quad A_1 = \pi x_1/L \ln(1/s). \tag{8.49}$$

At the last formula derivation, the Eq. (8.47) was resolved with regard to r_0 and the function arctan z is represented in the form of expansion over the argument. The upper cut-off frequency of the first pass-band is determined from Eq. (8.43) for $\gamma = \pi/L$.

8.6 Corrugated Slot Transmission Lines

8.6.1 Introduction, Problem Statement, Equivalent Boundary Condition

A class of waveguiding structures, which are referred as *corrugated slot transmission lines*, is an interesting example of transmission lines for super-fast processing systems for microwave and millimeter-wave ranges. This class of transmission lines was initially offered for utilization in missile space engineering. The thing is that the operation mode of super-fast processing systems under space conditions with regard to the temperature for external devices is very high-density: on the solar space vehicle side $+120°$, on the shadow side $-120°$ and no one dielectric can stand such a temperature drop. Professors V.A Neganov and E.I. Nefyodov and colleagues offered to replace the slowing dielectric plate in the slotted antennas with equivalent (in an electrodynamics sense) frequently periodic structure (see, for example, Neganov et al., Honl et al., Buduris and Shevenie [18, 19, 75, 109]).

In this section, we construct mathematical models of corrugated slot lines (some are shown in Fig. 8.7). These lines are related to the class of *periodic structures*, and for them, the method of analysis and calculation, described in Sects. 8.2 and 8.3, is true. However, in the case of small (or more precisely, rare: $kL \ll 1$, where L is the corrugation period) corrugation, at the calculation of such waveguiding structures, we can use *Weinstein–Sivov equivalent boundary conditions* (see Sects. 1.7, 1.8, and 1.9).

Creation of practical constructions of multi-functional modules on 3D-IC for microwave and millimeter-wave ranges, according the *optimality principle of the basing element*, requires a large set of regular and irregular transmission lines (predominantly of volumetric type), on the basis of which the specific functional units of 3D-IC are formed. In essence, the corrugated slot line, having a series of interesting properties, is of great interest. The basing element of the corrugated slot line is the slotted line (see Chap. 7) – the slot in the shielded metal layer over the corrugated impedance surface. Such a line has properties both of the slotted line and of the corrugated plane – the guiding structure of the *surface waves*, which significantly increase the functional possibilities of units based on such a line. On the other hand, for the corrugated slot lines, owing to the absence of the dielectric layer with $\varepsilon > 1$, the increased stability of electrodynamics characteristics to mechanical, thermal, and other influences, fewer losses, larger mechanical durability, are typical. In addition, strong dependence on the deceleration $\gamma/k > 1$ of eigenwaves of corrugated slot lines permits synthesis of the transmission lines and units based on them, with necessary *dispersion properties*. Finally, we would like to note that appearance itself of the corrugated slot line idea and the interest in it on the part of researchers is caused by the irrepressible desire to develop the class of transmission lines for

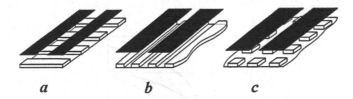

$$a \qquad b \qquad c$$

Fig. 8.7 Examples of corrugated slot transmission lines: with (**a**) transverse and (**b**) longitudinal grooves and based on (**c**) the two-dimensional corrugated plane

Fig. 8.8 The shielded model of the corrugated slot transmission line with (**a**) transverse grooves and its (**b**) transverse and (**c**) longitudinal sections of the corrugated surface

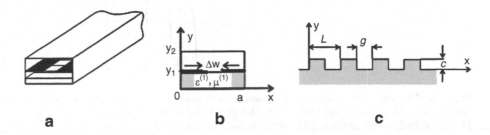

a   b   c

operation in equipment for space travel, where it is necessary to completely eliminate dielectric elements owing to their low reliability. Having in the mind the conception of super-fast processing systems on 3D-IC with electromagnetic couplings, which are currently being developed, we cannot ignore the availability, potential, aptitude, and usefulness of corrugated slot lines for application in 3D-IC.

Thus, we examine the guiding structures formed by a slot in the metallic plane over corrugated surfaces (Fig. 8.7). In general cases, such lines demonstrate properties of both the *slotted line* and the *corrugated plane* (*or, generally, the surface*). If the slot is far enough from the corrugation, the waves in the line propagate as in the slot structure.

With large enough slot dimensions, the impedance plane supports the directed propagation of the surface waves. Of course, the case in which properties of the slotted line and the corrugated plane are simultaneously demonstrated are of great interest.

The direct electrodynamics analysis of the structures shown in Fig. 8.7 is difficult; therefore, we proceed to examination of their shielded models (Figs. 8.8 and 8.9).

The properties of the eigenwaves of shielded transmission lines (Figs. 8.9 and 8.10) at large enough shield sizes will be close to properties of *non-emitted types* of eigenwaves of open structures.

Now we obtain expressions for elements of the *admittance matrix* of the region under the slot $\left[Y^-_{mij}\right]$. For this purpose, we describe the corrugated plane at $y = 0$ with the help of *equivalent boundary conditions* for periodic structures (see Sect. 1.7.3).

Small corrugations have the following form:

$$E_z = \eta^m H_x, \qquad E_x = 0 \tag{8.50}$$

for corrugation with transverse grooves (Fig. 8.8a) and

Fig. 8.9 The shielded model of the corrugated slot transmission line with (**a**) longitudinal grooves and (**b**) its transverse section

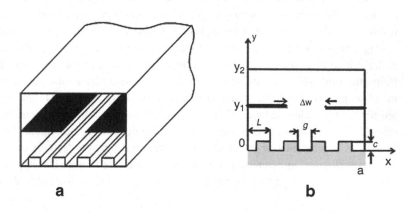

a   b

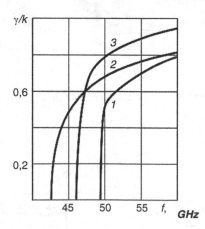

Fig. 8.10 The comparative analysis of dispersion properties of guiding structures. *1* The shielded slotted line. *2* The slot over the corrugated plane with longitudinal grooves. *3* The slot over the corrugated plane with transverse grooves: $a = 3.5$ mm, $y_1 = 0.1$ mm, $y_2 = 2$ mm, $\Delta w = 1$ mm, $i\eta^m = 0{,}5$

$$E_z = 0, \qquad E_x = \eta^m H_z \tag{8.51}$$

for corrugation with longitudinal grooves (Fig. 8.9a).

In the boundary conditions Eqs. (8.52 and 8.53), the η^m coefficient is defined by corrugation parameters:

$$\eta^m = i\,\xi_0\,(2g/L)\tan kc, \tag{8.52}$$

where g, L, c are geometrical parameters of corrugation (Figs. 8.8 and 8.9), $\xi_0 = \sqrt{\mu_0/\varepsilon_0}$ is the characteristic impedance of the vacuum. In this approximation, expressions for elements of $\left[Y_{mij}^-\right]$ for *corrugation with transverse grooves* (Fig. 8.8) are written as follows:

$$Y_{m11}^- = \frac{ic_m^2 \cot\left(r_m^{(1)}y_1 + l\right)}{\omega\mu^{(1)}r_m^{(1)}}, \quad Y_{m12}^- = Y_{m21}^- = \frac{\gamma\beta_m \cot\left(r_m^{(1)}y_1 + l\right)}{\omega\mu^{(1)}r_m^{(1)}},$$

$$Y_{m22}^- = -\frac{i\left[k_c^{(1)}\right]^2}{\omega\mu^{(1)}r_m^{(1)}} \cot\left(r_m^{(1)}y_1 + l\right) + \eta^m \frac{\varepsilon^{(1)}}{\mu^{(1)}} \frac{\cos ec\, r_m^{(1)}y_1}{\Delta_m^{(l)}}, \tag{8.53}$$

where $l = \arctan\left(i\eta^m c_m^2/\omega\mu^{(1)}r_m^{(1)}\right)$, $\Delta_m^{(l)} = \sin r_m^{(1)}y_1 + \tan l\cos r_m^{(1)}y_1$, and c_m, $k_c^{(1)}$, $r_m^{(1)}$ coincide with the similar expressions in admittance elements for the waveguide slotted line.

For corrugation with longitudinal grooves (Fig. 8.9) expressions for elements of $\left[Y_{mij}^-\right]$ have the form:

$$Y_{m11}^- = \frac{i\chi_m^2 \cot\left(r_m^{(1)}y_1 - \alpha\right)}{\omega\mu_0\mu^{(1)}r_m^{(1)}} + \eta^m \frac{\varepsilon^{(1)}}{\mu^{(1)}} \frac{\csc r_m^{(1)}y_1}{\Delta_m^{(\alpha)}},$$

$$Y_{m12}^- = Y_{m21}^- = \frac{\gamma\beta_m}{\omega\mu_0\mu^{(1)}r_m^{(1)}} \cot\left(r_m^{(1)}y_1 - \alpha\right), \tag{8.54}$$

$$Y_{m22}^- = -\frac{i\left[k_c^{(1)}\right]^2}{\omega\mu_0\mu^{(1)}r_m^{(1)}} \cot\left(r_m^{(1)}y_1 - \alpha\right),$$

where $\alpha = \arctan\dfrac{i\eta^m\left[k_c^{(1)}\right]^2}{\omega\mu_0\mu^{(1)}r_m^{(1)}}$, $\Delta_m^{(\alpha)} = \sin r_m^{(1)}y_1 - \tan\alpha\cos r_m^{(1)}y_1$.

At $\eta^m = 0$, Eqs. (8.53) and (8.54) coincide with formulas for elements for the input admittance tensor for the isotropic layer laying on the ideally conducting metallic plane (see Sect. 8.1).

8.6.2   Properties of the Symmetric Corrugated Slot Line

The properties of guiding slots over corrugated planes are defined from the dispersion equations obtained above.

A comparative analysis of dispersion properties of corrugated slot lines with different corrugations is given in Fig. 8.10. From these data, we can conclude that in the case of corrugation with longitudinal grooves, the cut-off frequency of the main wave begins to move toward zero with corrugation parameter growth (for instance, groove deepness), but does not achieve zero because in the case when $\eta^m > 1$, the boundary conditions Eq. (8.51) becomes inapplicable, and the corrugated slot line must be considered as the periodic structure. Reduction of the cut-off frequency with η^m growth can be explained by an equivalent increase in waveguide wide wall dimensions with growth of the depth of the corrugation grooves. In the case of corrugation with transverse grooves, the cut-off frequency shift effect is absent. These peculiarities of the dispersion dependence behavior of the corrugated slot line are caused by the view of the Y_{022} coefficient characterizing properties of the main wave. In the case of longitudinal corrugation, its parameter is included in Y_{022} as the phase shift $\cot\xi$, which, in turn, leads to zero shift of the dispersion equation. In the case of transverse corrugation, the Y_{022} dependence on the corrugation parameter disappears, and the form of Y_{022} coincides with the appropriate coefficient for the slotted line.

The interesting feature is the fact that the value and position of the opacity band depend on the corrugation parameters. This effect is especially interesting for the designing of controllable microwave devices [109].

Expression (8.53) gives the true results in the case of small corrugation because boundary conditions Eq. (8.50) describe the connection between average (over the period L) field values, which contains the zero harmonic only. If $\eta^m \cong 1$, the transmission line represents the periodic structure, the field that is, strictly speaking, unharmonious. The view of dispersion dependence is complicated, in which instead of the one opacity band, a set of bands appears, which corresponds to different spatial harmonics of expansion in the Fourier series.

Corrugated slot lines have a series of advantages over the usual slotted structures with dielectric layers: increased stability of line characteristics to various types of mechanical and thermal influences, fewer losses, greater rigidity, etc. Such transmission lines can be a base for the construction of various functional devices of microwave and millimeter-wave ranges.

8.7 Wire Lattices

8.7.1 Problem Statement, Two Polarizations

Frankly speaking, the very insignificant number of examples of periodic lattices shown in Fig. 1.13 does not, by any means, exhaust the manifold periodic and quasi-periodic structures that are used in practice (see, for example, Buduris and Shevenie [109]). Here, we give some information to help the reader to approach an analysis of specific systems in the future.

The analysis of the lattice behavior is essentially simplified if we use the two-side equivalent boundary conditions for the lattice of Eqs. (2.22 and 2.23) types, or, if we are speaking about the corrugated surface, conditions (2.18 and 2.19). Here, again, as for the analysis of the plane wave falling on the plane boundary of two media (Sect. 5.4), it is convenient to introduce consideration of two cases, H- and E-polarizations of the field, which fall on the lattice.

At H-polarization, the electric field is situated in the plane perpendicular to the conductors, and the magnetic field has a component along the lattice conductors (Fig. 8.11). If ξ designates the direction along the conductors, then at H-polarization $E_\xi = 0$. In this case, all field components are expressed (from Maxwell equations) by H_ξ, which is, as already mentioned, the Hertz vector. In the case of E-polarization, $H_\xi = 0$ and the field component E_ξ serves as the Hertz vector. Thus, Maxwell equations permit two types of independent solutions. The general case of arbitrary falling happens by superposition of solutions for both polarizations.

The general idea of formulas obtaining for the reflection coefficient (R) and transmission coefficient (T) consists of the utilization of the static solution (the Laplace equation $\Delta \begin{Bmatrix} H_\xi \\ E_\xi \end{Bmatrix} = 0$; Sect. 3.1) near conductors, and its continuation with the help of the Lorentz lemma (Sect. 1.6) to all remaining regions. Thus, the Lorentz lemma connects fields near the lattice, which have the L period imposed by the lattice geometry, and fields far from the lattice (in any case, at a distance much greater than

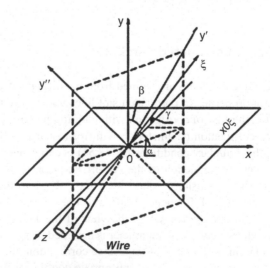

Fig. 8.11 The element of the wire lattice with the coordinate system

λ). Obviously, the field far from the lattice has not already had the L period, it is "smoothed." Coefficients of reflection (R) and transmission (T) satisfy the energy conservation law: $|R^2| + |T^2| = 1$.

8.7.2 Magnetic Polarization, the Maliuzhinets Effect

For distinctness, let the lattice be situated in the $x0z$ plane of the rectangular coordinate system, and direction ξ of the lattice conductors form the γ angle with the z-axis. The plane drawn in the ξ axis perpendicular to the $x0z$ plane is the plane of the plane wave falling on the lattice, and directing cosines are β (with the y axis), α (with the x axis). Satisfying these conditions, we obtain the final expressions for R and T:

$$\begin{cases} R \\ T \end{cases} = \frac{1}{2} \left\{ \frac{\beta + ik\left[(1-\gamma^2)l + \alpha^2 l_2\right]}{\beta - ik\left[(1-\gamma^2)l + \alpha^2 l_2\right]} \mp \frac{1 - ikL\Delta}{1 + ikL\Delta} \right\}. \tag{8.55}$$

Here, $l = S/2L$, S is an area of cross-section of the lattice conductor, $\Delta = \beta\, l_1/L - ikL(\beta^2\Delta_2 - \alpha^2\Delta_3)$. Undesignated quantities represent integrals over the conductor surface and they can be found as, for instance, in Nefyodov and Sivov, Buduris and Shevenie [60, 109].

When the normal wave falls on the lattice, we can obtain from Eq. (8.55) $R = (3/2)\, i\, k\, S/L$, $\quad T = 1 - (1/2)\, i\, k\, S/L$.

The result Eq. (8.55) for the H polarized wave reflects the unique property of frequent lattices – at the definite falling angle, the incident wave completely, without reflection ($R \equiv 0$), passes through the lattice. This effect was discovered by G.D. Maliuzhinets in his publications of 1937–1940 when designing decorative-absorbing coverings (in acoustic range). Substantiation of this effect of the complete transmission was fulfilled by him using hydro-dynamic analogies, allowing the statement that for the lattice of timbers of nonzero thickness, in the range $L/\lambda < 1/2$, the complete transmission is observed, when the incidence angle is $\varphi = \arccos q$, q is the relative slot width in the lattice. This discovery has attained world-wide recognition, and the effect itself is observed in many types of lattices (for instance, the jalousie type), in structures of artificial dielectrics, etc. In a certain sense, we can consider that this situation is similar to the known effect when the equivalent permittivity of the artificial dielectric has simultaneously negative values (see Sect. 1.5).

We note that the lattice with vertically located bands (Fig. 1.13g without a metal base, of course) at $L \approx q\ (\lambda/2)$, $q = 1$, 2, 3, ... represents the diffraction-coupled system of band open resonators (Sect. 11.2).

8.7.3 Electric Polarization

In the case of E-polarization ($H_\xi = 0$), the procedure of considerations remains the same and it results in:

$$\begin{cases} R \\ T \end{cases} = \mp \frac{1}{2} \left(\frac{1 - ik\beta l_2}{1 + ik\beta l_2} \pm \frac{1 - ik\beta l_3}{1 + ik\beta l_3} \right). \tag{8.56}$$

Quantities included in Eq. (8.56) should be chosen from Nefyodov and Sivov [60].

For horizontally located band conductors in the lattice we have: $l_2 = (L/\pi) \ln\left[1/\cos(\pi q/2)\right]$, $l_3 = (L/\pi) \ln\left[1/\sin(\pi q/2)\right]$, $l = l_2 = 0$.

Equations (8.55) and (8.56) permit the transfer to corrugated surfaces, when the filling coefficient of the L period by the metal is $q \to 1$.

8.7.4 Conclusions

The brief information presented here about frequently periodic structures, as equivalent boundary conditions of Eqs. (2.13, 2.14, 2.18, 2.19, 2.21, and 2.22) type and similar to them permitting the analysis to be conducted of many complicated transmission lines and basing elements of microwave and millimeter-wave engineering, containing lattices. However, the main physical conclusion from these formulas consists in the fact that the maximal reflection occurs in the case when the electric field of waves falling on the lattice is oriented parallel to its conductors, whereas the minimal reflection occurs when it is perpendicular to the conductors. The thing is that the incident field induces the conductor currents $\vec{j} = \sigma\, \vec{E}$ and in the case of $\vec{E}$ coincidence with direction of lattice wires, currents are maximal, and therefore, the reflected field is large. In reality, these considerations are useful only as a whole. Thus, for example, refusing the transverse current account in the lattice conductors may lead to noticeable errors. In acoustic lattices, we know about the *Lamb error* related to precisely mentioned circumstances (see, for instance, Nefyodov and Sivov [60]). Russian professor A.N. Sivov was the first to draw attention to this error.

Evidently, the boundary conditions on frequently periodic structures mentioned are the first step to obtaining appropriate boundary conditions on multi-layer lattices, which form the basis of nano- and meta-structures, for example, electromagnetic crystals [50, 60, 120].

8.8   Summary

Proceeding to studying of the chapter about periodic structures, we noticed their wide distribution in nature, science, and engineering and gave some examples starting from the Mendeleev periodic system of elements. Indeed, it is impossible to list those areas of science and engineering in which the remarkable properties of periodic structures are used to this or that extent. We saw that the Floquet theorem, which allows estimation of "integral" properties of considered structures, is a basis. But it is impossible to obtain the general system properties without the scrupulous analysis of parameters of each period of the actual structure. As we saw, the periodic structures can be divided to two large groups: either in the free space or in the waveguide. In addition, structures may be infinite or finite.

The integral equation with respect to the longitudinal component of the surface current density on metallic strips forms the basis of an analysis. This equation has a simpler form if we transfer to Pistohlkors–Schwinger variables. As a result, the Fredholm equation of the first kind is obtained, which is suitable for deducing the result.

In connection with needs of antenna structures for space applications, the new class of corrugated slot transmission lines was offered and developed. The description was based on utilization of equivalent Weinstein–Sivov boundary conditions (Sect. 2.1.5).

In practice, as before, great attention is given to systems with wire lattices. They are considered at the end of Chap. 8.

Checking Questions

1. Draw field pictures and their intensity distribution in structures of Fig. 8.1 assuming that the lowest wave has the main importance and the structure period $kL \ll 1$ and $kL\sqrt{\varepsilon} \ll 1$.
2. Give the physical representation of the $\omega - \beta$ diagram of the periodic transmission line (see Fig. 8.3).
3. Make sure that in fact in isotropic structures with a periodically metallized dielectric the odd waves cannot propagate (see Sect. 8.4.2) at small values of s parameter ($s \ll 1$).
4. Draw pictures of electric fields in the frequently periodic comb structure (with the period $kL \approx 0.1$) shown in Fig. 8.7c, for the case of shallow ($kc \leq 1$) and deep ($kc \gg 1$) grooves.
5. Show that in the periodic comb structure shown in Fig. 8.8c and having the characteristic equation in the form $p = k \tan kc$, the surface wave exists only under condition $c/\lambda \in (0, \quad 1/4)$.
6. The surface wave propagates over the $x = 0$ plane along the z axis so that it has the form $\exp\{-px + ihz\}$, $h = h' + i\, h''$, satisfying the characteristic equation $p = i\,\zeta\, k\,(\partial/\partial y \equiv 0)$. Assuming $\zeta = 1/\sqrt{\varepsilon} = |\zeta|\exp\{-i\delta/2\}$, when $|\zeta| = 1/\sqrt{|\varepsilon|} \ll 1$, and $\delta \leq \pi/2$ is the angle of electric losses, determine conditions at which the Zennek wave phase velocity $u = \omega/h'$ is greater than the light speed in vacuum.
7. Determine the wavelength propagating in a slowing structure on the frequency 10 GHz, if we know that at a distance of 5 cm the amplitude of the magnetic field vector decreases by an order compared with the amplitude of the structure surface.

Dielectric Waveguides and Light-Guides

9.1 Plane Dielectric Waveguide

Dielectric waveguides and lightguides occupy a great part of modern electrodynamics. The plane dielectric waveguide (Sect. 9.1) is the simplest representative of them. First wave investigations in such structures (the Goubau line type) are linked with the names Zenneck and Sommerfeld. A field in a layer satisfies the Helmholtz equation and the general boundary conditions (Sect. 2.1.2). Electrical and magnetic waves are examined with the help of the Hertz longitudinal vector. We obtain the transcendent dispersion equations for even and odd waves, which require a numerical solution. In particular, we note that the even electric surface wave E_{00} in the layer has zero critical frequency (the Katsenelenbaum effect). Many examples of the field component behavior in the layer and in the surrounding space are given. Using Programs P.9.1, P.9.2, the reader may be acquainted with field pictures of the plane dielectric waveguide.

Several physical consequences (Sect. 9.1.3) are offered to the reader. In particular, the Katsenelenbaum direction coupler is described. The wave expansion of the plane dielectric waveguide in plane waves is shown (Sect. 9.1.3). We examine the guiding structure from two layers (separated by an air) of the uniform dielectric, which is a model of many devices in the millimeter-wave range. The existence of fast waves is discussed. So-called anti-surface and leaky waves are described, which have a mathematical sense only, because they do not satisfy the emission conditions (Sect. 2.5). Nevertheless, we may apply the physical sense to the leaky wave and it can be used in practice. For instance, it allows reduction of complexity and difficulties at calculation.

This specific section of our lecture course is devoted to consideration of optical waveguides (lightguides) and dielectric integral structures (Sect. 9.2). Study of optical waveguides gave birth to integral optics. We show that in the optical region, the material dispersion can be quite significant. An outlet was found and it consists in the fact that the dielectric waveguides with permittivity of the free space are closed to the parameters of the film. Thus, there are E_{00} and H_{00} waves only, which do not have the lower frequency limit (Sect. 9.2.1).

Dielectric waveguides are extremely important for practice (Sect. 9.2.2). The solution of Helmholtz equations in the cylindrical coordinate system according to the Fourier method is represented in the form of expansions over cylindrical functions. We consider examples of various guiding systems. For instance, the metallic rod with the thin dielectric envelope (the Goubau line; Sect. 9.2.3). For comparison, we show fields of the EH_{11} wave of the dielectric rod and the main wave of the metallic wire with the thin dielectric (impedance) covering.

The tubular dielectric waveguide (Sect. 9.2.4) constitutes the lightguide fundamentals. The presence of a large parameter ($ka \gg 1$) in the tubular structures stipulates, on the one hand, the greater multi-wave operation of the transmission line, but on the other hand, it facilitates peculiar *selectivity* of the system, when "needless" wave types are simply highlighted.

During assimilation of new high-frequency ranges (millimeters, sub-millimeters), the role of quasi-optical waveguides and resonators based on it becomes stronger and stronger (Sect. 9.3). They have the capability to provide the "natural" selectivity of the necessary (operation) wave (Sect. 9.3.1). The lens line is another example of the quasi-optical waveguide (Sect. 9.3.2). The phase correction method was developed for its analysis. As an example of its application, the problem is reduced to the uniform Fredholm integral equation. For the first time, this approach was used by Russian academician Mandelstam; later by Fox and Lee. The solution is expressed through the Hermit polynomials. The diaphragm line is another example of the quasi-optical waveguides (Sect. 9.3.3).

© Springer International Publishing AG, part of Springer Nature 2019
E. I. Nefyodov, S. M. Smolskiy, *Electromagnetic Fields and Waves*, Textbooks in Telecommunication Engineering,
https://doi.org/10.1007/978-3-319-90847-2_9

There is a problem with the completeness of the eigenfunction system of the quasi-optical transmission lines (Sect. 9.3.4).

At the end of this chapter, the most interesting circle of problems is examined about the basing elements of the quasi-optical waveguides, of integral optics and lightguide engineering (Sect. 9.4). One of the problems is connected with excitation of quasi-optical waveguides. The main approaches to emission inlet (outlet) in optical micro-waveguides are excitation with the help of diffraction lattices, dielectric prisms, and the direct excitation "to the end surface," etc. We determine conditions and circuits for excitation. A great number of excitation methods are connected with dielectric prism utilization (Sect. 9.4.2). The method of phase synchronism is widely used (Sect. 8.2). Examples of thin-film elements of integral optics are presented.

As an example, we study the dielectric prism in the waveguide sharp bend (Sect. 9.4.3). The transformation coefficient of the prism is defined in the so-called Born approximation. As a result, we formulate the fundamental operation principle of quasi-optical devices – the principal of inter-cancellation of the spurious wave (the Nefyodov effect). Another example is the directional coupler based on two spaced plane dielectric layers. We obtain and discuss the approximate expressions for the propagation constant and the half-wavelength of beating.

This chapter is finished by utilization of MathCAD packet for analysis of the structures considered in this chapter.

9.1.1 Structure, Problem Statement

In Chap. 8 we became acquainted with the main properties of *slow electromagnetic waves* and structures, in which these waves can exist. Nevertheless, the area of surface waves does not become exhausted by the examples examined. One of the "oldest" examples of surface waves is the *Zenneck wave* existing in the space above Earth or the water surface. Later, much attention was paid to waves in the wire covered by the dielectric layer (the *Goubau line*) or, more generally, the wire with finite conductance. In recent years, great attention has been drawn to dielectric waveguide of the light range (the lightguides) [19, 37, 104–107, 110, 123].

Evidently, these examples can be continued, but for clarification of the physical sense we stop at the simpler two-dimensional system – the dielectric uniform layer (Fig. 9.1). As a matter of fact, in such a seemingly simple model, it was possible to reveal the most interesting features, about which, unfortunately, we can (owing to lack of volume) mention only very briefly. However, we note that *L.M. Brekhovskikh* made a serious contribution to the theory and practice of multi-layer structures for electromagnetic and acoustic waves.

The electromagnetic field must satisfy the Helmholtz Eq. (3.1) or (3.2), taking into consideration the task's two-dimensionality ($\partial/\partial y \equiv 0$) and the excluded dependence on the longitudinal z-coordinate ($\exp\{ihz\}$) at $\vec{j}_{\text{out}} = 0$, $\rho_{\text{out}} = 0$. Then, two of the same type of ordinary differential equations for the electric $\Pi^e(x)$ and magnetic $\Pi^m(x)$ Hertz vectors take place:

$$d^2\Pi/dx^2 + g^2\Pi(x) = 0, \quad g^2 = k^2\varepsilon\mu - h^2. \tag{9.1}$$

In Eq. (9.1), we understand under Π either $\Pi^e(x)$ or $\Pi^m(x)$.

The Eq. (9.1) has specific solutions: outside the plate ($|x| > a$):

$$\Pi(x,z) = A\,e^{-px+ihz} \tag{9.2}$$

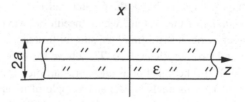

Fig. 9.1 The plane dielectric layer, dimensions, and the coordinate system

and in the plate ($|x| < a$):

$$\Pi(x, z) = [B \sin(gx) + C \cos(gx)] e^{ihx}. \tag{9.3}$$

The first term in Eq. (9.3) defines the odd solutions (with regard to the x coordinate), whereas the second term defines the even solution.

Components of the electromagnetic field are found by simple differentiation of Hertz vectors $\vec{\Pi}^{e,m}$ (9.2, 9.3) on (x, z) coordinates as:

$$\begin{aligned}
\vec{E} &= \operatorname{grad} \operatorname{div} \vec{\Pi}^e + k^2 \varepsilon \mu \vec{\Pi}^e + ik\mu \vec{\Pi}^m, \\
\vec{H} &= -ik\varepsilon \operatorname{rot} \vec{\Pi}^e + \operatorname{grad} \operatorname{div} \vec{\Pi}^m + k^2 \varepsilon \mu \vec{\Pi}^m.
\end{aligned} \tag{9.4}$$

9.1.2 Electric and Magnetic Waves

The electric Hertz vector has a form at $|x| > a$ defined by (9.2), and in the plate

$$\Pi_z^e(x, z) = \Pi^e(x) \, e^{ihz}. \tag{9.5}$$

Determining by Eq. (9.4) the field components E_x, E_z, H_y ($E_y = H_x = H_z \equiv 0$) in the plate, we have:

$$E_x(x, z) = ihgB \cos(gx) \exp\{ihz\},$$

$$E_z(x, z) = g^2 B \sin(gx) \exp\{ihz\}, \tag{9.5a}$$

$$H_y(x, z) = -ikg\varepsilon \cos(gx) \exp\{ihz\}$$

and outside of it

$$E_x = -ihpA \exp\{-px + ihz\},$$

$$E_z = -p^2 A \exp\{-px + ihz\}, \tag{9.5b}$$

$$H_y = -ikpA \exp\{-px + ihz\}.$$

Equating components E_z and H_y from Eq. (9.5a) and (9.5b) on the boundary "dielectric-air" ($x = a$), we obtain some algebraic relationships, and excluding from them the A, B constants, we come to the following dispersion equations for even *electric waves* (the upper line in Eq. (9.6)):

$$pa = \frac{ga}{\varepsilon} \left\{ \begin{array}{c} \tan \\ -\cot \end{array} \right\} (ga) \tag{9.6}$$

and, relatively, for odd electric waves (the lower line in Eq. (9.6)).

From the interchangeable duality principle (see Sect. 2.4; formula (2.40)), replacing ε with μ in (9.6), we obtain the dispersion equation for magnetic waves:

$$pa = \frac{1}{\mu} ga \left\{ \begin{array}{c} \tan \\ -\cot \end{array} \right\} (ga). \tag{9.7}$$

Fig. 9.2 The scheme of the
graphical solution of Eqs. (9.6,
9.7). (**a, b**) Curves corresponding
to the right parts of equations for
even and odd waves. (**c**) Graphical
explanation of the leaking wave
appearance

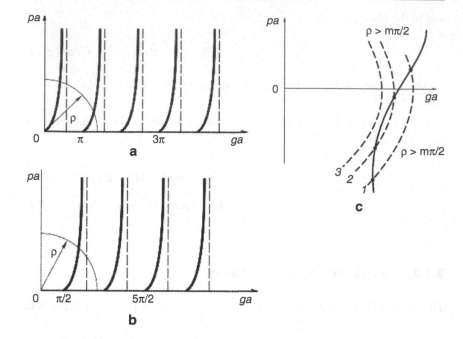

Transcendent dispersion Eqs. (9.6) and (9.7) are based on the analysis of the fundamental electric and magnetic waves of the dielectric layer. The transverse wave numbers p and g are connected between them as:

$$(ga)^2 + (pa)^2 = (ka)^2[\varepsilon(-1)]. \tag{9.8}$$

The latter equation can be compared with the appropriate formulas for metallic waveguides (compare Eq. (7.5)).

Dispersion Eqs. (9.6, 9.7) should be solved numerically, and now such a solution can be obtained with arbitrarily given accuracy with the help of our MathCAD programs (see Appendix B – http://extras.springer.com/2019/978-3-319-90847-2). The Program P.9.1 allows sequential determination of the roots of even electric waves from Eq. (9.6) together with the condition (9.8). Equation (9.8) is interesting because it gives the possibility of determining the accuracy of roots found from (9.7) and (9.8) (see Programs P.9.1, P.9.2, P.9.3, and P.9.4).

However, for the initial physical analysis and understanding of the merits of the case, it is sufficient to have a graphical picture of the solution behavior of these equations. The right parts of Eqs. (9.6) (at $\varepsilon = 1$) are presented in Fig. 9.2, which means that the scale variation does not influence the sense. They remind us the upper parts of dispersion characteristics of the multiport reactive impedance without losses (see Neganov et al. and Buduris and Shevenie [19, 109]).

To decrease the wave field outside the layer during moving away of the layer (at |x| > a; see Eq. (9.2)), it is necessary to satisfy the condition $p > 0$; therefore, the right parts of the equations should be positive. Obviously, at nonfulfillment of this condition, the surface wave in the examined structure is absent. Solutions of the dispersion equation are obtained as the intercept points of the curves in Fig. 9.2a, b with the circumference, whose radius is

$$\rho = ka\sqrt{\varepsilon\mu - 1}, \tag{9.9}$$

and its center coincides with the origin ($pa = ga = 0$).

From Fig. 9.2c we see that the circumference Eq. (9.9) at $\rho < \pi$ for even waves intercepts only one branch, which defines the *electric surface wave* E_{00} in the layer, the main wave for the dielectric waveguide. Its critical frequency is $f_{cr} = 0$. This effect was discovered by B.Z. Katsenelenbaum in due course. The sequential increase in ρ gives waves $E_{2m, 0}$ (Fig. 9.2c; see also Program P.9.1 in Appendix B – http://extras.springer.com/2019/978-3-319-90847-2). Results for odd waves E_{m0} of the dielectric waveguide can be obtained using Program P.9.2).

Longitudinal wave numbers of these slow surface waves are

$$h = \sqrt{k^2 + p^2}. \tag{9.10}$$

Fig. 9.3 Deceleration of (**a**) the main wave of the uniform dielectric layer E_{00} of ka and (**b**) even waves of ε. (**c**) Deceleration of E_{00} wave with respect to layer parameters. (**d**) Deceleration of E_{10}, E_{20}, E_{40} electric waves of ka at $\varepsilon = 10$. (**e**) Deceleration of E_{10} wave of ka. (**f**) Deceleration of odd E_{10}, E_{20}, E_{50} waves of ka at $\varepsilon = 10$.

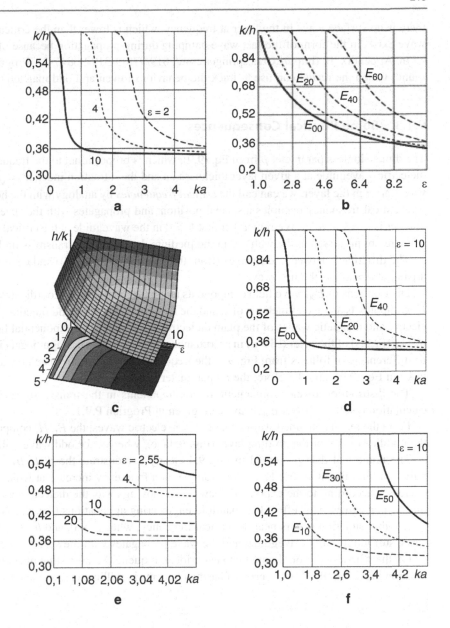

In practice and in the study of dielectric waveguide physics, the concept of *deceleration* of the definite wave is often used, for example, the slow even wave E_{00} defined as a ratio of its phase velocity (or any other chosen velocity) to the light speed in the vacuum: k/h. An example of such a dependence for the E_{00} wave is presented in Fig. 9.3a (see also, Appendix B – http://extras.springer.com/2019/978-3-319-90847-2), Program P.9.1, in which the root search is automatic and the accurate interval of root localization is simultaneously determined. When the layers are thick enough, for instance, at $ka \approx 6$, deceleration values are determined as $1/\sqrt{\varepsilon}$.

A complete enough picture of electric E_{00}, E_{20}, E_{40} wave decelerations with respect to ka at $\varepsilon = 10$ is seen from Fig. 9.3d. We clearly see that for "thick" layers ($ka \approx 5.7$), the deceleration of waves of a higher order (in the main) aspires to deceleration values in the space filled by the uniform dielectric, namely, to $\sqrt{\varepsilon}$.

The behavior of odd electric waves E_{m0} at $m = 1, 3, 5$ is shown in Fig. 9.3e, f calculated using the Program P.9.2 (Appendix B – http://extras.springer.com/2019/978-3-319-90847-2). The presence of the cut-off frequencies, at which these waves arise, is these wave peculiarities. In this way, they differ from E_{00} and H_{00} waves, which do not have cut-off frequencies.

It easy to see that there are many common points in the behavior of eigenwaves of metallic waveguides and waves of dielectric waveguides. However, the wave appearing in the dielectric layer (the waveguide) at its critical frequency ($p = 0$ и then $h^2 = k^2$) propagates with the speed of light. In metallic waveguides, on the contrary, $h = 0$. There is another difference:

there is no surface wave in the layer at frequency, which is lower than the critical frequency of the given surface wave. This wave exists in the form of the fast wave damping during propagation because of the emission.

In Appendix B (http://extras.springer.com/2019/978-3-319-90847-2), Programs P.9.1, P.9.2, P.9.3, and P.9.4 can be found, which the reader can use to track the behavior of even and odd magnetic waves.

9.1.3 Some Physical Consequences

The dimensionless parameter ρ from Eq. (9.9), which is proportional to the frequency, defines a number of eigenwaves of the dielectric waveguide at a given layer thickness $2a$ and the refraction index $n = \sqrt{\varepsilon\mu}$. Frequencies, at which new propagating waves arise in the layer, we can call the *critical frequencies* by analogy with the hollow transmission lines. The E_{00} wave with zero critical frequency occupies a special position, and propagates with the speed of light (at $pa = 0$, the longitudinal wave number is $h = k = \omega/c$). We remember that $h = 0$ in the waveguide at the critical frequency, and the wave does not propagate (if there are no losses in the walls or in the medium, by which the transmission line is filled).

The difference between odd waves (from the low line in Eq. (9.6)) and even waves consists in the fact that the E_{10} wave arising at $\rho > \pi/2$ is the first wave.

Distributions of E_{00} wave field components along the "vertical" $x-$coordinate are shown in Fig. 9.4. We clearly see that E_x, H_y components are even functions of x, and the E_z component is the odd function. Similarly, we can draw components of other electric and magnetic waves of the plain dielectric waveguide. The exponential field reduction when moving away from both planes (facets) of the layer ($|x| > a$) in accordance with Eq. (9.2) is the significant circumstance. The degree of this damping can be different. As it follows from Fig. 9.4, the frequency growth for the specific wave leads to ρ parameter growth proportional to ka in Eq. (9.9), and therefore, the pa parameter increases.

The distribution of electromagnetic field components in the transverse section of the dielectric layer is typical. Some calculation results for even E_{mn}-waves are given in Program P.9.1.

From the results obtained we see that for even electric waves, the E_x, H_y components are the even functions with respect to the middle ($x = 0$) of the guiding layer (Fig. 9.4a–c), whereas for odd waves, all components are the odd functions of x.

The numerical data presented in Fig. 9.4 clearly demonstrate the *dielectric effect* connected with electromagnetic field retraction into the dielectric layer, for example, at a frequency increase at which wave deceleration aspires to $\sqrt{\varepsilon}$. The field decreases according to the exponential law when moving away the dielectric–air boundary, i.e., waves have the character of the *surface waves*. This effect has manifold engineering applications (see also Nefyodov [52]).

Another situation occurs near the critical frequency, when $p \approx 0$, and $h \to k$. Thus, the field is mainly concentrated in the space surrounding the waveguide, and the wave propagates with the speed close to c.

On the whole, we should note that near critical frequencies, interesting phenomena are observed that serve as the basis for creation in IC and 3D-IC of a series of basing elements in millimeter-wave and sub-millimeter-wave ranges, say, in particular,

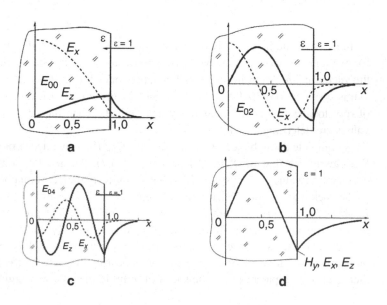

Fig. 9.4 Distribution of field components of even electric (**a**) E_{00}, (**b**) E_{02}, (**c**) E_{04} waves and (**d**) odd E_{10} waves in the transverse section of the dielectric layer (at $x = 1$ the air–dielectric boundary is located; $x = 0$ the layer middle)

the *BZ directional coupler* and others. The idea of this coupler (*BZ coupler*) was offered by *B.Z. Katsenelenbaum* (see, for instance, Gvozdev and Nefyodov [14], p.175–178).

9.1.4   Wave Expansion of the Planar Dielectric Waveguide into Plane Waves

We have already used similar procedures more than once (see, for example, Sects. 6.8 and 6.10). Let us take for example the odd electric wave of the dielectric waveguide with the longitudinal component of the Π_z^e Hertz vector, and write it with the help of *Euler formulas* (without insignificant multipliers) as: $\Pi_z^e = \exp\{i(gx + hz)\} - \exp\{i(-gx + hz)\}$.

Taking into account that $h^2 = k^2\varepsilon\mu - g^2$ and introducing, as earlier (see Sect. 6.8, Eq. (6.34)), the angle of each these waves with respect to the *z*-axis as ϑ, we obtain

$$g = k_n \sin\vartheta, \quad h = k_n \cos\vartheta, \quad k_n = k\sqrt{\varepsilon\mu}. \tag{9.11}$$

Thus, the situation reminds us of the metallic plain and rectangular waveguides. However, there is an essential difference, already mentioned by us, and we now look at the geometrical–optical picture shown in Fig. 9.5.

From Fig. 9.5a, b we see that the field inside the dielectric waveguide is a sum of two plane waves, whose propagation direction forms the angle ϑ with *z*-axis. The incidence angle of these waves onto the dielectric boundary (the plane $|x| = a$) is evidently equal to $\varphi = \pi/2 - \vartheta$ (Fig. 9.5a). The waves in the dielectric waveguides studied above were *slow waves*, i.e., they satisfied the condition

$$h = k_n \cos\vartheta = k_n \sin\varphi > k \tag{9.12}$$

and the incidence angle φ satisfied the inequality

$$\sin\varphi > 1/n, \quad n = \sqrt{\varepsilon\mu}. \tag{9.13}$$

Therefore, these plane waves experience the *complete internal reflection* on the dielectric–air boundary.

Thus, the waveguide properties of the dielectric waveguides are caused by the *complete internal reflection*, owing to which the electromagnetic energy keeps within the limits of the dielectric layer. The electromagnetic field in the space surrounding the dielectric arises owing to energy transpiration at complete reflection into another medium; thus, the effluent field damps at moving away from the boundary plane according to the *exponential law*.

At frequencies essentially exceeding the critical frequency of each wave, the ϑ angle is small and plane waves propagate almost on the *z*-axis, to which $h \approx k_n$ corresponds. At frequency decrease, the ϑ angle increases, and the φ angle decreases,

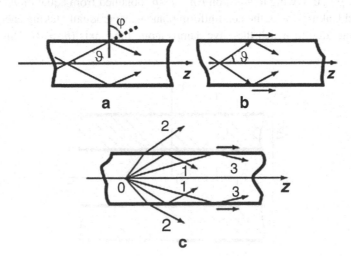

Fig. 9.5 The geometric–optical picture of waves in the planar dielectric waveguide. (**a, b**) Expansion of waveguide waves into plain waves. (**c**) Waves from the point source. (Taken from Nefyodov and Kliuev [73], p. 300)

which occurs in the usual waveguide (Sect. 7.7). However, the ϑ angle cannot grow to the $\pi/2$ value, as in the metallic waveguide, namely, at the critical frequency

$$\cos\vartheta = \sin\varphi = 1/n \qquad (9.14)$$

and in the surrounding space, the refracted beam appears, which passes along the boundary (Fig. 9.5b). The plane wave propagating in the vacuum parallel to the z-axis with c speed corresponds to this beam. For lower frequencies, this wave cannot exist as the surface wave.

Surface waves E_{00} and H_{00}, which have no critical frequency and exist at arbitrarily low frequencies, take a special position. If the condition $\rho \ll 1$ if fulfilled, then, according to Eq. (9.7), it causes inequalities $pa \ll 1$ and $ga \ll 1$. This means that, as both in the dielectric and over it, the field becomes practically transverse, along the plate, the plane wave propagates slightly perturbed by the plate.

9.1.5 Note About Leaking Waves

Up to now, we have considered the only slow (surface) waves in the plane dielectric waveguide. However, nature happens to be more complicated than its geometrical–optical description. As we see from Fig. 9.5a, b, slow waves in the plate correspond to plane waves experienced on the dielectric–air boundary at *complete reflection*. Earlier experts considered that there are no other waves besides these waves.

At the same time, the *fast waves*, reflection from which is not complete, may propagate in the dielectric plate and in the multi-layer plane-parallel structure from dielectric layers (in Fig. 9.6 the case is shown of two air-separated layers). Therefore, these waves propagate along the plate damping because of emission in the surrounding space (Fig. 9.6).

We already noted above that for $\rho < m\pi/2$ there are no surface waves E_{m0} and H_{m0}. Nevertheless, these waves actually exist at $\rho < m\pi/2$, not as slow waves, but as *fast waves*; the *complex roots* of the dispersion Eqs. (9.6) and (9.7) (upper lines) at even m and lower lines (9.6) and (9.7) at odd m correspond to them. More precisely, the fast wave arrives from the surface wave not right away for $\rho < m\pi/2$, but passed the stage of the slow "anti-surface" wave (Fig. 11.7 in Nefyodov [26]).

The thing is that these equations having the form $pa = F(ga)$ formally permit at $\rho > m\pi/2$ negative values of p corresponding to negative parts of curves shown in Fig. 9.2b: their intersection with the circumference Eq. (9.7) (the circumference 1 in Fig. 9.2c) gives the anti-surface wave; slow, not damping with the field, which *grows exponentially* outside the plate with $|x|$ growth. Such p values were not examined by us as not having the physical sense; however, we must take them into consideration to provide a complete picture. When the ρ parameter proportional to the frequency decreases and passes through the $\rho = m\pi/2$ value, then, at first, the surface wave becomes the second anti-surface wave (circumference 2), at a further reduction of ρ, points in the plane ga, pa (corresponding to two such waves) approaches and, finally, the circumference 3 does not already intersect with the curve $pa = F(ga)$. Then, the dispersion equation has the complex-conjugated roots $p = p' + ip''$, $p' < 0$. Owing to equation $h'h'' = p'p''$ obtained from equation $h^2 = k^2 + p^2$, whose root $p'' < 0$ defines for $h' > 0$ the so-called leaking wave: the non-uniform plane wave gradually taking energy from the field in the plate passed from the dielectric plate, due to which the wave damps along the z-axis ($h'' > 0$). This wave may be referred to as inflowing.

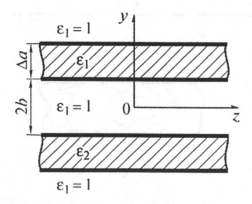

Fig. 9.6 The guiding structure from two air-separated layers of the uniform dielectric

Anti-surface and *leaking (outflowing) waves* do not factually excite and have only a mathematical sense. Strictly speaking, the leaking wave belongs to the so-called *non-eigenwaves* because, owing to exponential growth of its field during moving away of the layer, it does not satisfy the emission conditions (see Sect. 2.5; eq. (2.41)). And, nevertheless, we can give the leaking wave a physical sense, and it is used in practice [19, 52]. Thus, at analysis of rod dielectric antennas, representation of emission field formation due to the leaking wave allows the essential complexity and labor expenditure of calculations to be reduced [52].

At wave excitation in the dielectric plate or in the plates system, in the dielectric rod and in many other systems, the leaking waves do not play any noticeable role as they damp rather quickly when moving away from the source, and at a large distance, the surface waves predominate. However, this is not concerned with the dielectric *rod antenna*, *BZ coupler*, the *prism exciter* of the light-guide, and a series of other elements of the waveguide section mentioned earlier. Because of inevitable losses in the dielectric, the surface waves also damp, but these losses increase the damping of leaking waves. Nevertheless, this is not always the case: there are systems, in which damping of leaking (fast) waves is lower than damping of surface waves, if taking into consideration losses in the dielectric. These are dielectric tubes, which are briefly examined in Sect. 10.2.

9.2 Optical Waveguides, Dielectric Integrated Structures

9.2.1 Introduction

The task of the dielectric plate considered in the previous section and the task about the plates system mentioned find a new life with the appearance of *plain (planar) optical waveguides* (light-guides) in *integral optics*. Such waveguides consist of the thin dielectric film on a substrate having a slightly lower refraction index than for film itself. Films with variable refraction index on thickness are applied. In practice, the completely dielectric strip structures are used.

Not stopping on all these complicated systems, we note only the following. In the optical range, where the wavelength is small, the dielectric layer supports propagation of many surface waves, i.e., the *multi-wave mode* in the transmission line may occur. However, if in the microwave and partially in the millimeter-wave ranges, the material dispersion $\varepsilon = \varepsilon(\omega)$, $\mu = \mu(\omega)$ does not practically manifest, then in the optical area the position is not the same, and the material dispersion may be entirely significant. If so, the multi-wave operation mode becomes extremely unprofitable, because radio signal components (especially, wide-band signal, which is frequently used in radio engineering, radio physics, radar technology, and other areas of engineering) propagate with various velocities, which leads to sharp distortions of transmitted signals. Thus, contradictions are present: on the one hand, the dielectric waveguide should be wide to provide a large channel capacity, but, on the other hand, this leads to the multi-wave mode with its mentioned properties. The outlet was found out and it consists in the fact that the dielectric waveguide with ε, μ permeability is surrounded by the medium (or film) with ε_0, μ_0, which are closed to ε, μ. Then, the parameter (9.9) takes the form

$$\rho = ka\sqrt{\varepsilon\mu - \varepsilon_0\mu_0} \tag{9.15}$$

and it is not difficult to achieve condition $\rho < \pi/2$, at which the surface waves E_{00} and H_{00} only exist with a small difference in refraction indices of the dielectric waveguides and the environment ($n = n_0$). If the medium with permittivity and permeability closed to ε, μ joins to one side of the layer, and the other side is air, the surface waves of higher numbers are absent for $\rho < 3\pi/4$, but the surface waves E_{00} and H_{00} themselves have critical frequencies, they exist as surface waves for $\rho > \pi/4$ only.

9.2.2 Dielectric Waveguides

In the general case, the dielectric waveguide is a multi-layer dielectric structure (Fig. 9.7). The specific cases of this structure are shown in this figure: the guiding dielectric rod (b), the dielectric rod with metallic cover – the circular waveguide with solid filling (c), the same but with laminated filling (d), the simplest tube dielectric waveguide (e). Evidently, there may be a large number of such structures. The common geometrical property unites them – they have rotation symmetry.

The scheme of dispersion equation obtained for structures in Fig. 9.7 are approximately the same, as we saw when considering the coaxial transmission line (Sect. 6.9). In all these cases, the electromagnetic field satisfies the *Helmholtz equation* in the cylindrical coordinate system (r, φ, z) of Eq. (6.48) type. After extraction of the function of longitudinal

Fig. 9.7 Dielectric guiding structures. (**a**) The multi-layer dielectric waveguide. (**b**) The dielectric rod. (**c**) The dielectric rod in a metallic cover. (**d**) The two-layer dielectric rod in a metallic cover. (**e**) The tube dielectric waveguide

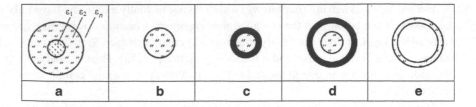

coordinate z (namely, $\exp\{ihz\}$) and representation of the required solution by the Fourier method in the form of the product $\Psi(r, \varphi) = R(r)\Phi(\varphi)$, we obtain the known solutions (compare Eq. (6.9)):

$$R(r) = AJ_\nu(\chi\, r) + BY_\nu(\chi r),$$
$$R(r) = \tilde{A}H_\nu^{(1)}(\chi r) + \tilde{B}H_\nu^{(2)}(\chi r). \tag{9.16}$$

and for azimuth functions

$$\Phi(\varphi) = C\cos(\nu\varphi) + D\sin(\nu\varphi) \text{ or } \Phi(\varphi) = \tilde{C}e^{i\nu\varphi} + \tilde{D}e^{-i\nu\varphi}. \tag{9.17}$$

The selection of necessary functions from Eqs. (9.2) and (9.3) is defined by the task geometry (Fig. 9.7). Thus, for the rod waveguide (Fig. 9.7b) in the internal area ($r < R$) in Eq. (9.2), it is necessary to assume $B = 0$ (because the $Y_\nu(r)$ function for $r \to 0$ has a singularity). For an external area ($r > R$) it should be $\tilde{A} = 0$, because we must keep the decreasing (at $r \to \infty$) solution only (thus, the decrease speed is larger than $1/\sqrt{r}$). On the dielectric–external medium boundary, the tangent continuation to the boundary conditions should be satisfied for electric (E_z, E_φ) and magnetic (H_z, H_φ) fields (Sect. 2.1.2; formula (2.9)).

Similar conditions should be fulfilled for all other guiding structures in Fig. 9.7. On metallic surfaces, as, for example, for the dielectric rod in a metallic cover (Fig. 9.7c) or the two-layer dielectric rod in the metallic cover (Fig. 9.7d), we must satisfy the ideal Dirichlet boundary condition (2.11)) or Neumann condition (2.12).

The derivations are simple but bulky enough; thus, we omit them and give only some of the results.

9.2.3 Metallic Rod with a Thin Dielectric Cover

This structure (Fig. 9.7c) has been known for a long time; it is referred to sometimes as the *Goubau line*. Investigation of the guiding properties of this structure, as a whole, is conducted in the same scenario as in the previous case. We note only some of the peculiarities. Obviously, for symmetric waves ($\partial/\partial\varphi \equiv 0$) at enough low frequencies, the picture differs a little from the planar waveguide on the metallic basis (see Sect. 10.2). In essence, the symmetric wave (in the absence of dielectric or some slowing, for instance, frequently periodic structure) transfers into the transverse wave of the single-wire line (Fig. 9.8a).

The field structure of the main nonsymmetric wave of the metallic rod with the thin dielectric cover or the impedance cover (Fig. 9.8b) reminds us of the EH_{11} wave field of the dielectric rod (Fig. 9.8).

9.2.4 The Tube Dielectric Waveguide (The Light-Guide)

Dielectric tubes, as we already mentioned, are light-guides and transmission lines in the millimeter-range (Fig. 9.7d) that are widely used in practice. The main difference between this class of transmission lines and, say, hollow waveguides, consists in the fact that the former, as a rule, are wide, i.e., in the overwhelming majority of cases $ka \gg 1$, where a is a radius of the internal tube cavity. Transmission lines with simplest single-layer walls are shown in Fig. 9.9. As a matter of fact, walls can be multi-layer, contain layers with air gaps, or the frequently periodic metallic lattices, etc. The internal tube cavity can be filled by a gas that allows, for instance, performing of the fast deployment of the transmission lines or antennas in the field conditions.

Analysis of the tube structure *in corpore* is bulky enough. Different approaches [52, 107] to its simplification were offered; however, the approach developed in Weinstein [125] is probably most physically well-founded. As waves effectively

Fig. 9.8 (**a**) The EH_{11} wave field of the dielectric rod and (**b**) the main wave field of the metallic wire with thin dielectric (impedance) cover. (Taken from Nefyodov and Kliuev [73], p. 310)

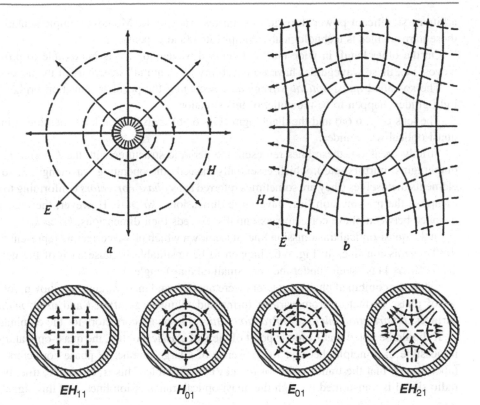

Fig. 9.9 Field structures of some of the waves in the transverse section of the tube waveguide (E *lines* solid lines, H *lines* dotted lines). (Taken from Nefyodov and Kliuev [73], p. 311)

EH_{11} H_{01} E_{01} EH_{21}

reflected from walls (the reflection coefficient of the plain wave comprising this waveguide wave is closed, in any case, to 1 modulo) present the main interest, we, naturally, start from the known solutions for the circular metallic waveguide.

The presence of the large parameter ($ka \gg 1$) in tube structures causes, on the one hand, the multi-wave operation mode in the transmission line, but on the other hand, promotes the peculiar *selectivity* of the system, when "not-used" wave types are shone through (see beams of *2* types in Fig. 9.5c). The similar principle, as we shall see later (Sect. 11.1), was laid by A.M. Prokhorov in the fundamental explanation of the *open resonator* operation, which opened a new page of modern radio physics and electronics.

At a small wave incidence angle on the tube wall ($\vartheta \ll 1$, $\varphi \approx \pi/2$), the reflection coefficient for waves of both polarizations $|R_{1,2}| \to 1$, and therefore, we can use in the perturbation method (see Sect. 10.2), as the first approximation, the ideal Dirichlet conditions (2.11) and Neumann conditions on the internal tube side. This allows precise enough values of wave linear damping of the tube waveguide to be obtained.

Pictures of electric (solid lines) and magnetic fields of several waves are shown in Fig. 9.9. The EH_{11} wave occupies a special position. It reminds us of the plane transverse wave in the field structure. It has minimal damping (see Task 1 in "Checking Questions").

9.3 Quasi-optical Waveguides

9.3.1 Introduction

During assimilation of new frequency ranges (millimeter and sub-millimeter ranges), the small convenience of usual shielded (closed) waveguides (for the wave, and the closed resonators, see Sect. 11.1) for practical systems becomes increasingly clear to researchers. The thing is not only in the sharp loss of growth in metals and dielectrics, which, for example, for metallic waveguides (practically for all waves excluding the H_{01} wave of the circular waveguide) increase proportionally to $\sqrt{f}$. The usual damping of Russian dielectrics is $\tan\delta \approx 10^{-4}$; the best foreign samples have $\tan\delta \approx 10^{-6}$. The requirement of the single-wave operation mode, which is "natural" for decimeter and centimeter ranges, is difficult to implement in millimeter-waves, because the average transverse size a of the closed waveguide is approximately equal to λ, and therefore in millimeter-waves it happens to be so small that, first, it cannot be performed technologically, and second, it is impossible to transmit the arbitrary significant power through this narrow waveguide. Moreover, implementation of basing elements and devices on these narrow lines is technologically complicated (and expensive).

On the other hand, in wide ($ka \gg 1$) closed waveguides, it is impossible to provide effective suppression of spurious waves, i.e., these waveguides have no capability for "natural" *selectivity* of the necessary (operating) wave.

The so-called *quasi-optical waveguides*, becoming the peculiar transition bridge between metallic waveguides and the light-guides, happen to be the outlet of this situation.

The lens (Fig. 6.6a) and the diaphragm (Fig. 6.6b) transmission lines mentioned in Chap. 6 are the simplest examples of quasi-optical waveguides.

Strictly speaking, these lines represent the periodic structures with the L period (distance between line elements; lenses, diaphragms, mirrors, etc.), which essentially exceeds the operating wavelength λ, so that $kL \gg 1$. On the other hand, the elements themselves (they are sometimes referred to as *phase correctors* conforming to the simplest calculating approach) that comprise the transmission line, have large dimensions: $ka \gg 1$. However, the main condition consists in the fact that the distance between phase correctors essentially exceeds their dimensions: $kL \gg ka$.

In the quasi-optical transmission lines it is known which phase correctors represent the plane or focusing mirrors (Fig. 6.6c, d). Transmission lines in Fig. 6.6c happen to be unsuitable because each of the next phase correctors (due to conditions $kL \gg ka \gg 1$) is "seen" under the very small (sliding) angle.

Later, the combined phase correctors were applied, and their variants are shown, for instance, in Fig. 11.4c in [26]. A great set of phase correctors from various forms and fillings was offered and used: *multi-layer dielectric covers* of reflecting surfaces, *polarization mirrors* (*diffraction lattices*) in phase correctors for the elimination of unnecessary polarization, etc.

It should be noted that from the point of view of classical optics, the quasi-optical transmission lines of the lens line type is impossible in principle, because the presence of a large number of phase correctors introduces such noticeable distortions (aberrations) that the transmission of *images* is impossible. This is, of course, true, but the thing is that not an image, but a radio signal is transmitted through the quasi-optical transmission lines, and this signal is insensitive to optical aberrations.

9.3.2 Lens Line, Fundamentals of the Theory, Method of Phase Correction

We examine the scheme of quasi-optical transmission lines on an example of the *lens line* (Fig. 6.6a). This scheme is simple enough and clear: the field at the input of the next lens (for $z = L - 2\delta$) is defined according to the unknown field at the output of any lens (for example, the left lens at $z = 0$). After that, according to the input field, we determine the field at the output of this lens ($z = L$). Then, and this is the main idea of the scheme, we write a *condition* that the field at the output of the second lens differs by the *constant multiplier only* from the field at the output of the first lens.

Thus, we obtain an equation for the wave field, which propagates along the transmission line. For some, for instance, the transverse component of the magnetic field $H_y \equiv u(\vec{r}) = u(x, y, z)$, it has the form [66, 106, 107]:

$$\frac{ik}{2\pi L} e^{ikL} e^{i\varphi(\xi,\eta)} \int\limits_S u(x,y,0) \exp\left\{ \frac{ik}{2L}\left[(x-\xi)^2 + (y-\eta)^2 \right] \right\} dx\, dy = \chi\ u(\xi,\eta,0). \tag{9.18}$$

At derivation of the *uniform integral Fredholm* Eq. (9.18), which was used in 1912 by Russian scientist *L.I. Mandelstam*, the condition of field recurrence in apertures of the series lens with accuracy up to constant multiplier χ:

$$u(\xi,\eta,L) = \chi u(\xi,\eta,0). \tag{9.19}$$

We would like to note, by the way, that the Eq. (9.18) is often referred to as the *Fox* and *Lee* equation, which, as we see, does not correspond to reality.

Another important condition was the condition of the wave passage through the lens: the *condition of phase correction*, according to which the geometric–optical beam refraction on the lens edges is not taken into account, but the beam phase $k\sqrt{\varepsilon\mu}\, l(x)$ only accepts a fact, where $l(x)$ is the lens thickness at a given value of the "vertical" x coordinate.

Solutions of three-dimensional Eq. (9.18) define eigenwaves of the lens line. Eigenwaves of the two-dimensional lens line are especially clear, i.e., the line consisting of cylindrical lenses. For the so-called *confocal lens system,* when its *focal distance* f is $f = L/2$, the focuses of the adjacent lens coincide. Thus, the m-th eigenfunction has the form

$$w_m(x) = \exp\{-(kx^2/2L)\} \ H_m\left(x\sqrt{k/L}\right), \tag{9.20}$$

where functions $H_m(\zeta)$ are *Hermite polynomials*: $H_0 = 1, H_1 = \zeta, \ldots$.

From (9.20), we see that the field on the vertical axis (in the two-dimensional case) decreases according to Gaussian law. We introduce a concept of a zone half-width, in which the field has the same order as in the center [65, 66, 106]:

$$x_s = \sqrt{L/k} \ (2\nu - \nu^2)^{-1/4}, \quad \nu = L/2f. \tag{9.21}$$

Taking Eq. (9.21) into account, the first eigenfunction from Eq. (9.20) has the form:

$$w_1(x) = \exp\left\{-x^2/2x_s^2\right\}. \tag{9.22}$$

The confocal system provides the maximal field concentration on lenses, and therefore, minimal transverse dimensions of the beam, whose width increases with growth of the wave m number in the line.

In more general three-dimensional cases, when the phase corrector is limited by a circumference, eigenfunctions [66] are solutions of the equation of (9.18) type for the confocal line ($\nu = 1, f = L/2$) with infinitely lengthy correctors:

$$w_{mq}(r, \varphi) = (r/r_s)^m \exp\{-r^2/2r_s^2\} L_q^m (r^2/r_s^2) \cos(m\varphi), \tag{9.23}$$

where $L_q^m(r)$ *Laguerre adjoint polynomials* are "replacing" Hermite polynomials in the two-dimensional line. The first of them is $L_0^m = 1, L_1^m = 1 + m - r$. In Program P.9.4 (Appendix B – http://extras.springer.com/2019/978-3-319-90847-2) for calculations of eigenfunctions (9.23), it happens to be simpler to use its definition [29]: $L_q^m(r) = (e^r r^m/q!) d^q/dr^q(e^{-r} r^{m+q})$. Pictures of some eigenfunction distribution (9.23) in the transverse section of the lens line are presented in P.9.4.

In our lecture course we repeatedly mentioned that at radio signal transmission through the transmission line, the single-wave mode is preferable, when the only one operating wave is used. Usually, the lowest wave of the specific transmission line is such a wave. At transmission of power, it is usually useful to apply some combined wave type. Thus, Russian professor E.P. Korshunov and colleagues proved that at power transmission, if instead of the main wave with field intensity (on cross-section) (see P.9.4) we use the wave with some "truncated" Gaussian distribution, we observe that four-times more power acts in the rectenna (rectifying antenna).

The program P.9.5 permits distributions for waves to be constructed with arbitrary indices.

9.3.3 Diaphragm Line

The diaphragm line (Fig. 6.6b) is an interesting example of the quasi-optical transmission line. In it, the phase correction is absent ($\phi \equiv 0$) and the wave in such a line is supported owing to wave diffraction on a hole in the infinite (or finite) shield. Thus, the shield material, in any case, is indifferent within the limits of *physical optics* estimations. In fact, focusing properties of a hole have been known and described a long time ago. Obviously, these focusing properties in the diaphragm line are expressed more poorly than, for instance, in the lens line. For the nonfocal case, the concept of the effective parameter can be introduced

$$c_{\text{eff}} = (ka/L) \ (2\nu - \nu^2)^{1/2}, \nu = L/2f. $$

9.3.4 Completeness of the Eigenfunction System of Quasi-optical Transmission Lines, Three-Dimensional Case

If $w(x, y)$ is a field on some phase corrector (a lens, a diaphragm etc.) and $w_m(x, y)$ is the eigenfunction, then the field $w(x, y)$ can be expanded into fields of eigenwaves:

$$w(x,y) = \sum_m A_m w_m(x,y), \quad A_m = \frac{\int w(x,y)\; w_m(x,y)\; dxdy}{\int w_m^2(x,y)\; dxdy}. \tag{9.24}$$

The latter equation is a consequence of the orthogonality property of the line eigenfunctions:

$$\int_S w_m(x,y)\; w_n(x,y)\; dxdy = 0, \quad m \neq n. \tag{9.25}$$

We considered here (for obvious simplicity) the two-dimensional case. In the three-dimensional case, the result includes the adjoint *Laguerre polynomials* instead of Hermite polynomials.

9.4   Basing Elements of the Quasi-optical Waveguide, Integrated Optics and Light-Guide Engineering

9.4.1   Excitation of Quasi-optical Waveguides

The dielectric prism in the integrated optics devices and in light-guiding engineering happens to be rather suitable constructively and technologically and, chiefly, the effective device from an electrodynamics point of view. The main methods of emission inlet (outlet) into (out of) the optical micro-waveguides are their excitation with the help of *diffraction lattices, dielectric prisms,* and direct *excitation "in the end surface"* [52], the system of some distributed diodes, etc. The ratio between the effective thickness of the micro-waveguide h_{eff} and the wavelength λ affects, mainly, on selection of each of these excitation methods. Under h_{eff} we have in mind the transverse (with respect to propagation direction) dimensions of the light-guide and the surrounding environment, in which all or almost all energy of the propagating wave (or a sum of energies of propagating wave types) is concentrated.

Usually, one can distinguish two specific cases: $h_{eff}/\lambda \gg 1$ and $h_{eff}/\lambda \approx 1$. For $h_{eff}/\lambda \gg 1$, we deal with so-called *single-wave light-guide*, in which the difference between refraction indices of the carrying layer and the surrounding environment is small. For this reason, the transverse dimensions of the carrying layer essentially exceed λ, and therefore, the end-surface of this waveguide represents a large area (aperture) convenient for an energy inlet (outlet). For effective excitation of the necessary wave type in the layer, it is necessary to reconstruct the transverse field distribution of the given wave type on the end-surface with maximal possible accuracy with the help of an excitation device. This requirement is well-known in the theory of radio waveguide excitation in longer wavelength ranges (see Sect. 12.1). The situation of $h_{eff}/\lambda \gg 1$ occurs in the case of small deceleration, when the surface wave field has the larger length than the thickness of the guiding layer. Effectiveness of excitation in this case requires the larger aperture from the exciter.

9.4.2   Excitation Circuits for Dielectric Waveguide

A group of tasks concerning irregularities of the *cross-section jumps*, *break*, *bend*, *disruption*, etc. type joins the task of optical dielectric waveguide excitation to the end-surface. The group of tasks concerning *emission* (excitation) from the *open end* of the *plain or circular waveguides* with transparent walls can be attributed to these problems.

If $h_{eff}/\lambda \approx 1$, excitation in the end-surface becomes ineffective and complicated to implement. The inlet (outlet) method through *the side surface of the light-guide* happens to be the competitive approach compared with excitation in the end-surface. In essence, this is a task (see Sect. 6.10) about power branching in two coupled transmission lines.

In this connection, the circuit of surface wave excitation in the open line *1* through a slot in the metallic waveguide wall *2* (Fig. 9.10a) and the circuit of two film light-guides *1* and *2* coupling through the wide window (Fig. 9.10b) are the simplest examples.

Both circuits (*a*) and (*b*), in principle, are absolutely identical. The coupling portion between transmission lines is defined by the slot size *L*. At surface wave propagation along films at the portion *L*, we have a system of two *coupled film waveguides* (for a structure in Fig. 9.11a we can consider the mirror representation of the film *1* in the ideally conducting lower wall of the waveguide *2* as the second film waveguide) with an air gap between the dielectric layers.

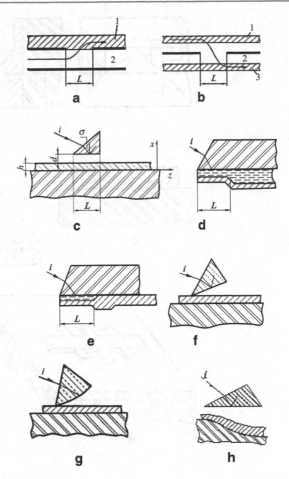

Fig. 9.10 Excitation circuits of dielectric waveguides (light-guides). (**a**) Slot: through the coupling slots between dielectric (*1*) and metallic (*3*) waveguides. (**b**) Through the slot between two light-guides (*1* and *2*). (**c**) Prism-like: with constant gap value. (**d**, **e**) Through a substrate. (**f**) The linear gap. (**g**) The quadratic law of variation of the gap value. (**h**) The more complicated law of variation of the gap value

Beyond the L portion, coupling is eliminated owing to shields. It is clear that there is some transition area ΔL, when coupling is nonzero because of the surface wave diffraction of waveguide waves at the edges of the L slot. Usually, $\Delta L/L \ll 1$. The mutual exchange of wave energy of the one waveguide into other at portion L is referred to as beats. The beats wavelength is defined as:

$$\Lambda = 2\pi/(h_{\text{even}} - h_{\text{odd}}),\qquad(9.26)$$

where h_{even}, h_{odd} are longitudinal wave numbers of even and odd waves of the three-layer symmetric waveguide (in L). Having selected the length L of the interaction portion divisible by integer (more exactly, almost integer) number Λ, we can achieve effective energy transmission from one transmission line into another.

Excitation of the *m-th* surface wave in the film waveguide with the help of the dielectric prism (Fig. 9.11f), in principle, has no difference from the just considered beat process. Really, the monochromic excitation (*i*) wave $\exp\left\{i\,\vec{k}\,\vec{r}\right\}$ should have z-component of the wave vector equal to kn_m^* ($k_z = n^*k$), which exceeds the kn value for the film and the gap (between the prism and the film). The field, which has the form of the surface wave in the gap, i.e., the *i* beam should fall on the lower prism facet under the angle of the complete internal reflection, has the necessary projection k_z. If the refraction index of the prism material $n_{\text{refr}} > n_m^*$, utilization in *i* wave, the suitable polarization and selection of θ angle allow fulfillment of the *phase synchronism* condition of the excitation wave and the *m-th* surface eigenwave of the film. This condition has the form $n_{\text{refr}} \sin\theta = n_m^*$. The

Fig. 9.11 Thin-film elements of integrated optics. (**a**) The refraction beam rotation in the film with varying thickness. (**b**) The thin-film prism of the surface wave. (**c**) Introduction of foreign inserts into the film and the substrate. (**d**) Rotation of the film beam-guide with the "mirror". (**e**) A filter. (**f**) The prism-like wave separator of E and H waves in the planar light-guide. (Taken from Nefyodov and Kliuev [73], p. 320)

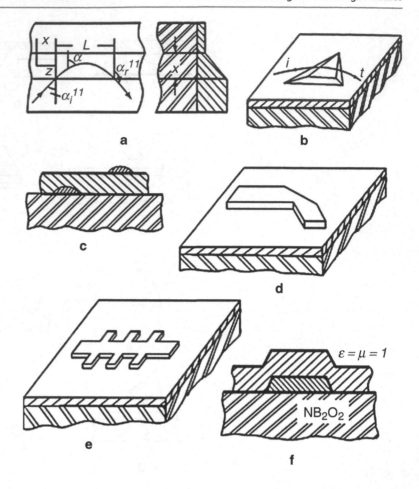

phase synchronism condition is widely used in microwave electronics, when it is necessary to equate the electron speed v_{el} and the phase velocity of the one from the surface (slow) wave v_{ph}, so that $v_{ph} \approx v_{el}$, to realize the effective interaction of the field with the electron flow. In electronics, surface waves are often referred to as *spatial harmonics* (see also Sovetov, and Wood [94, 103]).

The mechanism of energy inlet into the light-guide through its side surface in Fig. 9.11c has the name *tunnel inlet*; sometimes, it is also referred to as the *prism inlet*.

At prism emission inlet into the film, it is also necessary to provide the maximal energy transit of the excitation wave i into the film. Here, we have the same mechanism as in the structures with a slot (Fig. 9.11a, b): it is necessary to provide shielding of wave-guiding film coupling one from another beyond the interaction L portion. We present two variants with the so-called *emission inlet through the substrate* (Fig. 9.11d, e). Here, as the air gap (compare Fig. 9.11c), the dielectric with the small refraction index is used. Emission enters into the film through the beveled edge in the substrate and, thus, in these circuits there is no separate prism, as there was, for instance, in the excitation circuit presented in 9.11c. Such constructions (without the separate prism) are convenient for integral optics. The interaction L portion in Fig. 9.11d on the right is restricted by the sharp increase in the gap value, which makes the re-emission into the prism negligible.

On the contrary, in the circuit in Fig. 9.11e, the interaction L portion on the right is restricted by the sharp increase in the film thickness. This increase is such that the wave from the L portion may pass through the irregular portion, and higher wave types do not excite in the regular waveguide.

In contrast to circuits with sharp variations of properties at the end of the interaction portion (as, for example, in Fig. 9.11d, e), the whole group of energy inlets was developed with the *gap varying according to linear* (Fig. 9.11f), *quadratic* (Fig. 9.11g), or the *more complicated law* (Fig. 9.11h). In any case, we can theoretically achieve 100% of the beam energy inlet into a waveguide by selection of the graded junction construction. Practically, this forms values of the order of 60–80%.

9.4.3 Dielectric Prism in the Waveguide Break, Principle of Mutual Compensation

The dielectric prism (Fig. 6.5a, c), with whose help the breaks in the transmission line on relatively small angles $\vartheta(\vartheta/\pi \ll 1)$ are overcome, is one of the widespread elements of the quasi-optical transmission line. The elementary prism calculation by *methods of geometrical optics* does not give the possibility of estimating its effectiveness in utilization as the rotation basing element, because in this case, we cannot determine the spurious wave amplitudes that inevitably arise when falling on the prism of the one from eigenwaves of the wide waveguide, quasi-optical (for instance, lens) transmission line (Fig. 6.8a), etc. On the other hand, application of electrodynamics approaches, for example, the *method of cross-sections*, is also ineffective because of the large number of eigenwaves that may exist in the wide waveguide. At the same time, the geometric–optical approach allows determination of a connection between the break angle ϑ of the line, the angle 2θ at prism apex, and ε of the prism material. This connection is expressed as follows:

$$\vartheta = 2\arcsin\left(\sqrt{\varepsilon}\sin\theta\right) - 2\theta. \tag{9.27}$$

Let the H_{10} wave of the plain wide ($ka \gg 1$) waveguide ($i \equiv 10$) fall onto the prism (for distinctness). We want to understand the operation principle of the prism in the waveguide break on a small ϑ angle. Hence, in the task statement, the small parameter $\vartheta/\pi \ll 1$ is present. Besides, from Eq. (9.27) another condition follows directly; namely, the small value of ε, or more precisely, an insignificant difference between prism ε and 1: $\varepsilon - 1 \ll 1$. Thus, there are two smallness parameters ϑ and $\varepsilon - 1$ in this task. These peculiarities allow "division" of the initial task into two tasks: the break of the plain waveguide on the small angle ϑ and the prism from material with $\varepsilon - 1 \ll 1$, but located in the rectilinear waveguide.

On the small break, the spurious waves j arise with amplitudes

$$F_{ij}^{(u)} = -Aka\vartheta + O(\vartheta^2), \tag{9.28}$$

where $A = (i8/\pi^2)[j/(j^2 - 1)^2], j = 2m, m = 1, 2, \ldots$.

On the other hand, amplitudes of spurious waves j from the prism in the straight waveguide are expressed with the help of the integral over the prism volume:

$$F_{ij}^{(\varepsilon)} = \left[i(\varepsilon - 1)/2h_j\right]\int_V E_i E_j dV. \tag{9.29}$$

Here h_i is the propagation constant of the wave falling onto the prism, which is $h_i \approx k$ owing to the condition $ka \gg 1$.

The integral in Eq. (9.29) is easily calculated in the so-called *Born approximation*, when the field E_j inside the prism can be changed by the field E_i of the incident wave because of the condition $\varepsilon - 1 \ll 1$. This results in:

$$F_{ij}^{(u)} = -Aka(\varepsilon - 1)\theta + O\left((\varepsilon - 1)^2\right), \tag{9.30}$$

Thus, two considered irregularities – the break of the empty waveguide and the prism in the rectilinear waveguide – result in only even waves of $H_{2j, 0}$ type with the same amplitudes but different signs, i.e., these waves happen to be shifted in phase by π. Hence, they compensate for each other (in higher orders of powers of task smallness parameters ϑ and $\varepsilon - 1$). Thus, it is necessary for this fulfillment of the condition:

$$\vartheta = (\varepsilon - 1)\theta, \tag{9.31}$$

which provides the mutual compensation for spurious waves caused by these two irregularities

$$F_{ij}^{(u)} = -F_{ij}^{(\varepsilon)}. \tag{9.32}$$

The condition (9.32) expresses the main operation principle of quasi-optical devices – the *principle of mutual compensation of spurious waves* (*the Nefyodov effect*) [52].

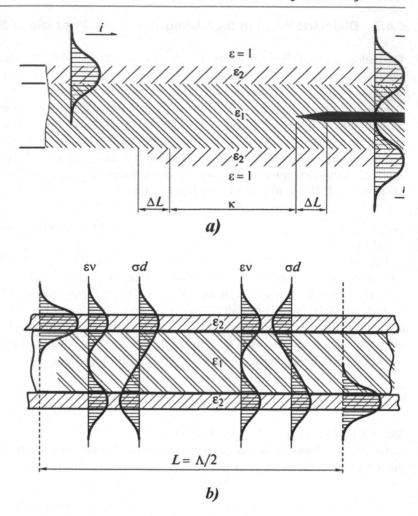

Fig. 9.12 The directional (controllable) coupler. (a) The switcher based on two plain dielectric waveguides, (b) separated by the dielectric layer

9.4.4 Directional Coupler

The circuit of the directional coupler based on two plain dielectric layers (ε_2) separated by the layer with ε_1 permittivity and located in the air ($\varepsilon = 1$) is shown in Fig. 9.12.

We have already mentioned above that waves in this waveguide are divided into even and odd. Obviously, for implementation of the directional coupler, the mode in which there are one even and one odd wave is the most profitable. For this, the condition $k \cdot \Delta b \sqrt{\varepsilon_2 - 1} < \pi$ must be fulfilled.

The operation principle of the directional coupler consists in the following. Let the incoming wave (i) fall on the left end (Fig. 9.12a) of the selected part l. It excites both waves: even and odd (their distribution along the vertical coordinate is shown in the figure). Phase velocities and, therefore, their longitudinal wave numbers, are different: $h_{\text{even}} > h_{\text{odd}}$. Thus, at the distance l, the odd wave (arisen simultaneously with the even one) "overtakes" the even wave in the phase by π. At this moment, the field summation occurs in the lower layer, i.e., the i wave transfers into the t wave of the lower layer. Thus, the transition region l represents a half of a beat wave of even and odd waves in the coupled line: $l = \Lambda/2 = \pi/(h_{\text{even}} - h_{\text{odd}})$. Strictly speaking, determination of wave numbers requires the solution of bulky transcendent equations of (9.6), (9.7) type (see Nefyodov [52]).

In the optical range, the structure usually represents two thin enough films located at a large distance one from another, i.e., the following conditions are fulfilled: $\varepsilon_1 = 1$, $g \cdot \Delta b \ll 1$, $\alpha\, b_1 \gg 1$. Wave numbers are connected by equations: $g^2 = k^2 (\varepsilon_2 - 1) - \alpha^2$, $h^2 = k^2 + \alpha^2$. Now we can obtain approximate expressions for propagation constants:

$$h_{\text{even,odd}} \approx k + \left(\alpha_0^2/2k\right)\left(1 \pm 2\exp\{-2\alpha_0 b_1\}\right). \tag{9.33}$$

Hence, half of the beat wavelength is

$$l = \Delta/2 \approx (\pi k/2\alpha_0)\exp\{2\alpha_0 b_1\}. \tag{9.34}$$

The analysis fulfilled, as the accurate reader, of course, notices, everywhere, to a great extent, is based on the general principles of wave behavior, and on utilization of clear enough physical considerations. Such an approach seems to us to be the most complete and simple.

9.5   Examples of MathCAD Packet Utilization

We have already repeatedly mentioned in our lecture course the wide development and modernization of information transmission systems in quasi-optical, optical, and terahertz ranges of electromagnetic waves during the last decade. This is a huge field of activity for young and older scientific researchers and engineers. There is an enormous number of literature sources from all over the world devoted to this wide field. In our lecture course, several programs (P.9.1, P.9.2, P.9.3, and P.9.4) are dedicated to the examination of the physical content of the main concepts and numerical data for electric and magnetic waves of the main determining class of transmission lines of this range; namely, the planar uniform dielectric layer.

The attached programs P.9.1, P.9.2, P.9.3, and P.9.4 are devoted to the examination of the properties of electric even waves E_{00}, E_{20}, E_{40} and odd waves E_{11}, E_{21}, E_{31} of the simplest models of the dielectric waveguides, namely: the plane uniform dielectric layer, which is a convenient model for consideration of the wide class of transmission lines.

The programs examine the behavior of the main characteristics of the specific wave deceleration and distribution of the electromagnetic field in transverse section of the line depending on the normalized wave number and the permittivity of the layer. The procedure of the finding of the appropriate root of dispersion Eqs. (9.6, 9.7) of the layer is described preliminarily.

9.5.1   Dielectric Layer, Electric Even Waves

The geometry, dimensions, and coordinate system are shown in Fig. 9.13.

After defining the initial data (ε, μ, ka), program P.9.1 forms the dispersion equation, determines the interval for root search, deceleration, and two- and three-dimensional pictures are constructed for electric even waves with checking of the accuracy.

9.5.2   Dielectric Layer, Electric Odd Waves

Program P.9.2 acts in a similar way, but for electric odd waves.

9.5.3   Dielectric Layer, Magnetic Even Waves

Program P.9.3 acts in a similar way, but for magnetic even waves.

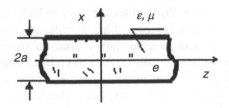

Fig. 9.13 Geometry, dimensions, and coordinate system

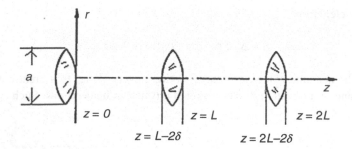

Fig. 9.14 Geometry, dimensions, and coordinate system

9.5.4 Dielectric Layer, Magnetic Odd Waves

Program P.9.4 acts in a similar way, but for magnetic odd waves.

9.5.5 Quasi-optical Dielectric Lens Transmission Line

The lens transmission line is one of the widespread quasi-optical transmission lines. Program P.9.5 is devoted to analysis of characteristics of the main waves of the regular quasi-optical dielectric lens line. It is possible to track pictures of field components on the separate lens (the phase corrector). In some applications, the internal structure of the inherent wave field in the line is of interest. There is such a possibility in Program P.9.5.

The geometry, dimensions, and coordinate system are shown in Fig. 9.14.

Program P.9.5 for quasi-optical conditions ($ka \gg 1, kL \gg 1, L \gg a$), for the given confocal parameter, for the permittivity ε of the lens material, and for the focus distance defines initial parameters and determines the calculation scheme. Then, the Laguerre polynomials are calculated and then the eigenvalues of the lens line. The program constructs the three-dimensional field pictures. After that, the Program calculates the internal structure of the field.

9.6 Summary

In connection with rapid assimilation of the millimeter-wave range and frequencies of the light spectrum part, the dielectric waveguides and lightguides have attracted great interest. The plane dielectric waveguide is one of the attractive models of the simplest line, which supports the surface waves. Sect. 9.1 is devoted to its analysis. We describe the numerical investigation of both polarizations and show that a solution is represented by the infinite series of eigenwaves. All of them have a cut-off frequency except for the E_{00} wave (the Katsenelenbaum effect). The appropriate field distribution pictures in the transverse section of the line are presented.

The class of leaky waves is described, which allows essential extension of our information about the physics of dielectric guiding structures. In particular, these waves determine the physics of the dielectric rod antenna, the Katsenelenbaum directional coupler, the plasma exciter of the lightguider, and a series of other elements of the waveguide section.

The optical waveguides and dielectric integrated structures form a large part of modern engineering and physics. The future millimeter-wave and terahertz-wave ranges depend on them. We have considered the various types of lines in this chapter. Among them are the tubular dielectric waveguide (lightguide), the Goubau line, etc.

The gap between millimeter-wave and optical ranges is filled by radio physics and radio engineering of quasi-optical waveguides and their elements. Here, we met with the rather rare boundary of waveguide engineering and quasi-optics itself. Transmission lines and devices of quasi-optics are normally taken from the usual optical considerations (the geometric optics), but we include amendments according to the presence of the finite sizes of the obstacle. Obtaining eigenwaves is usually reduced to the homogeneous Fredholm integral equation. Apparently, Mandelstam used this approach for the first time. The principle of mutual compensation (the Nefyodov principle) was suggested to explain the operational essence of quasi-optical devices.

This ideology was extended in Weinstein's publications (Chap. 11). Strict solutions of crucial problems and the effect of wave reflection from the open end at the approach of the incident wave's frequency to the critical one (the Weinstein effect) form the appropriate basis.

The problem of the eigenfunction system completeness of the quasi-optical transmission lines is given special consideration (Sect. 9.3.4).

The excitation schemes of dielectric waveguides are shown. The essence of the beating phenomenon and its influence on the degree of wave energy transmission from one waveguide to another one is explained (the phase synchronism condition).

The material in this chapter is comparatively new and has a great future, to which young researchers can make a great contribution.

Checking Questions

1. Why are field pictures of the tube waveguide (Fig. 6.11) similar to appropriate field pictures in the circular waveguide (see Sect. 6.9; Figs. 6.20 and 6.21)?

2. Over the plain dielectric layer from polystyrene with a 12 mm-thickness laying on the ideally conducting surface, the surface wave propagates, and has a wavelength of 24 mm. At which distance from the dielectric surface will the magnetic strength amplitude decrease by e^2 times (see, for instance, Nefyodov and Fialkovsky [58], p.313)?

3. How much are the critical wavelength of the lowest electric E_{10} and magnetic H_{10} waves in the electrodynamics structure of the previous task?

4. In the lens line, lens dimensions are $ka = 10\,\pi$ and the distance between them is $kL = 100\,\pi$. How much is the beam width and which distribution of its intensity is in the plane located in the middle between adjacent lenses?

5. For the plane electromagnetic wave of $1 \cdot \exp\{i\,(kz - \omega tt)\}$, the Maxwell Sect. (1.3.3) are transferred into two following equations: $[\vec{k}\,\vec{E}] = (\omega/c)\mu\,\vec{H}$ and $[\vec{k}\,\vec{H}] = -(\omega/c)\varepsilon\,\vec{E}$. From these equations, we directly see that if $\varepsilon > 0$, $\mu > 0$, vectors $\vec{E}$, $\vec{H}$, $\vec{k}$ form the *right triplet*. What will the triplet be if $\varepsilon < 0$, $\mu < 0$?

6. In Fig. 5.12, the geometric–optical picture of plane electromagnetic wave (i) on the plane boundary ($z = 0$) of air-some medium ($z > 0$), which is characterized by ε, μ permeability, is presented. Show that the refracted beam t^{refr} actually corresponds to the case $\varepsilon > 0$, $\mu > 0$, and the beam $t^{\text{inc}} - \varepsilon < 0$, $\mu < 0$.

7. What will be with the reflected beam r if the beam i comes from the medium, in which $\varepsilon_1 > 0$, $\mu_1 > 0$, into the medium with $\varepsilon_2 = -\varepsilon_1$, $\mu_2 = -\mu_1$? Compare this case with the case of the absence of full reflection for waves.

8. What is the sense of the principle of mutual compensation of spurious waves?

9. Using Eq. (9.34), calculate the length of the directional coupler – light switcher, for instance, at the red light for $\lambda = 0.63$ μm, $\Delta b = 0.3$ μm, $b_1 = 0.9$ μm, $\varepsilon_2 = 1.2$.

10. Using Programs P.9.1, P.9.2, P.9.3, and P.9.4, calculate two- and three-dimensional pictures of electrical and magnetic even and odd waves for $\varepsilon = 4$; 6, $\mu = 0.8$; 1.2, $ka = 3$; 7. Explain the results.

11. Using Program P.9.5, the calculated lens line for $L_z = 8$; 12, $\lambda = 0$, 8; 1, 3. Explain the results.

Volumetric Closed Resonators

The electromagnetic oscillation occurring in finite guiding structures and resonators is studied in this chapter.

Again, we return (see Sect. 2.1.9) to the important concept to electrodynamics of finite structure, i.e., the structure that is restricted in all three coordinates (Sect. 10.1). For instance, the half-plane, the open end of the waveguide, etc. are concerned with the class of crucial or semi-infinite structures. Having restricted such a system in length, we obtain a structure of finite length, to which we can regard a band (a strip), the open resonator etc. A great interest in the abstract, the semi-infinite, crucial structure is caused, first of all, by the fact than many crucial structures permit a strict analytical solution.

Construction of the solution for the finite structure based on this strict solution for the crucial structure is rather simple. For transition from the solution of the crucial problem to the "actual" solution (finite), there are several known approaches, which allow the effective approximate (for instance, asymptotic $2kl \gg 1$) solution to be obtained. The Haskind–Weinstein method and the method of successive diffractions are such approaches.

The typical physical phenomenon in the finite strictures is (possible but not obligatory) the resonance state. A wide class of problems regarding the properties of resonant structures is based on consideration of the finite segment of the regular (or irregular) transmission line.

Section 10.2 contains a discussion of the general properties of cavity resonators. At first, we consider the closed resonance systems, the simplest of which is the conventional resonant single circuit (Sect. 10.2.1). The resonance frequency (the only frequency!) of this single circuit is defined by the known Thomson formula. It is easy to track the gradual transition from the single circuit with lumped elements to the closed or open cavity resonator with distributed parameters and, in the end, to the elementary radiator – the Hertz dipole.

At present, the so-called open resonators form the special class of resonators (Chap. 11). The idea of open resonator application in the quantum generators, which in the end led to quantum electronics, was suggested by A. Prokhorov. N. Basov also actively participated in this research. American physicist Ch. Townes worked independently. The dielectric resonators are connected to this class of open resonators.

Section 10.2 is devoted to discussion of general properties of cavity resonators. Usually, the close resonator based on the regular transmission line is built on restriction of this line segment with the help of metallic planes (partitions). However, we may ensure the resonance in the restricted segment when the role of these actual partitions is played by so-called critical sections. They represent some of the imaginary planes from which the eigenwave is almost fully reflected. This occurs when the wave propagation constant goes back to zero (fully or almost). This is called the Weinstein effect.

The main characteristics of the resonator are: the oscillations type, the electromagnetic field structure, the fundamental frequency, and the Q-factor of operation oscillation (Sect. 10.2.2). As a whole, the significant resonator characteristic is the spectrum of degrees of rarefaction of its natural frequencies one after another.

The general theory of resonators is discussed in Sect. 10.3. The discussion begins from the simple case, namely, we analyze the resonator as the finite transmission line segment (Sect. 10.3.2). The simplest oscillation type consists of superposition of direct and backward waves of the same type. The resonant length of the resonator is defined as the integer number of half-wavelengths in the line. Thus, the fundamental frequency of this type of oscillation is obtained. In the simplest case, the resonator is built on the transverse *TEM* wave (Sect. 10.3.3). Then, using eigenfunctions of appropriate waveguides (Sects. 6.4, 6.5, 6.6, and 6.7), we construct eigenfunctions of the appropriate oscillation of resonators. In the presence of losses, the fundamental frequencies can be conveniently considered complex.

© Springer International Publishing AG, part of Springer Nature 2019
E. I. Nefyodov, S. M. Smolskiy, *Electromagnetic Fields and Waves*, Textbooks in Telecommunication Engineering,
https://doi.org/10.1007/978-3-319-90847-2_10

As examples, we examine the rectangular, cylindrical, and spherical resonators (Sect. 10.4). For each type of resonator, we construct eigenfunctions and resonance frequencies of the main types of oscillations. Field structures of these oscillations are presented. The quasi-stationary resonators are separately considered, which are used, for example, in electronic microwave devices.

An analysis of most types of resonators is conducted by the method of partial regions, which is already known by us (Sects. 6.3, 6.4, 6.5, 6.6, and 6.7). The case is specifically studied, when we use the integral equation for a solution joining the boundary of partial regions (Sect. 10.5.3). The approximate solution of the integral equation can be obtained using various numerical methods, in particular, the Bubnov–Galerkin method.

In the whole series of cases, the perturbations method is sufficient for resonator calculation (Sect. 10.6). As the basis for the perturbation method application for resonators of the "wrong" shape, it is often enough to know a solution for the regular shape resonator.

This chapter finishes by showing the MathCAD packet application of electromagnetic oscillation analysis of rectangular (Program P.10.1) and cylindrical (Programs P.10.2 and P.10.3) closed resonators.

10.1 Finite Structures in Electrodynamics

10.1.1 Definition of Finite Structure

A concept of the finite (or key) structure is widely used in classical electrodynamics [71, 99, 111, 125]. The finite structures presented, for instance, in Fig. 2.3, are the usual circuits of the key (or semi-infinite) structures.

The half-plane (Fig. 2.3a; on the right), the open end of a waveguide (Fig. 2.3b), the open end of the planar waveguide with dielectric filling (Fig. 2.3c, d) are key structures. The finite structures – a band (a strip), an open resonator or the system of diffraction-coupled open resonators (Fig. 2.3a–c), an asymmetric strip- or slot-lines (Fig. 2.3c, d) – correspond to key structures. Of course, the possible variants of finite and key structures, that we can meet in practice, are not exhausted by these examples [71, 99, 124].

However deep we can penetrate into electrodynamics, it is always useful to remember mechanics, in which resonant and oscillation systems do not occupy the last place.

A great interest in seemingly abstract semi-infinite key structures (Fig. 2.3, on the right) is caused, first of all, by the fact that many key structures allow a *strict solution* [57, 98, 111]. Expansion of the strict solution results for the key structure on the finite structure does not cause the specific principal difficulties, although sometimes it happens to be excessively complicated in "arithmetic." Appropriate methods have been developed comprehensively enough in Russian and world-wide literature.

In the very general view, a transition procedure from the key structure to the finite one is presented, say, in the nonstationary case, as follows. Let the nonstationary wave in the form of the single Heaviside step [29]) be: $1(t) = 0$ at $t < 0$, but $1(t) = 1$ at $t > 0$ and it falls under some angle on the band located in a free space (Fig. 2.3a; on the left).

At the time moment $t = 0$, the wave achieves the left edge of the band and experiences diffraction on this edge. Before the time moment $t = 2l/c$, the diffracted field does not "feel" an influence of the second edge of the band, and the scattering picture is *exactly described* in terms of the semi-infinite structure (Fig. 2.3a, on the right). At $t = 2l/c$, the diffraction field achieves the second edge and the *second* scattering of the wave occurs, which was diffracted at the left edge, and the *primary* scattering of the incident field on the second edge occurs. Then the picture repeats. The *nonstationary* case of diffraction is typical because of the finite number of diffracted waves for each time moment of observation.

The stationary case causes another picture. Here, the diffraction picture should be considered right away at full volume: we should add up to a whole chain of]*edging waves*. Such a task is rather complicated and it can hardly be solved effectively within the limits of modern methods of mathematical physics. This is connected, in particular, with the fact that Mathieu functions, by which the *accurate expansion* of the solution is possible (the Fourier method – variable separation; see Sect. 6.3), have no satisfying integral representation.

In this connection, strict methods of solution for finite structures, which permit the *effective approximate* (asymptotic $2kl \gg 1$) *solution* to be obtained, represent a special interest. In our opinion, the *Haskind–Weinstein method* and the *method of sequential diffractions* [57, 111, 126] are such suitable methods.

The typical physical phenomenon in finite structures is (possible but not necessarily) the resonance condition. It is known that the resonance condition was described for the first time by Galileo Galilei in 1602 in a publication devoted to investigation of pendulums and musical strings.

A wide class of tasks about the properties of resonance structures is based on consideration of the finite segment of the regular (or irregular) transmission line (Fig. 10.1). Thus, it is clear that the transmission line itself can be of any type, and its restriction in length (in the z coordinate) is provided with the help of vertical electric (Fig. 10.1a) or magnetic (Fig. 10.1b) walls. In other words, in the selected segment ($2l$ in length) of the transmission line, in essence, the mode of standing wave is established owing to superposition of the line's eigenwave reflected sequentially from the first and the second walls ($z = 0$ and $z = 2l$).

Therefore, if, say, some wave of the i-th number of the regular line, for example, H_{in}, was a base, then the oscillation of the resonator obtained is referred to as H_{inq}, $q = 1, 2, \ldots$.

The wave can be reflected not only from the real walls, but also from imaginary walls, as in Fig. 10.1e, when the critical sections for the given wave type serve as such "reflectors." Thus, the longitudinal wave number of this wave in the critical section is $h = 0$ and the reflection coefficient model $|R_{i, -i}| \rightarrow 0$, i.e., the wave, is almost completely reflected from the critical section (evidently, with accuracy up to transformation of the wave number i into waves of other numbers and types due to line irregularity along the z-axis).

In future, we shall see that critical sections may be located not only in the convergent waveguide part (as in Fig. 10.1d), but also in its expanding part (Fig. 10.1c–f). This effect allows construction of open resonators with the uniquely rare spectrum of natural oscillations (see Nefyodov [23]).

10.2 General Properties of the Cavity Resonators

10.2.1 Definition, General Properties, Evolution of Resonance Structures, Closed and Open Resonators

In electrical engineering lectures for university students, the resonance systems are studied in the form of series and parallel *oscillating circuits* (Fig. 10.2). These systems are concerned with structures with *lumped parameters* (L, C, R).

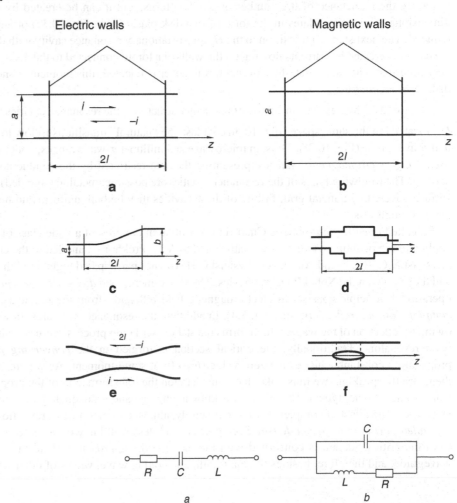

Fig. 10.1 Formation of a resonator from the finite segment of (**a**, **b**) regular and (**c–f**) irregular transmission lines

Fig. 10.2 (**a**) Series and (**b**) parallel oscillating circuits of the lumped L, C, and R elements

Fig. 10.3 The gradual transition from the parallel oscillating circuit with (**a**) lumped elements to (**d**) the open resonance cavity with distributed parameters and (**e**) to the Hertz dipole through three intermediate stages: (**b**) the circuit from a capacitor + distributed half-turns of an inductance and (**c**) close resonance cavity, (**d**) the open disk resonator, (**e**) the Hertz dipole, (**f**) the Popov pin radiator

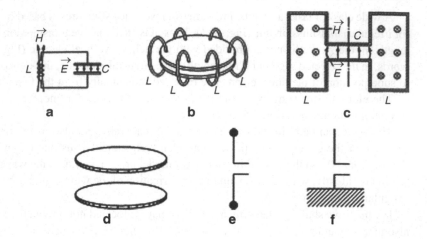

Strictly speaking, any lumped element, for example, the inductance L, during frequency growth, obtains parameters of other elements. For instance, in the inductance circuit (a coil), the capacitive couplings appear between windings.

If to add the natural *ohmic losses* and *losses for emission*, this circuit reflects the inductance properties at high frequencies. The resonance frequency ω_{res} (the only frequency!) for the oscillating circuit is defined by the known *Thomson formula*: $\omega_0 = 1/\sqrt{LC}$.

The transition from the conventional parallel oscillating circuit with lumped parameters L, C, R (Fig. 10.3a) to the disk-shaped *open resonance cavity* (Fig. 10.3d) during growth of the operating frequency f is shown in Fig. 10.3. In the oscillating circuit (Fig. 10.3a) the electric field $\vec{E}$ of a capacitor is mainly concentrated in the capacitor, whereas the magnetic field $\vec{B}$ is concentrated in the inductance, and they are supposedly "uncoupled" in a space. During frequency growth, the necessity arises to reduce the inductance of the number of winding turns, and it can be created by wide half-turns (Fig. 10.3b). Thus, the dimensions of capacitor plates in the form of two disk plates (as in Fig. 10.3c) or any other (rectangular, elliptical, etc.) are reduced. The next step is a transition to the H- quasi-stationary resonance cavity with distributed parameters: the inductance in the cavity is created by currents flowing on the walls of a toroid connected to the disk-shaped capacitor (Fig. 10.3c). Finally, at very high frequencies, at first, the external toroid cover is rejected (this moment is shown in Fig. 10.3d), and we come to the disk *open resonator*.

At the end of this "evolution" chain of resonance structures, the resonance cavities have the united electromagnetic field $\vec{E}$, $\vec{H}$ "located" in the same space (Fig. 10.3d). At last, the "natural" transitions follow to the Hertz dipole (Fig. 10.3e) and to the half-wave dipole (Fig. 10.3f). Thus, in microwave and millimeter-wave ranges, under the resonance cavity, we understand the *oscillating electromagnetic system* representing the area restricted by the conducting or other (fully or partially reflecting) surface. The following types of the resonance cavities are possible: metallic (shielded), dielectric, and resonators with the side surface covered by a metal grid. Filling of these cavities may be both uniform and non-uniform, with different isotropic and anisotropic inserts.

So-called *open resonators* (see Chap. 11) constitute in the present a wide class of resonators. The idea of open resonator application in quantum oscillators was attributed to A.M. Prokhorov, and led to the creation of quantum electronics. Russian physicist N.G. Basov and American physicist C.H. Townes participated together with A.M. Prokhorov in these investigations and they had won the Nobel Prize in physics. The above-mentioned *dielectric resonators* can be included in this group, whose operational principle consists in electromagnetic field reflection from the boundary of two dielectric media (the effect of *complete internal reflection*; see Sect. 5.4). In addition, the resonance structures, in which a field inside the cavity is retained owing to reflection of the wave, which forms the stable oscillating process, from so-called *critical sections*, can be attributed to open resonators. Traditionally, the critical section is formed in the *converging part of the waveguide*, and the wave propagation constant achieves the zero value ($h \to 0$) at some moment. As we are speaking about an irregular waveguide, then, strictly speaking, we must take into consideration the transformation of the propagating wave into waves of other types (for instance, higher types). This effect was known long ago and is adequately described in the literature, including students' textbooks. The effect of complete (or more precisely, almost complete) wave reflection from the critical section located in the *expanded part of the waveguide* (see Sect. 6.9) is much less well known. Specification about the *almost complete reflection* from the critical section is concerned with the fact that the approach itself to this section acts in the irregular part of the waveguide, and therefore, besides the main falling operating wave, waves of other types and number occur. They should be taken into account; however, this irregularity is usually weak, and transformation into other types of waves is small, for example, if α is the angle of generatrix slope to the z-axis, then the transformation constitutes the value of α^2 order, i.e., $\alpha^2 \ll 1$.

On the basis of this effect, in Russia (*for the first time in the world*), open resonators *with a uniform spectrum of natural frequencies*, such as for a string, were offered and developed. This allowed development of a wide range of resonators and electronic devices based on these resonators with the uniquely rare spectrum of fundamental frequencies, in other words, with a huge *spectrum of rarefaction*, which is impossible in the *open two-dimensional resonators*, and especially, in *closed resonators*.

In microwave engineering, cavities are used as *oscillating systems* of the electronic filters for various purposes, *measuring instruments*, etc. The wide application of resonance structures takes place in *volumetric integrated circuits (or 3D-IC)* in *microwave and millimeter-wave* ranges, in *antenna engineering,* etc.

When frequencies of exciting oscillations in the cavity vary, the very different *structures* of the electromagnetic fields may be formed. In general, an infinite number of resonance oscillations is possible in the cavity.

The main *type* of oscillations has a minimal resonance frequency. If resonance frequencies of two and more oscillations with various field structures coincide, such oscillations are referred to as *degenerated*. In this case, within the limits of excitation of this type of oscillation, the equivalent circuit represents the conventional *resonance circuit* characterized by the equivalent active resistance, the capacitance, and the inductance (Fig. 10.3a).

10.2.2 Main Characteristics of the Resonator: Oscillation Type, Electromagnetic Field Structure, and Fundamental Frequency

At selection of this or that type of resonance cavity, the following electrical characteristics and parameters are of critical importance. These characteristics are defined from a solution of appropriate boundary electrodynamics tasks concerning fundamental oscillations of the resonator. Classification of oscillation types is similar to classification of transmission lines: E-(electric E_{mnp}) and H-oscillations, LE- and LM-oscillations, and *hybrid* oscillations. The frequency at which the oscillation amplitude achieves a maximum, is referred to as the *resonance frequency* f_{res}.

In general, the *resonator Q-factor* (Q) is defined by the Sovetov formula [94]:

$$Q = -2\pi / \ln\left(1 - W_{dis}/W_{acc}\right), \tag{10.1}$$

where W_{acc} is the energy accumulated in the resonator; W_{dis} is the energy dissipated over the T period on the resonance frequencies.

In the case of small losses over the oscillation period T: $W_{dis}/W_{acc} \ll 1$, the Eq. 10.1 transforms into a more widespread convenient equation:

$$Q = \omega W_{acc}/P_{los}, \tag{10.1a}$$

where P_{los} is the loss power.

From a physical point of view, the definition of Q through energy and frequency is independent (invariant) of the equivalent circuit used.

The energy dissipated during the T period can be presented as

$$W_{dis} = W_{dis}^{(res)} + W_{dis}^{(ext)}, \tag{10.2}$$

where $W_{dis}^{(res)}$ is energy dissipated in the resonator during the period on the resonance frequency (in indices *nom* $\equiv$ *n*); $W_{dis}^{(ext)}$ is energy dissipated in the external circuit during the period on the resonance frequency.

Accordingly, the following kinds of resonator Q-factors are distinguished:

$$Q_0 = 2\pi W_{acc}/W_{dis}^{(res)} \tag{10.3}$$

is the *unloaded (proper) Q*-factor of the resonator,

$$Q_{\text{ext}} = 2\pi\, W_{\text{acc}} / W_{\text{dis}}^{(\text{ext})} \tag{10.4}$$

is the *external Q*-factor of the resonator,

$$Q_{\text{load}} = 2\pi\, W_{\text{acc}} / \left(W_{\text{dis}}^{(\text{res})} + W_{\text{dis}}^{(\text{ext})} \right) \tag{10.5}$$

is the *loaded Q*-factor of the resonator.

In open resonance structures, we use the *emitted Q*-factor Q_{emit}, i.e., connected to the emission of the energy part of the operation oscillation type from the resonator cavity (see Chap. 11).

There is a simple connection between resonator Q-factors, such as, by the way, the waveguide:

$$1/Q_{\text{load}} = 1/Q_0 + 1/Q_{\text{ext}}. \tag{10.6}$$

The *unloaded (proper) Q*-factor Q_0 is connected with the *bandwidth* $2\Delta\omega$ of the resonator as:

$$Q_0 = \omega_0 / 2\,\Delta\omega. \tag{10.7}$$

The coupling coefficient (K_{coupl}) is the ratio of the power transmitted by the resonator into the external circuit (P_{load}) to the power lost in the resonator on the resonance frequency $\left(P_{\text{dis0}}^{(\text{res})} \right)$:

$$K_{\text{coupl}} = P_{\text{load}} / P_{\text{load0}}^{(\text{res})}. \tag{10.8}$$

10.2.3 Spectrum of Rarefaction of Fundamental Frequency

In a series of other resonator characteristics, the *spectrum of rarefaction* (or more precisely, the degree of the spectrum of rarefaction) is its most important feature. Sometimes, it can be changed to a concept of the frequency interval defining the latter as a distance between the resonance frequency of operating oscillation and the nearest, most intensive non-operating type of oscillation. However, a concept of a *spectrum of rarefaction of fundamental frequencies* is more general. Let us examine briefly the merits of the case.

As we see from Eq. (10.1), the *resonator Q-factor* directly depends on energy *accumulated* by the resonator. Obviously, the accumulated energy W_{acc} is defined by its volume V. Let us introduce a concept of the *spectrum of rarefaction of fundamental frequencies* as the number of resonator oscillations ΔN, which occupy the frequency interval $\Delta\omega$. Then, in accordance with the Courant theorem (or the *Rayleigh–Jeans formula*; see Sect. 5.10), the *spectrum of rarefaction of fundamental frequencies* of the *closed* resonance cavity (three-dimensional) is determined as:

$$N_{\text{raref}} = \Delta N / \Delta\omega = \left(\omega^2 / 2\pi^2 c^2 \right)\, V. \tag{10.9}$$

Here, ω is the radian frequency of the oscillations, c is the light speed.

From Eq. (10.9) we see that in the case of a closed resonator, the *spectrum of rarefaction of fundamental frequencies* is directly proportional to a cube of some linear size L: $N_{\text{raref}} \approx L^3$.

The *Rayleigh–Jeans formula* has an *asymptotic character*, i.e., its accuracy increases with the frequency ω growth, and a specific form depends on the surface shape of the S area restricting the V volume. Most simply, Eq. (10.9) looks for volumes of the "*correct*" shape (a rectangular parallelepiped, the finite segment of the cylindrical tube, a sphere, etc.; Fig. 10.4).

The Courant formula defines the spectrum of rarefaction for volumes of an *arbitrary shape* and because of this, contains some (correction) terms in addition to Eq. (10.9) [23, 126].

The spectrum of rarefaction of fundamental frequencies for two-dimensional *open resonators* is determined as

Fig. 10.4 Closed resonance cavities. (**a**) Rectangular. (**b**) Cylindrical. (**c**) Coaxial. (**d**) Microstrip (almost closed). (Taken from Nefyodov and Kliuev [73], p. 331)

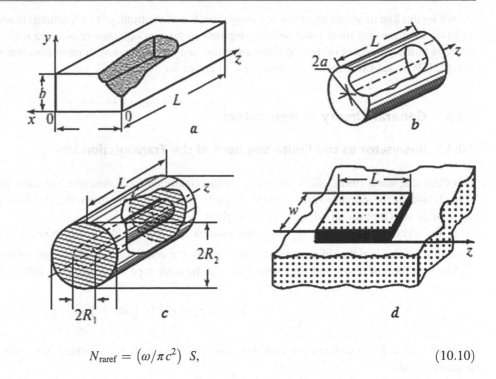

$$N_{\text{raref}} = (\omega/\pi c^2) \, S, \qquad (10.10)$$

where S designates the area of the two-dimensional region occupied by the resonator's electromagnetic field. Thus, the *spectrum of rarefaction* in the two-dimensional cavity is proportional to the *square* of the linear size $N_{\text{raref}} \approx L^2$ and the spectrum of *open cavities* is significantly rarer than that for closed ones (compare with Eq. (10.9)).

We note that precisely owing to a transition to *open cavities* (offered by A.M. Prokhorov and independently by Ch. Townes) the development of *lasers, masers,* and many other devices of *quantum electronics* became possible.

At last, for a one-dimensional region with the length interval L the *spectrum of rarefaction of fundamental frequencies* has the form:

$$N_{\text{raref}} = (2/\pi c) \, L. \qquad (10.11)$$

Thus, $N_{\text{raref}} \sim L$. It is known that the *one-dimensional structure*, for example, the well-known *string*, has the rarest spectrum of fundamental frequencies. In essence, the *spectrum concentration* for a string is *absent* and it has a *uniform (equidistant) spectrum*.

In due course, the evident question arose: is it possible to create open resonance cavities in which the *spectrum of fundamental frequencies would* be determined, as in Eq. (10.11) instead of, say, Eq. (10.10)? As it happens, such oscillating structures with the string spectrum (10.11) may be developed and they were suggested in Russia nearly 40 years ago.[1] They had acquired the name *open resonators with nonfocusing mirrors*. A large number of *passive* and *active* instruments and microwave, millimeter-wave, and optical-range devices were based on this class of resonance structures. Among them: *cavities, wave-meters, lasers, klystrons, magnetrons, orothrons* (also referred as oscillators of diffraction emission), and many others. Evidently, a great future is expected for this direction in science.

Initially, for implementation of this class of cavities properties of *near-surface waves* (see Sect. 6.9) of the *coaxial waveguide* were applied. Later, it was proved that *near-surface waves* (*oscillations*) take place in a whole series of other guiding systems, in particular, in the *strip-slot lines* [75, 126, 127].

Estimation results of proper Q-factors allow predetermination of a large range of Q-factors, which are peculiar to different oscillating systems. Of course, these data should be supplemented by an analysis of specific systems, taking into consideration their geometric–physical parameters.

---

[1] This was research by E.I. Nefyodov and I.M. Rossiyskiy. It is remarkable that at that time I. Rossiyskiy was a student in the 3rd academic year of the Bachelor cycle.

We would like to attract attention of young (and, to tell the truth, not only young) readers to the circumstance that the first stage of the discovery (in Russia) and development of the class of *open resonators with nonfocusing mirrors* was a detailed analysis of the properties of the *coaxial waveguide*, i.e., the structure with properties that were seemingly studied a very time ago [23]. Who says that there is nothing new to invent for bicycles?

10.3   General Theory of Resonators

10.3.1 Resonator as the Finite Segment of the Transmission Line

It is clear that some (finite) segment of any longitudinally uniform structure truncated by two transverse conduction planes (Fig. 2.3a–d) is the simplest *resonator*. All further considerations are true with regard to any truncated segment of the regular waveguide structure (of an arbitrary cross-section).

In the truncated area (Fig. 10.1a–c), the existence of such fields is only possible, which, in addition to the boundary conditions of the initial guiding structure, also satisfy the condition $\vec{E}_\tau = 0$ (or any other) on the "transverse" planes.

Superposition of direct and backward waves of the same type is given by the following equation:

$$\vec{E}_\tau(x,y,z) = \left(Ae^{i\gamma z} + Be^{-i\gamma z}\right)\vec{e}_\tau(x,y),$$

(10.12)

in which A and B are some of the complex constants. To satisfy the boundary conditions $\vec{E}_\tau = 0$ at $z = 0$ and $z = L$, it is necessary that

$$\vec{E}_\tau = E_0 \vec{e}_z \sin\left(l\pi z/L\right); \qquad l = 0,1,2,\ldots\infty,$$

$$E_0 = -i2A, \qquad \gamma = l\pi/L.$$

(10.13)

As $\gamma = 2\pi/\lambda$, then

$$L = \lambda/2.$$

(10.14)

The condition (10.14) means that the length of the guiding structure segment L should be a multiple of the *half-wavelength* of some type of regular waveguide, from which the given type of resonator oscillations is formed (see also the resonance conditions (6.7), (6.8)).

Because for the uniformly filled cavity with parameters ε and μ, the value of $\gamma^2 = (\omega/c)^2 \varepsilon\mu - g^2$, where g is the transverse wave number, then *fundamental frequencies* of the resonator are determined as

$$\omega = \left(c/\sqrt{\varepsilon\mu}\right)\sqrt{g^2 + (l\pi/L)^2}.$$

(10.15)

Equation 10.15 determines in an explicit form the *resonator's fundamental frequencies*. In general, there is an infinite set of such frequencies in the cavity. Fundamental frequencies for real ε and μ form the sequence of real values $0 < \omega_1 \leq \omega_2 \leq \ldots \leq \omega_l \leq \ldots < \infty$.

10.3.2 Resonator on the Transverse Wave

In the case of *T-oscillations*, $g = 0$ and eigenvalues depend on the longitudinal size L only:

$$\omega_l = \left(c/\sqrt{\varepsilon\mu}\right)\ (l\pi/L), l \neq 0.$$

(10.16)

As in the case $l = 0$, the resonator fundamental frequency is defined as

$$\omega = \left(c/\sqrt{\varepsilon\mu}\right) g. \tag{10.17}$$

10.3.3 Boundary Problems for an "Arbitrary" Resonator

In general, for an arbitrary hollow cavity with the uniform isotropic medium, one of the following tasks can be formulated as:

$$\nabla^2 \vec{E} + (\omega/c)^2 \varepsilon\mu \vec{E} = 0 \text{ in } V, \vec{E}_\tau = 0 \text{ on } S \text{ or} \tag{10.18}$$

$$\nabla^2 \vec{H} + (c/\omega)^2 \varepsilon\mu \vec{H} = 0 \text{ in } V, \left(\text{rot } \vec{H}\right)_\tau = 0 \text{ on } S, \tag{10.19}$$

where V is the cavity volume and S is the boundary surface.

If the hollow cavity is concerned with the class shown in Fig. 10.5, it is suitable to project field vectors on the z-axis. This leads to *scalar tasks* for the E_z, H_z components.

10.3.4 Calculation of Losses, Q-Factor, Concept of the "Complex" Frequency

When analyzing and designing cavities, it is necessary to take into account *losses* caused by absorption into dielectric and metallic elements, and also by emission into external space. Thus, the total energy of the electromagnetic field W in the cavity can be determined from the following ordinary differential equation:

$$dW/dt + (\omega/Q_0) W = 0, \tag{10.20}$$

where Q_0 is the proper (unloaded) Q-factor. Its solution

$$W(t) = W(0)\exp\left\{-(\omega/Q_0)t\right\} \tag{10.21}$$

Fig. 10.5 Structure of the oscillation field of (**a**) H_{110}, (**b**) E_{110}, (**c**) H_{101}, (**d**) H_{011} in the rectangular resonator. (Taken from Nefyodov and Kliuev [73], p. 338)

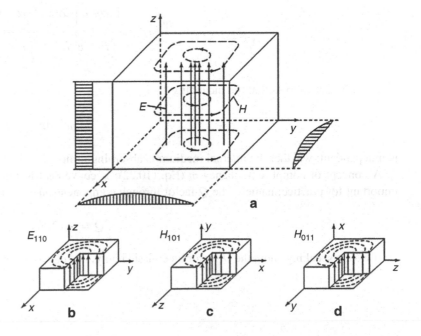

shows that the energy reserve of proper oscillations exponentially decreases with time. Because energy is connected with the field by quadratic law, amplitudes of $\vec{E}$ and $\vec{H}$ field components vary according to the law: $\exp\{-(\omega/2Q_0)t\}$.

If losses are present, the resonator *fundamental frequencies* become *complex*:

$$\omega = \omega' + i\omega''. \tag{10.22}$$

In the case of the hollow cavity with ideally conducting cover, which is filled by the absorbing medium, we obtain:

$$\omega = \left(c/\sqrt{|\varepsilon||\mu|}\right)\sqrt{g^2 + (l\pi/L)^2}\exp\{i(\Delta + \Delta^{\text{met}})\}, \tag{10.23}$$

where $\tan\Delta = \varepsilon''/\varepsilon'$, $\tan\Delta^{\text{met}} = \mu''/\mu'$, and index met designates losses in the metal.

At calculation of the Q-factor Q_0 of resonators, different factors defining losses can be taken into account separately. Let losses in the resonator be

$$P_{\text{loss}} = P_{\text{loss}}^{\text{diel}} + P_{\text{loss}}^{\text{met}} + P_{\text{loss}}^{\Sigma}, \tag{10.24}$$

where $P_{\text{loss}}^{\text{diel}}$ are losses in dielectric, $P_{\text{loss}}^{\text{met}}$ are losses in metal, P_{loss}^{Σ} are losses in emission. Then

$$1/Q_0 = 1/Q_{\text{diel}} + 1/Q_{\text{met}} + 1/Q_{\Sigma}, \tag{10.25}$$

where the partial Q-factors are

$$Q_{\text{diel}} = \omega W/P_{\text{loss}}^{\text{diel}}, Q_{\text{met}} = \omega W/P_{\text{loss}}^{\text{met}}, Q_{\Sigma} = \omega W/P_{\text{loss}}^{\Sigma} \tag{10.26}$$

In the case of the energy-isolated resonator with absorbing medium (losses on emission are absent) we obtain

$$1/Q_0 = 2\tan[(\Delta + \Delta^{\text{met}})/2]. \tag{10.27}$$

For hollow resonators of V volume with ideally conducting walls and the uniform isotropic medium of ε and μ parameters, the following equations are true. (Here the asterisk means the complex-conjugated quantity).

$$\frac{\omega^2}{c^2}\varepsilon\mu = \frac{\int_V \left|rot\,\vec{E}\right|^2 dv}{\int_V \vec{E} * \vec{E}\,dv} = \frac{\int_V \left|rot\,\vec{H}\right|^2 dv}{\int_V \vec{H} * \vec{H}dv}. \tag{10.28}$$

From here it follows that the inequality

$$\omega^2\varepsilon\mu > 0 \tag{10.29}$$

is independent: whether the internal medium is absorbing or not.

A concept of complex frequency ω (Eq. (10.22)) is convenient in the concern that the following resonator property is important for practice; namely, the Q-factor through components $\omega = \omega' + i\omega''$ is expressed as

$$Q = \omega'/2\omega''. \tag{10.30}$$

If we are speaking about the Q-factor on emission only, i.e., the Q-factor caused by losses on emission only, then we obtain

$$Q = q/4p'', \quad q \gg 1. \tag{10.31}$$

The p'' parameter is defined from the resonance condition (Sect. 6.2.6) and similar to that, and, obviously, usually $|p''| \ll 1$.

The total Q-factor of the resonator is formed, as mentioned in Eq. (10.3), from losses on emission, losses in the metal of walls, mirrors, and magnetic–dielectric filling.

10.4 Rectangular, Cylindrical, and Spherical Resonators with Metallic Walls

10.4.1 Variable Separation Method, Rectangular Resonator

On examination of resonators with "correct" shapes, the universal *variable separation method* remains the principal one, as in the waveguide theory. We use this method, but do not repeat, of course, the material from Chap. 6. The three-dimensional *Helmholtz equations* are the theoretical fundament. Below, in examples of the resonator with coordinate boundaries and uniform filling, we illustrate the *variable separation method* (the *Fourier method*; Fig. 10.4a).

The three-dimensional *Helmholtz equation* in Cartesian coordinates has the form

$$\frac{\partial^2 \psi_i}{\partial x^2} + \frac{\partial^2 \psi_i}{\partial y^2} + \frac{\partial^2 \psi_i}{\partial z^2} + k^2 \varepsilon \mu \psi_i = 0, \tag{10.32}$$

where through ψ_i one of the components of the resonator field vectors $\vec{E}$ or $\vec{H}$ is designated.

We should attract attention to the fact that in the resonator theory the *k parameter is unknown*. In the waveguide theory, the wave number of a free space was known in advance ($k = \omega/c = 2\pi/\lambda$), and we looked for the longitudinal wave number as: $h = \sqrt{k^2 \varepsilon \mu - g^2}$. As earlier, in accordance with the variable separation method, the unknown solution ψ_i is represented in the form of a product of various coordinate functions:

$$\psi_i = X(x) Y(y) Z(z). \tag{10.33}$$

After substitution of this representation in the *Helmholtz Equation* (10.32) and dividing all terms by a product XYZ, we obtain three ordinary differential equations:

$$\frac{d^2 X}{dx^2} + \chi_x^2 X = 0, \frac{d^2 Y}{dy^2} + \chi_y^2 Y = 0, \frac{d^2 Z}{dz^2} + \chi_z^2 Z = 0 \tag{10.34}$$

and an additional condition

$$\chi_x^2 + \chi_y^2 + \chi_z^2 = k^2 \mu \varepsilon. \tag{10.35}$$

Solutions of first two equations were obtained in Sects. 6.4 and 6.5. The solution of the third equation (10.34) has the same form:

$$Z = \begin{cases} A_1 \cos \chi_z z + A_2 \sin \chi_z z \\ A_3 e^{i\chi_z z} + A_4 e^{-i\chi_z z}, \end{cases} \tag{10.36}$$

where $A_i (i = \overline{1,4})$ are unknown constants.

We note that the third equation in (10.34) can be obtained for more general cases of variable separation. For instance, the general solution for structures that are uniform along the z-axis (Fig. 10.1), can be searched in the form $\psi_i = T(\xi_1, \xi_2) Z(z)$, where the function $T(\xi_1, \xi_2)$ depends on transverse (with respect to the z-axis) coordinates ξ_1 and ξ_2.

For the area in the parallelepiped shape (Fig. 10.4a), we may state various boundary tasks. As an example, we give the boundary task solution for boundary conditions (the first mixed problem):

$$\psi_i = 0 \text{ for } x = 0, a; \psi_i = 0 \text{ for } y = 0, b; \partial \psi_i / \partial z = 0 \text{ for } z = 0, L. \tag{10.37}$$

Solutions of the boundary task (10.36) form the system of eigenfunctions ψ_{imnl} at eigenvalues k_{mnl}^2:

$$\psi_{imnl} = N_{mnl} \sin(\pi m x / a) \sin(\pi n y / b) \cos(\pi l z / L), \tag{10.38}$$

$$k_{mnl}^2 \varepsilon \mu = (m\pi/a)^2 + (n\pi/b)^2 + (l\pi/L)^2,$$

$m = \overline{1, \infty}$; $n = \overline{1, \infty}$; $l = \overline{0, \infty}$, N_{mnl} are coefficients, which are still indefinite.

Oscillations E_{110} and H_{110} are the lowest oscillation types of the rectangular hollow resonator. Figure 10.5 shows the picture of fields and intensity distribution (on axes) of some of the oscillations (see also Program P.10.1).

10.4.2 Cylindrical Resonator

The Helmholtz equation in the cylindrical coordinate system is written as

$$\frac{1}{r}\frac{\partial}{\partial r}\left(r\frac{\partial \Psi}{\partial r}\right) + \frac{1}{r^2}\frac{\partial^2 \Psi}{\partial \phi^2} + \chi^2 \Psi(r, \phi) = 0.$$

Representing of the Ψ function as a product: $\Psi = R(r)\ \Phi(\phi)\ Z(z)$, we obtain three differential equations, whose solutions are expressed as a set of functions of this equation (depending on the boundary condition type on the S surface), and besides, an additional condition occurs:

$$\chi^2 + \chi_z^2 = k^2 \varepsilon \mu. \tag{10.39}$$

Solutions to the first two equations were studied in detail in Sect. 7.9. For the third equation, the solution can be written in the usual form (10.41).

As an example, we give the boundary task solution for the cylindrical resonator (Fig. 10.4b) at boundary conditions (the first fixed problem):

$$\psi_i = 0 \text{ for } r = R, \partial \psi_i / \partial z = 0 \text{ for } z = 0, L. \tag{10.40}$$

Eigenfunctions and *eigenvalues* are defined by the following equations:

$$\psi_{nml} = N_{nml} J_n\left(\frac{\mu_{nm}}{R}r\right)\begin{Bmatrix} \cos n\varphi \\ \sin n\varphi \end{Bmatrix} \cos\frac{l\pi z}{L}, \; k_{nml}^2 \varepsilon \mu = \left(\frac{\mu_{nm}}{R}\right)^2 + \left(\frac{l\pi}{L}\right)^2, \tag{10.41}$$

μ_{nm} is the m-th root of equation $J_n(\mu) = 0$, N_{nml} are indefinite coefficients; μ_{nm} roots can be found by Program P.6.12 and ν_{mn} roots of equation $J_o'(\nu_{mn}) = 0$ by Program P.6.13.

Thus, in the cylindrical resonator, the oscillations H_{nml} and E_{nml} can exist. We note that electric oscillations with zero third index E_{nm0} exist, the magnetic at $l = 0$, i.e., H_{nm0}, do not exist. The oscillation with the lowest fundamental frequency is usually used as the main operating type of oscillation. Here, there are definite features related to the resonator length L. If we have a "short" resonator, the oscillation of electric type with $l = 0$ is the main oscillation. For the "long" resonator, the magnetic oscillation with $l = 1$ is the main oscillation. Therefore, there is a definite choice between oscillations E_{010} (the wave number $k = 2.405/R$) and H_{111} $\left(k = \sqrt{(1.841/R)^2 + (\pi/l)^2}\right)$. Fundamental frequencies of these oscillations coincide, as we see, at $l/2R \approx 1.015..$ At $l/2R > 1.015$, H_{111} is the main oscillation. Otherwise, E_{101} is the main oscillation. It is interesting that its frequency does not depend on the resonator length L and usually $L < 2R$. The field pictures of these oscillations are shown in P.10.1 and P.10.2 (Appendix B – http://extras.springer.com/2019/978-3-319-90847-2).

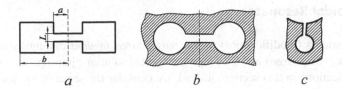

Fig. 10.6 Quasi-stationary resonators in microwave electronics. (**a**) Toroidal. (**b**) Calculation model for circuit analysis. (**c**) the separate resonator of the magnetron

10.4.3 Quasi-Stationary Resonators Used in Microwave Electronic Devices

The basic resonator shapes in microwave electronics are the *toroid* (Fig. 10.6a, b) and the separate *resonator of the magnetron* (Fig. 10.6c).

10.4.4 Q-Factors of some Oscillation Types of Closed Resonators

We already discussed that the resonator Q-factor (on the operating oscillation type) is one of the main resonator parameters, and, at the same time, its strict calculation, on the one hand, is complicated enough, and on the other, in practice, it is frequently sufficient to have estimation values with low accuracy. Therefore, sometimes, the Q value can be determined based on approximate theory, for example, according to the quasi-static approach, to the perturbation method etc. Let us give some data for the main classes of resonators.

1. *Rectangular resonator. E_{110} oscillation:*

$$Q = \frac{1}{d} \; \frac{ab(a^2 + b^2)L}{2(a^3 + b^3)L + ab(a^2 + b^2)} \quad \text{or for } a = b : Q = \frac{1}{d}\frac{aL}{a + 2L}. \tag{10.42}$$

Hereinafter, d represents the depth of the surface layer (the skin layer).

2. *Circular cylindrical resonator. E_{010} oscillation:*

$$Q = (1/d) \; (aL/(a + L)). \tag{10.43}$$

3. *Toroidal resonator. The main type of oscillations.* Let the toroid cross-section be represented by the rectangle with an external radius of b, internal radius of a, and the height h. Then, its Q-factor is determined as

$$Q = 2D/d, \quad D = [h \ln (b/a)] \; / \; [h(1/a + 1/b) + 2 \ln (b/a)]. \tag{10.44}$$

In Sect. 10.5 we describe the toroidal resonator in detail.

10.5 Resonators of Other Shapes, Concept of the Partial Area Method

10.5.1 Introduction

In practice, there is a huge number of resonators and their directly coupled or diffraction-coupled systems. The single resonator itself can be of many various shapes. In particular, many such structures are used in microwave electronics, measuring technologies, and other areas of radio engineering. We become acquainted in the next sections with two of the probably most widespread and universal methods of resonance structure analysis; namely, with the *partial area method*, in the development of which Russian scientists played the main roles. First of all, G.V. Kisun'ko [112], and the *perturbation method* [79, 90, 107, 125].

10.5.2 An Example: Toroidal Resonator

At present, there are many variants (modifications) of the *partial area method* distinguished by approaches of resonator volume fragmentations into the *partial areas* and the procedure of field solution "joining" on boundaries (or areas). We do not try to describe all these modifications in this section; instead, we consider the sense of the partial area method in the simple *example* of calculation of azimuth-uniform kinds of fundamental oscillations in the toroidal resonator with cylindrical casing and bushing (Fig. 10.6b). For distinctness, we are limited by E-type oscillations.

Azimuthal–uniform oscillations of E-type suppose an absence of field variations along the ϕ coordinate. In this case, they are described by the following system of Maxwell equations ($\partial/\partial\phi \equiv 0$):

$$\begin{cases} E_r = (i/\omega\varepsilon_o\varepsilon)\partial H_\varphi/\partial z, \\ E_z = -(i/\omega\varepsilon_0\varepsilon r) \cdot \partial(rH_\varphi)/\partial r, \\ \partial E_r/\partial z - \partial E_z/\partial r = -i\omega\mu_0\mu H_\varphi \end{cases} \tag{10.45}$$

and boundary conditions on the ideally conducting resonator walls:

$$\begin{aligned} E_z &= 0 \text{ for } r = 0; \\ E_r &= 0 \text{ for } z = z_1 \ (0 \le r \le r_1); \\ E_r &= 0 \text{ for } z = z_2 \ (r_1 \le r \le r_2). \end{aligned} \tag{10.46}$$

Owing to resonator geometric symmetry, it is enough to consider the electrodynamics task for a half-structure at $z \ge 0$ (Fig. 10.6b).

Let us divide the meridional resonator section (Fig. 10.6a) by the line AA into two *partial areas* of the "correct" (rectangular in section) shape. The general solution (10.45) satisfying the boundary conditions (10.46) in each of the areas has the form (compare Sect. 7.3 from Paul [126]):

$$H_\varphi^{(1)} = \sum_{n=0}^{\infty} A_n J_0(c_n r) \cos\frac{n\pi z}{z_1}, \tag{10.47}$$

$$H_\varphi^{(2)} = \sum_{n=0}^{\infty} B_n F_0(\eta_n r) \cos\frac{n\pi z}{z_2}, \tag{10.48}$$

where $F_v(\eta_n r) = Y_0(\eta_n r)J_1(\eta_n r_2) - J_v(\eta_n r)Y_1(\eta_n r_2)$,

$$c_n^2 = k^2\varepsilon\mu - (n\pi/z_1)^2, \eta_n^2 = k^2\varepsilon\mu - (n\pi/z_2)^2.$$

At the boundary of the areas, at $r = r_1$, the "joining" (equating, matching, fitting) condition is true ($0 \le z \le z_1$):

$$H_\varphi^{(1)} = H_\varphi^{(2)}, \tag{10.49}$$

$$\partial H_\varphi^{(1)}/\partial r = \partial H_\varphi^{(2)}/\partial r. \tag{10.50}$$

10.5.3 "Joining" Condition on the Boundary of Partial Areas, Integral Equation

One of the most widespread approaches to satisfying joining conditions is specification of the field distribution function on the boundary of the areas. Introducing the designation ($r = r_1, 0 \le z \le z_1$) $\partial H_\varphi^{(1)}/\partial r = \partial H_\varphi^{(2)}/\partial r = f(z)$, differentiating series in (10.45) and (10.46) on r and equating the result of the $f(z)$ function expansion into the Fourier series on the segment $[0, z_1]$, we have

$$A_n = \frac{2\zeta_n}{z_1 c_n J_1(c_n r_1)} \int\limits_0^{z_1} f(z) \cos\frac{n\pi z}{z_1} dz, \tag{10.51}$$

$$B_n = \frac{2\zeta_n}{z_2 \eta_n F_1(\eta_n r_1)} \int\limits_0^{z_1} f(z) \cos\frac{n\pi z}{z_2} dz, \tag{10.52}$$

where $\zeta_n = 1/2$ if $n = 0$ and $\zeta_n = 1$ if $n \neq 0$.

Having substituted Eqs. (10.47) and (10.48) in Eq. (10.49), we obtain the integral equation with respect to $f(z)$ function:

$$\sum_{n=0}^{\infty} \frac{2\zeta_n}{c_n z_1} \cdot \frac{J_0(c_n r_1)}{J_1(c_n r_1)} \cos\frac{n\pi z}{z_1} \int\limits_0^{z_1} f(z') \cos\frac{n\pi z'}{z_1} dz' -$$

$$- \sum_{n=0}^{\infty} \frac{2\zeta_n}{\eta_n z_2} \cdot \frac{F_0(\eta_n r_1)}{F_1(\eta_n r_1)} \cos\frac{n\pi z}{z_2} \int\limits_0^{z_1} f(z') \cos\frac{n\pi z'}{z_2} dz' = 0 \tag{10.53}$$

or in operator form

$$Lf(z) = 0. \tag{10.54}$$

10.5.4 Approximate Solution

The integral Eq. 10.53 can be solved using various numerical methods, in particular, by the *Bubnov–Galerkin method*. We search for the *approximated* solution of Eq. 10.53 as

$$f^{(N)}(z) = \sum_{i=1}^{N} a_n \psi_n(z), \tag{10.55}$$

where $\{\psi_n\}$ is the complete system of basic functions. Substituting this expansion into Eq. (10.53) and imposing the following *orthogonality conditions*

$$\int\limits_0^{z_1} \psi_n^* L f^{(N)} dz = 0, n = \overline{1, N} \tag{10.56}$$

we obtain a system of algebraic equations $[D(k)]\,\vec{A} = 0$, where $\vec{A} = [a_1, a_2, \ldots a_N]$ is the vector of expansion coefficients (10.56), $[D(k)]$ is the square matrix of N order with elements

$$d_{ij} = \int\limits_0^{z_1} \psi_i^* L \psi_j dz. \tag{10.57}$$

Eigenwave numbers are defined as roots of the equation *Det* $[D] = 0$, after which, with the help of the D matrix, we calculate eigenvectors $\vec{A}_i$ $(i = \overline{1, N})$, which define the field distribution on the boundary $(r = r_1)$. The field expansion coefficients in the partial areas can be found according to the known $f(z)$ function with the help of Eqs. (10.51) and (10.52). We should note that calculation of each matrix element requires summation of infinite series, which converges rather *slowly*. It is expedient to *improve their convergence* similar to the way it is done in the theory of *strip-slot structures*.

As *basic functions*, we can use polynomials, trigonometric functions, etc. The significant growth of the solution convergence speed at an increase of N number in series (10.53) can be achieved owing to the introduction of basic functions that *have the same singularity near the bushing edge*, as the required solution. Note that for bushing with sharp edges, we have near the $f(z) \sim [1 - (z/z_1)^2]^{\tau - 1}$ edge, where $0 < \tau < 1$. Therefore, as basic functions $\{\psi_n\}$, it is expedient to use *Gegenbauer polynomials*, which are orthogonal in the segment $[0, z_1]$ with a weight $\rho(z) = [1 - (z/z_1)^2]^{-1/3}$, or *Chebyshev polynomials* of the first kind, which are orthogonal in the segment $[-z_1, z_1]$ with a weight $\rho(z) = [1 - (z/z_1)^2]^{-1/2}$.

The integral Eq. 10.53 can be effectively solved by the *method of almost full operator transformation* [17–19].

10.6 Resonator Calculation, Perturbations Method

10.6.1 General Approach and Assuming Estimations

The perturbation method is the most simple and universal approach, and allows an analysis of resonators with various perturbations of the" correct" area V_0, S_0 boundary by some irregularity V_1, S_1 (Fig. 10.7a) or by some isotropic and anisotropic insertions (Fig. 10.7b).

In the task statement itself, the requirement of small perturbation is contained. For instance, for the resonance (generally speaking, two-dimensional, say, the regular waveguide, or three-dimensional) region with the S_0 boundary, the perturbation S_1 should be insignificant (for instance, $a_0 \gg a_1$, where a_0 is some average linear size of the "correct" area, a_1 is the linear size of the perturbation area). Thus, under the term "correct area" we understand an area for which fundamental frequencies (for resonators) or propagation constants (for waveguides) are known.

Similarly, in the case of perturbation of the area filling (Fig. 10.7b): either the perturbation area V_1, S_1 is small enough, or its permittivity and permeability differ slightly from material parameters of the nonperturbed area. Of course, the small size of the perturbation area is not a guarantee of application possibility of the perturbation method itself. We can say that for inserted perturbations in both cases shown in Fig. 10.7, the perturbing body should not (significantly) change the field structure of the nonperturbed area.

Thus, in both of the cases considered, we must enter the *small parameter*. If there is not such a parameter, it can be "invented," but this approach is far outside the scope of our lecture course. We are limited by references only to the sources in which these issues are considered in detail.[2]

10.6.2 Summary

Using fields of a nonperturbed resonator, we obtained in the outcome rather simple "engineering" expressions, permitting, on the one hand, the physical content of the element under consideration to be clarified, and on the other hand, estimation

Fig. 10.7 Illustration of the perturbation method for the hollow "correct" resonator. (**a**) Perturbation of the S_0 area boundary. (**b**)Perturbation of filling. (Taken from Nefyodov and Kliuev [73], p. 347)

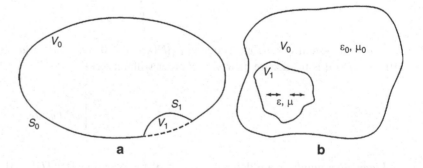

---

[2] The method of *small perturbations* and *variation methods* are described in many books (see, for instance, Neganov et al., and Mitra and Lee [17–19, 124]).

expressions for calculations to be obtained. Such expressions were given earlier (see Sect. 12.5 in Nefyodov [26]). By the way, it is most often sufficient in practice. As a result of wide experience in the design of different electrodynamic structures in microwave and millimeter-wave ranges, we can consider it acceptable if Q-factor estimation is made with an accuracy of 10%. As slightly different situation is in 3D-IC, where problems of the dissipated power account and heat dissipation issues may become decisive (this actually occurs).

There is another approach to the utilization of the perturbation procedure. Thus, for instance, during analysis of an elliptical waveguide (see Sect. 7.10 in Nefyodov [26]), instead of the application of unpopular elliptical functions, we can use the perturbation method to obtain the wave propagation constants, ohmic losses, and others. There are other numerous examples of the application of the perturbations method to solutions of practical tasks, so to speak, with few casualties.

10.7   Examples of MathCAD Packet Utilization

In Appendix B (http://extras.springer.com/2019/978-3-319-90847-2) there are three programs of electromagnetic oscillation analysis of the rectangular (P.10.1) and cylindrical (P.10.2 and P.10.3) closed resonators.

10.7.1 Rectangular Resonator with E-Oscillations

Program P.10.1 is devoted to the analysis of electric oscillations E_{mnq} of the rectangular closed resonator at different values of numbers m, n, and q representing the number of field maxima on coordinates x, y, and z. There is the possibility of studying all three components of the total field: $E_{x, y, z}$. There is also the possibility of looking into the dynamics of the animation picture of the main determining picture component of E_z, the Hertz vector.

The geometry, dimensions, and coordinates are shown in Fig. 10.8.

For initial values (a, b, L, ε, μ) and variable indices m, n, q, Program P.10.1 calculates the electric field and constructs the three-dimensional pictures of different components of the electric field.

10.7.2 Cylindrical Resonator with E-Oscillations

Programs P.10.2 and P.10.3 allow an analysis of electromagnetic oscillations of the cylindrical closed resonator. Thus, Program P.10.2 permits the behavior analysis of electromagnetic oscillation components of electric $E_{m, n, q}$ oscillations. Program P.10.3 examines the magnetic $H_{m, n, q}$ oscillations. There is the possibility of looking into the dynamics of the animation picture of the main determining components of the H_z, the Hertz vector.

For initial values (R, φ, η) Programs P.10.2 and P.10.3 calculates the left part of the equation, determining the roots, the matrix of roots, performs the analysis of the magnetic field (longitudinal component), and constructs the three-dimensional pictures of different components of the magnetic field.

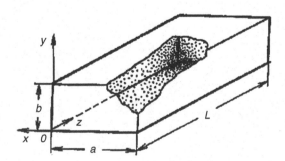

Fig. 10.8 Geometry, dimensions, and coordinates of the rectangular resonator

10.8 Summary

We completed the discussion of the most important problem of electrodynamics – closed cavity resonators. As we showed, many of them were built as parts (segments) of the finite length of the regular or nonregular transmission lines (Chaps. 6 and 7). The knowledge of eigenwaves of the mentioned regular and nonregular lines essentially helped us to examine their great variety. We showed the gradual transfer from the classical resonance single circuit with lumped elements to the open-cavity resonator with distributed parameters (Chap. 11) and to the elementary radiator – the Hertz dipole. We discussed in detail the main resonator characteristics: the oscillation type, the electromagnetic field structure, and the fundamental frequency (Sect. 10.2.2).

We draw the reader's attention to an important point for a general understanding of the properties and features of resonant structures; namely, the spectrum of rarefaction of the fundamental frequency (Sect. 10.2.3). This resonator property has special importance for open systems (Chap. 11).

On the base of the general theory, we considered various forms of resonators, revealed its main properties, and showed the electromagnetic field structure's features. Expressions for the Q-factors of some oscillation types of close resonators are obtained and discussed.

If the resonator shape does not coincide with any from 12 known coordinate systems, we suggested to use the method of partial regions and the method of small perturbation, which is well-known in physics.

Checking Questions

1. Using the variables separation method, determine eigenvalues and eigenfunctions of the rectangular resonator, whose geometry and boundary conditions are shown in Fig. 10.4a; goundary conditions $E_\tau = 0$ on all walls.
2. Using the variables separation method, determine eigenvalues and eigenfunctions of the cylindrical resonator (Fig. 10.4b), which is cut into two equal parts by the ideally conducting plane along the z-axis.
3. Using the variables separation method, determine eigenvalues and eigenfunctions of the cylindrical resonator, whose geometry and boundary conditions are shown in Fig. 10.4b; boundary conditions $E_\tau = 0$ on all walls.
4. Using the variables separation method, determine eigenvalues and eigenfunctions of the spherical resonator with ε and μ parameters, on the surface of which the boundary condition $\vec{H}_\tau = 0$ is true.
5. Write expressions for *Fourier expansions* in the sectorial–cylindrical area.
6. Obtain the integral equation with respect to the tangent electric field $y = y_1$ for E_{nm0}-oscillations of the T-resonator, whose geometry is shown in Fig. 10.5 for $k\rho_1 \ll 1$, $\partial/\partial z \equiv 0$.
7. Using Eq. (10.28), obtain expressions for the frequency of the main oscillation of the rectangular resonator with the dielectric rod ($\partial/\partial z \equiv 0$) using the method of small perturbations (Sect. 10.6).
8. Using Program P.10.1 for parameters $a = 0.8$; 1.3; $b = 1.5$; 2.0; $\varepsilon = 3$; 7; $\mu = 0.8$; 1.5 calculate the three-dimensional pictures of the electric field in the rectangular waveguide. Explain the results.

Now we proceed to studying of one of the most interesting issues of our lecture course: the open resonance structures. The description begins from the discussion of the main idea and main properties of open cavity resonators (Sect. 11.1). One of the greatest discoveries of the twentieth century – the creation of a laser – became possible owing to the usage of open-system (a waveguide, a resonator) properties. The main feature of the open structure is the capability to "spontaneously" become free from the spurious types of waves (oscillations). The honor of discovering these properties of open systems belongs to Prokhorov. The evolution chain of oscillating systems at operating frequency increases from the resonant single-circuit to the open resonator we have tracked in Sect. 10.2.1 (Chap. 10). Here, we note some of the main classes of open resonators, which are formed as a part of both the waveguide with open ends and waveguides from the focusing mirrors. In particular, we describe the class of open cylindrical coaxial resonators. Based on these, systems with an extremely rare spectrum of fundamental frequencies were developed (the Nefyodov–Rossiyskiy effect). This resonator can serve, for example, as a model of the plasma-beam discharge discovered and investigated by Russian academician P. Kapitsa.

Section 11.1.2 is devoted to a description of the operation principles of open resonance structures. One of these operation principles consists of high-Q oscillation, kept in some finite waveguide segment. It is based on the property of almost full reflection of the wave, which falls on the open end of the structure (the Weinstein effect). Thus, if the wave frequency approaches a critical one for this wave type, the reflection coefficient modulus tends to be 1. The second principle – the electromagnetic field "keeping" some volume at the expense of wave reflection from the crucial section or from the caustic surface. These two cases have nonprincipal differences from an engineering point of view.

We study various types of open resonance systems (Sect. 11.2), for instance, the open resonator with rectangular mirrors (Sect. 11.2.1). In the class of open resonance structures, we also include their diffraction-coupled resonators, for instance, with band and rectangular mirrors (Sect. 11.2.3). As examples, we additionally examine open resonators with circular mirrors and contiguous axis symmetric structures, systems of diffraction-coupled disk open resonators (Sect. 11.2.4).

The large section on open systems (Sect. 11.3) is devoted to open resonators with nonplanar mirrors. They have, as a rule, an essentially larger Q-factor, better stability of the operation oscillation type and other useful properties. An essential difference between open resonators and closed type resonators is the large volume (greater stored energy) and the rarefied spectrum of fundamental oscillations. The latter property relates to the self-filtering of high-order oscillations (Sect. 11.3.1). The resonance condition form is given. We specially consider the open coaxial cylindrical resonator with dielectric inserts. Based on this, open resonators with the uniquely rare spectrum of fundamental frequency can be performed (Sect. 11.3.3).

Separately, we give data on open coaxial cylindrical resonators with dielectric inserts. In general, specialists pay a great deal of attention to dielectric resonators and filters. They are particularly interested in the design of millimeter-wave structures (Sect. 11.3.4).

© Springer International Publishing AG, part of Springer Nature 2019
E. I. Nefyodov, S. M. Smolskiy, *Electromagnetic Fields and Waves*, Textbooks in Telecommunication Engineering,
https://doi.org/10.1007/978-3-319-90847-2_11

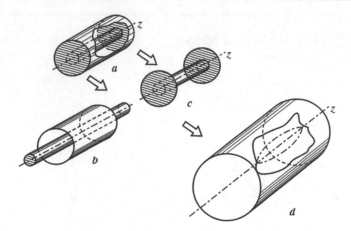

Fig. 11.1 Transition from (**a**) the closed coaxial resonator to three variants of open resonators: (**b**) the *open coaxial cylindrical resonator* to (**c**) the *open disk-shaped resonator* of the coaxial type and to (**d**) the *open coaxial cylindrical resonator*. (Taken from Nefyodov and Kliuev [73], p. 353)

11.1 Main Idea and Properties of Open Resonators

11.1.1 Some History

One of the greatest discoveries of the twentieth century – the creation of a *laser* – became possible thanks to application of the property of the *open system* (a waveguide, a resonator) to "spontaneously" become free from unnecessary wave (oscillation) types [23, 73, 106, 107, 128, 130]. In other words, in the open system, the "spontaneous" *spectrum of rarefaction of eigenwaves* (for the waveguide) or *oscillations* (for the resonator). We had already tracked in Fig. 10.2 (from *a* to *e*) the chain of transitions (evolution) of oscillating systems at the operation frequency growth in systems from the *single oscillating circuit* to the *open resonator*. A similar picture can be seen at transition from the closed coaxial resonator (Fig. 11.1a) to three variants of resonators: the *open coaxial resonator* (Fig. 11.1b), to the *open disk-shaped resonator* of the coaxial type (compare Fig. 11.1c) and, finally, to the *open coaxial cylindrical resonator* (Fig. 11.1d).

In the first two cases, transitions are obtained by means of, first, dropping from the main coaxial resonator (Fig. 11.1a) of the transverse walls (Fig. 11.1b), and in the other case, of the side surface (Fig. 11.1c). The *open coaxial cylindrical resonator* is situated slightly aside in the sense that it is implemented based on the weakly-irregular portion of the internal conductor of the coaxial – the spindle-shaped axis-symmetrical insert (Fig. 10.1f and Fig. 11.1d).

The open coaxial cylindrical resonator shown in Fig. 11.1d results in the large class of congeneric structures (see Sect. 11.2.4). The main point here is in the transition from the closed resonator to the open one with a *uniquely rare spectrum of fundamental frequencies*: within the framework of one figure, we show the transition on the chain of the *Rayleigh–Jeans* formulas:

$$N_{\text{raref}} = \frac{\Delta N}{\Delta \omega} = \frac{\omega^2}{2\pi^2 c^2} V \quad \rightarrow \quad \frac{\omega}{\pi c^2} S \quad \rightarrow \quad \frac{2}{\pi c} L. \tag{11.1}$$

This process is shown more obviously in Fig. 11.2. Thus, the frequency response of the closed resonator (Fig. 11.2a) "locks" many resonance frequencies of the resonator, i.e., each of them has a small amount of power from the active body, and, hence, generation does not happen. Another situation is in the open resonator (Fig. 11.2b), when the power "falls" on one resonance oscillation and generation becomes possible. Accordingly, in the third case (Fig. 11.2c), we meet the significantly greater spectrum of rarefaction; therefore, many practical problems may be solved. So, for example, at creation of powerful microwave oscillators, the degree of the spectrum of rarefaction of the open resonator is not sufficient (Fig. 11.2b), and other (non-operating) oscillation types fall into the system pass-band. Utilization of resonators with a uniquely rare spectrum of natural oscillation (Fig. 11.2c) is the effective approach to offset from non-operating modes.

In other words, this is a transition from the "dense" spectrum proportional to the volume of the closed resonator V, through the spectrum of the open "classical" resonator with the spectrum defined by its *cross-section area S* to a *uniform* (sometimes

Fig. 11.2 Picture of the spectrum of "density" for (**a**) closed, (**b**) opened, and (**c**) open coaxial cylindrical resonators

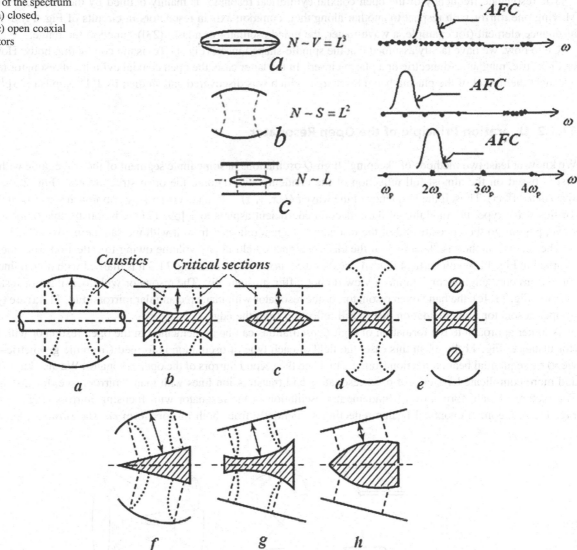

Fig. 11.3 The open coaxial cylindrical resonator with focusing mirrors and its "natural" variants. (**a**)- With barrel-shaped external and cylindrical internal mirrors. (**b**) the internal mirror in the form of a rotation paraboloid, the external mirror is cylindrical. (**c**) The internal mirror in the form of a rotation ellipsoid. (**d, e**) Toroidal resonance structures. (**f–h**) Tuned resonators made based on resonators in the circuits in **a–c** relatively. Caustic surfaces are shown in figures by dotted lines, surfaces of critical sections are shaded. (Taken from Nefyodov and Kliuev [73], p. 355)

called equidistant) spectrum, defined by the same resonator *linear size*. In the specific case of the open coaxial cylindrical resonator, this size is defined by the minimal distance between plates of the coaxial structure (Fig. 11.1d). This type of resonator, in its operation principle, can be based either on properties of a weakly irregular waveguide (resonators in Fig. 11.3a–c), or on the focusing properties of one (Fig. 11.3a, b, f, g) or both (Fig. 11.3d, e) mirrors. Thus, resonators in Fig. 11.3c and h can be realized only because of reflections of quasi-eigenwave from critical sections (in Fig. 11.3 they are shown by straight dashed lines). At first glance, resonators from Fig. 11.3c and h should not have focusing properties (at least, from the geometric–optical point of view). However, the operation principle of these resonators operating on magnetic oscillations of H_{m1q} type consists, as for resonators based on a weakly irregular waveguide (Fig. 11.3a, b), in almost full reflection of H_{m1} wave from *critical section* (see Nefyodov [23]).

At pronounced mirror focusing properties, oscillations inside the resonator are "kept" by *caustic surfaces* (in Fig. 11.3b, c they are shown by shaded lines). Note that mirror curvature and dimensions in resonators of *toroidal* type (Fig. 11.3d, e) can be various for each of mirrors that form the resonator.

The resonance frequency of the open coaxial cylindrical resonator is mainly defined by the distance l between mirrors. Moving one mirror with respect to another along their common axis in resonators in circuits of Fig. 11.3f–h, we obtain the resonance element (for example, a wave-meter, the tuned oscillator or a laser [23]), tuned in the range.

Up to now, we have tacitly assumed that the spindle-shaped insert (Fig. 11.1c) is the conducting body. However, this may be dielectric, magnetic–dielectric or a plasma insert. In the latter case, the open coaxial cylindrical resonator may serve as, for instance, the model of the plasma-beam discharge, which was discovered and studied by P.L. Kapitsa [23].[1]

11.1.2 Operation Principle of the Open Resonator

We know at least two methods of "keeping" high-Q oscillations in some finite segment of the waveguide with V volume. The first is based on the almost full reflection of the incident wave (i) onto the open structure end (Fig. 2.3b; right side) (the *Weinstein* effect). Thus, if the frequency of the wave i $U_i(x, y, z) = A_i$ $u_i(x, y)$ exp $\{ih_iz\}$ approaches a critical value (at $h_i \rightarrow 0$) for this wave type, the modulus of the reflection coefficient aspires to 1 [69, 111]. Obviously, this (almost fully) reflected wave, passing to the opposite end of the resonator, is again reflected from it with the same effectiveness.

The second method is "keeping" of the electromagnetic field in any volume owing to reflection from the *critical section* (marked in Fig. 11.3 with vertical shaded lines) or the *caustic surface* (in Fig. 11.3 it is marked with dotted lines). The last two cases from our "engineering" point of view do not differ substantially. The resonator with focusing (for instance, spherical) mirrors (Fig. 11.4c), the band open resonator, or the resonator with plane rectangular mirrors (Fig. 11.4a) are typical examples of open resonator implementation due to full reflection from the edge.

Another approach to the formation of high-Q oscillations can be implemented in the open resonator with focusing mirror (for instance, Fig. 11.4c, e). In this case, the field of each type of oscillation is formed (from the geometric–optical point of view) by sequential beam reflection from the first and the second mirrors of the open resonator. We also know the three-mirror and more complicated systems of open resonators and transmission lines with many mirrors (see, for instance, Fig. 11.3b). The picture of field formation of fundamental oscillation in the resonator with focusing mirrors (Fig. 11.4c) is shown in Fig. 11.4e. Geometric–optical beams reflecting sequentially from both mirrors form *caustic surfaces*, beyond which they

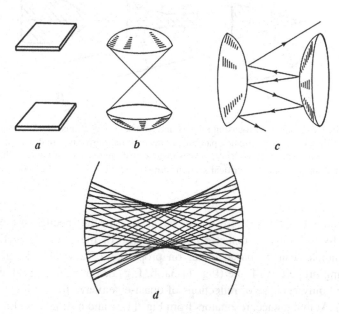

Fig. 11.4 Examples of open resonators. The resonator with (**a**) plain rectangular and (**b**) disk-shaped mirrors. (**c**) The resonator with focusing mirrors. (**d**) The "unstable" resonator. (**e**) The geometric–optical picture of field formation of the fundamental oscillation in the resonator with focusing mirrors. (Taken from Nefyodov and Kliuev [73], p. 356)

---

[1] *Kapitsa P.L.* The free plasma cord in the high-frequency field at high pressure (in Russian) // Journal of experimental and theoretical physics, 1969, vol.57, № 6, p.1801–1866.

Fig. 11.5 Distribution of currents on the rectangular mirror of the open resonator; the same for the system of diffraction-coupled resonators and the distribution of field intensities between mirrors for the following oscillations (shown by vertical or horizontal shading): (**a**) - $E_{11q}^{(x)}$, (**b**) - $E_{11q}^{(y)}$, (**c**) - $E_{21q}^{(y)}$, (**d**) - $E_{21q}^{(y)}$. (Taken from Nefyodov and Kliuev [73], p. 358)

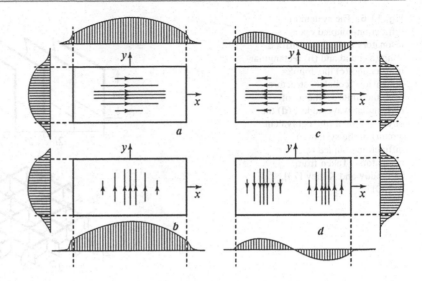

cannot come out, i.e., as though the field is "kept" by the caustics. The field near the caustic is described by *Airy functions* and it decreases beyond the caustics according to exponential law (compare Fig. 11.4e; beyond the shaded lines).

It is clear that mirror dimensions (they can be different) must exceed the transverse dimensions of the forming oscillation. Otherwise, oscillations of this type are "highlighted," i.e., are not kept by mirrors of these dimensions.

Sometimes, in high-power electronics, as we have already mentioned, the unstable low-Q open resonators are used (Fig. 11.4c). The Sovetov formula (10.1) is useful for estimation of the Q-factor.

Some examples of open resonators with different mirrors are shown in Figs. 11.6 and 11.7. Sometimes, open resonators are referred to as Fabry–Perot resonators. We think that we can call them Prokhorov resonators or, at worst, the Fabry–Perot–Prokhorov–Shavlov–Townes resonators, which correspond better to historical truth. If we use modern terminology, the open resonator from the well-reflected plane mirrors *1,2* located at the distance $2l$ is the basis for the Fabry–Perot interferometer. The interested reader can find a full description of the Fabry–Perot interferometer in lecture courses on optics (see, for instance, Born and Wolf, and Sivukhin [46, 99]).

Q-factors of closed (hollow) cavities and open resonators are varied in their very wide limits. So, for example, unstable resonators have $Q \approx 10$, microstrip resonators have $Q \approx 10^2$, hollow cavities in the microwave range have $Q \approx 10^3...10^5$, open resonators in the light range have $Q \approx 10^5...10^6$, etc. The presented data are largely estimated. We may improve the Q-factor of the operating oscillation by 1–2 orders and effectively decrease the level of spurious oscillations by using some of the measures, for example, the use of noncontinuous (broken) mirrors (with application of frequently periodic structures [60, 109]).

Nevertheless, we return to examination of the simplest case of the open resonator with plane mirrors. The *equivalent impedance boundary conditions of the resonance type*, with which we became acquainted in Sect. 2.1, comprise the basis of oscillations analysis in this open resonator. Precisely these conditions define the main properties of the resonance in the open area V with plane S mirrors. Among them there are: the *key problem* – the semi-infinite planar waveguide (Fig. 2.3b) with transition to the *finite structure* – the *band open resonator* (Fig. 2.3a). The sense consisted in the fact that if the wave frequency ($\omega \to \omega_{cr}$), which comes to the open end of the waveguide, is closed to the critical value (i.e., the longitudinal wave number $h \to 0$; the *Brillouin wave* falls onto the plane mirror almost along the normal), then the i wave is almost fully reflected ($|r| \to 1$) from this open end, i.e., the modulus of the wave reflection coefficient is closed to 1.[2] This process is usually referred to as the *Weinstein effect*. In other words, the almost integer number of $q=1,2,3...$ of half-waves keeps within the distance $2l$ between mirrors, i.e., the following *resonance conditions* are fulfilled:

---

[2] This result was obtained by Russian physicist L.A. Weinstein in 1947. However, he at that time he was unable to look at the *finite structure* and come to the seemingly natural result –the *open resonator*. Much later he developed the excellent theory of open resonators and waveguides [111].

Fig. 11.6 The system of diffraction-coupled open resonators with plane mirrors. With (**a**) band and (**b**) rectangular mirrors and (**c**) the double-dimension "cellular" structure from the finite segments of rectangular waveguides, (**d**) the physical model (the equivalent circuit) of the system of diffraction-coupled open resonators. (Taken from Nefyodov and Kliuev [73], p. 260)

Fig. 11.7 Open resonance systems with plane mirrors. (**a**) The disk-shaped resonator. (**b**) The circular resonator. (**c**) The ring resonator on the asymmetric stripline (*1* the metallic ring, *2* the dielectric substrate, *3* the metallic conducting shield). (**d**) The system of ring diffraction-coupled open resonators. (**e**) The metallic rod with ribs. (**f**) The radial open resonator. (**g**) Diffraction-coupled disk-shaped and ring open resonators. (**h**) Two diffraction-coupled ring open resonators. (**i**) The system of ring resonators with the axis rod. (**j**) The diaphragm structure inside the metallic tube. (Taken from Nefyodov and Kliuev [73], p. 361)

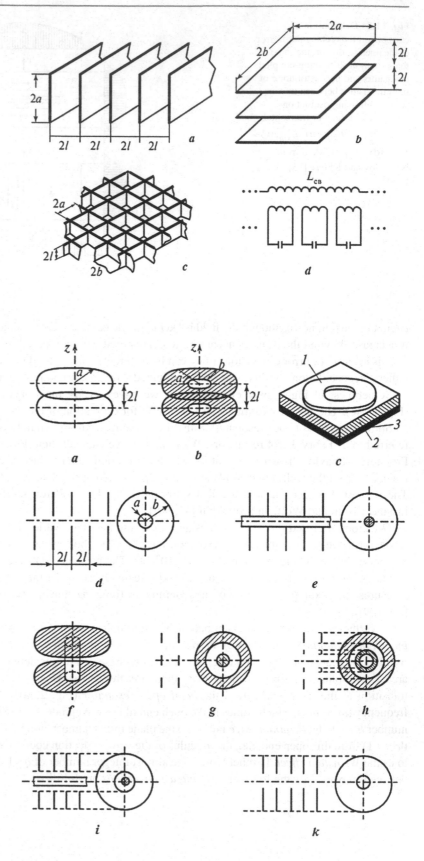

$$2kl = \pi q + 2\pi p, \quad |p| \ll 1. \tag{11.2}$$

It is sufficiently evident that in the same approximation, we can transfer from the band open resonator to the open resonator with rectangular (Fig. 11.4a) or disk-shaped (Fig. 11.4b) mirrors. The approximation mentioned is referred to as the *parabolic equation approximation*, when the Helmholtz equation $\Delta F + k^2 F = 0$ is replaced approximately by the *parabolic equation* $\frac{\partial^2 W}{\partial x^2} + \frac{\partial^2 W}{\partial y^2} + 2ik \frac{\partial W}{\partial z} = 0$, whose solutions are well-known. The great contribution to the development of the *parabolic equation method* was made by Russian scientist G.D. Maliuzhinets (see, for instance, Maliuzhinets [68], p. 606–609).

Another operation principle of the open resonator consists in using mirrors with *focusing properties* (Fig. 11.4c, e), which we mentioned just now. In contrast, in the powerful electronic devices, so-called "unstable" resonators are used (Fig. 11.4d).

11.2 Open Resonators with Plane Mirrors

11.2.1 Open Resonator with Rectangular Mirrors

For the first time, such an open resonator (Fig. 11.4a) was applied by A.M. Prokhorov in quantum electronics. The theory of this open resonator was constructed for the case of the large distance between mirrors ($kl \gg 1$ [111]); "*optical*" *approximation*, based on the strict problem solution regarding diffraction of the plane wave on the open end of the plane (Fig. 2.3b) and also circular or coaxial waveguides [23, 69]. In the optical range, owing to the small size of the wavelength (with respect to dimensions of the electrodynamics structures used) and the necessity of high Q-factor obtaining (i.e., the large volume V), the polarization "indifference" of oscillations is observed or, in other words, the *degeneration of oscillations* of $E_{m0q}^{(x)}$ and $H_{m0q}^{(x)}$ types. It consists, as usual, in the fact that at completely different field structures, these oscillations have the same *resonance frequency*. In shorter area (millimeter- and sub-millimeter waves) the polarization degeneration is eliminated at resonance condition (11.2) fulfillment and for finite values of the transverse wave number $q = 1,2,...$ Development of the *generalized theory of open resonators with plane mirrors* (at $q = 1,2,..$) becomes possible after development of the strict theory of wave diffraction, which runs against the edge of the semi-infinite waveguide under the "oblique" angle (see Nefyodov and Fialkovsky [58]). We now briefly give the main results of this theory. Unfortunately, a complete discussion would take up an impermissibly large amount of our lecture course [111]. Up to now, nobody has been able to obtain strict results for wave diffraction on the open end of the rectangular waveguide, although such structures find application in antenna engineering, microwave and millimeter-wave electronics, in periodic structures (see, for instance, Fig. 11.6) and in many other areas.

The p parameter from Eq. (11.2) for the two-dimensional case happens to be equal to $p = p_a + p_b$, where its components are relatively:

$$\begin{Bmatrix} p_a \\ p_b \end{Bmatrix} = \frac{\pi \begin{Bmatrix} m^2 \\ n^2 \end{Bmatrix}}{4\left(\begin{Bmatrix} M_a \\ M_b \end{Bmatrix} + \beta' + i\beta''_{E,H}\right)^2}, \quad \begin{Bmatrix} M_a \\ M_b \end{Bmatrix} = \sqrt{\frac{2k}{l} \begin{Bmatrix} a^2 \\ b^2 \end{Bmatrix}}.$$

Distribution of currents $j(x, y)$ on rectangular mirrors with sizes $a \times b$ for oscillations $H_{mnq}^{(x)} = E_{mnq}^{(y)}$ and $E_{mnq}^{(x)} = H_{mnq}^{(y)}$, and distribution of their intensities are shown in Fig. 11.5. The distribution of currents is:

$$j(x, y) = \begin{Bmatrix} \cos \\ \sin \end{Bmatrix} \frac{\pi m x}{2a\left(1 + \beta' \frac{1+i}{M_a}\right)} \begin{Bmatrix} \cos \\ \sin \end{Bmatrix} \frac{\pi n y}{2b\left(1 + \beta' \frac{1+i}{M_b}\right)}. \tag{11.3}$$

Obviously, outside the mirrors, conductance is absent: $j(x, y) = 0$. Nevertheless, fields at mirror ends are not zeros, but moving away from the mirror end its intensities decrease according to exponential law (i.e., quite fast).

The "identity" of magnetic and electric oscillations is, of course, conditional. As we already mentioned, they differ by $\beta''_{E,H}$ in (11.3); however, this difference has a small effect on the current distribution on mirrors and on the fields between them. Another thing is the resonance frequencies: they may be spaced apart from one another by the significant "distance" on the frequency scale.

At large enough distances between mirrors ($kl \gg 1$), but at fulfillment of resonance conditions (11.2), parameters are:
$\beta' = \beta''_{E,H} \to \beta_0 = 0.824$.

11.2.2 Emission *Q*-Factor of Open Resonators

The general expressions for emission Q-factor $Q_{emis} = \omega'/2\omega'' = q/4p''$ require the knowledge of values included in these formulas to determine this. If obtaining q is evident enough because usually $\lambda_{res} \approx 4l/q$, for the determination of p'' the definite efforts are necessarily related to the necessity (accurate or at least approximate) of the solution to the diffraction problem. This note equally concerns both single open resonators and their systems, which are coupled between them.

11.2.3 The System of Diffraction-Coupled Open Resonators with Band and Rectangular Mirrors

In practice, the periodic structures, for example, the structure of diffraction-coupled band open resonators (Fig. 11.6a) are widely used. In essence, in the same approximation of the *parabolic equation*[3] with *polarization supplements*, we can save the same formulas for fields that were written in the previous section. However, β, β', $\beta''_{E,H}$ parameters are different.

The equivalent circuit of the diffraction-coupled structure in the form of the coupled oscillating circuits is presented in Fig. 11.6d. The distributed inductance L_{coupl} defines the degree of mutual coupling of adjacent (and not only adjacent) cells of the structure [33, 126].

11.2.4 Open Resonator with Circular Mirrors and Congeneric Axisymmetric Structures, Systems of Diffraction-Coupled Disk-Shaped Open Resonators

Axisymmetric structures with plane mirrors constitute a wide circle of resonance systems and are widely used in practice. Some examples of these systems are shown in Figures 11.4b and 11.7. Earlier, we met with a series of "similar" structures. So, say, the disk-shaped open resonator (Fig. 11.7a) is the modified version of the closed cylindrical resonator (Fig. 11.1b) with the eliminated side wall (compare Sect. 2.3). The *resonance character* is the common point for structures in Fig. 11.7, i.e., an almost integer number of half-waves keeps within at the distance $2l$, and the equivalent boundary impedance conditions (11.2) of the resonance type are fulfilled on the open edges.

At the same time, in spite of external differences between these structures, they can be analyzed according to an already known scheme. At first, the electromagnetic field components are written in the cylindrical coordinate system, as we did for closed resonators (see Sect. 10.3). Then, satisfying the field to boundary conditions of the resonance type (say, in the form (11.2)) on the open edge of the structure, for instance, at $r = a$ for the disk-shaped resonator (Fig. 11.7a) or at $r = a$ and $r = b$ for the ring open resonator (Fig. 11.7b), we come to a highly complicated transcendent dispersion equation, which in the general form can be written as

$$\det \|d(\kappa)\| = 0. \tag{11.4}$$

In Eq. (11.4), for the disk-shaped open resonator (Fig. 11.7a), $\det \|d(\kappa)\|$ is the determinant of the fourth order for the normalized complex wave number $k = k' + ik''$.

In general, the dispersion Eq. (11.4) should be solved using numerical methods. However, taking the resonance character of oscillations into consideration and using the equivalent impedance boundary conditions (11.2) of the resonance type, we can obtain approximate expressions for resonance frequencies and diffraction losses. For the disk-shaped open resonator with mirrors of the a radius, the fundamental oscillation frequency of E_{mnq} is [111]:

---

[3] The transition from the wave equation to the parabolic one is substantiated (from a physical point of view) by the fact that the speed of field variation in the direction of wave propagation is much greater than in the transverse direction; hence, we can neglect the second derivation on x [111].

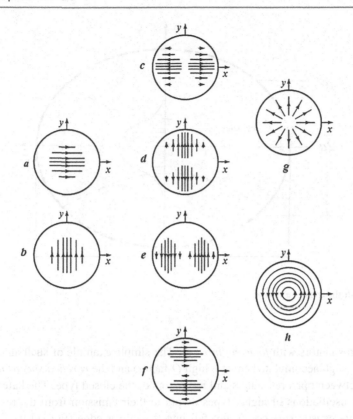

Fig. 11.8 Pictures of current distribution on the one of mirrors of the disk-shaped resonator for electric oscillations E_{01q} and E_{11q}. (a) the $E_{01q}^{(x)}$ oscillation, the $j_x = j_0(s_0, \rho)$ current. (b) The $E_{01q}^{(y)}$ oscillation, the $j_y = j_0(s_0, \rho)$ current. (c) The $E_{11q}^{(x)}$ oscillation, the $j_x = j_1(s_0, \rho)\cos\varphi$ current. (d) - the $E_{11q}^{(y)}$ oscillation, the $j_y = j_1(s_0)\sin\varphi$ current. (e) The $E_{11q}^{(y)}$ oscillation, the $j_y = j_1(s_0)\cos\varphi$ current. (f) The $E_{11q}^{(x)}$ oscillation, the $j_x = j_1(s_0, \rho)$ current. (g) The $E_{11q}^{(r)}$ oscillation, the $j_x = j_1(s_0, \rho)$ current, (h) the $E_{11q}^{(\varphi)}$ oscillation, the $j_x = j_1(s_0, \rho)$ current. (Taken from Nefyodov and Kliuev [73], p. 362)

$$kl \approx \pi q + \left(l\nu_{mn}^2/2ka^2\right) \qquad (11.5)$$

where ν_{mn} is the n-th positive root of the equation $J_m(\nu) = 0$ (see Nefyodov [26], Section 7.9, Eq. (7.56); Program P.6.12; see the same).

Distributions of current on mirrors of the disk-shaped open resonator are shown in Fig. 11.8 (see also Nefyodov [26]).

In the captions, the following designations are used: $j_{0.1}(y)$ are Bessel functions, $\gamma = s_0$, ρ is their argument, where $\zeta = s_0\rho$, where $s_0 = \dfrac{2\nu_{mn}}{\left(M + \beta' + \beta''_{E,H}\right)}$, $M = \sqrt{\dfrac{2ka^2}{l}}$ is the n-th positive root of the equation $I_m(\nu) = 0$.

Note that the application of β', $\beta''_{E,H}$ coefficients (see, for instance, Nefyodov and Fialkovsky [58]) makes the written expression suitable, first of all, not only for large values $q \gg 1$, but for $q = 1, 2, \ldots$, and second, eliminates degeneration between the electric and magnetic with the same indices. Elimination of degeneration between waves H_{01} and E_{11} of the circular waveguide was discussed in Sect. 6.10.

11.3   Open Resonators with Nonplanar Mirrors

11.3.1   Definitions, Examples, Brief Characteristics

Some types of axisymmetric open resonators with nonplanar mirrors have already been seen, for instance, in Fig. 11.4c–e. Compared with open resonators with the planar mirrors, these resonators have, as a rule, the essentially higher Q-factor, more stability of the operating oscillation type, etc. The unstable resonators (Fig. 11.4d) are excluded. In this section, we examine

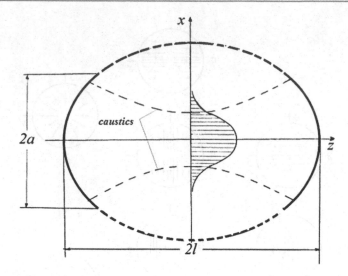

Fig. 11.9 The open resonator with the focusing mirror

almost exclusively the open resonators with *focusing mirrors*. The simple example of such an open resonator was shown in Fig. 11.4c. The *great volume* (high accumulated energy, high Q-factor) and the *rarefied spectrum* of fundamental oscillations are the essential difference between open resonators and resonators of the closed type. The latter property of open resonators relates to the self-filtering of oscillations of higher types owing to their emission from the resonator: the oscillation keeps inside the open resonator if its caustic surfaces do not fall into the mirror edge (Fig. 11.9). Otherwise, the energy of this oscillation is emitted outside the resonator.

Mirrors of the open resonator can be entire, semi-transparent (with transmission coefficient equal to, for instance, 0.99), may have one or several holes in one or both mirrors of the circular or ring type. Energy enters or exits through the holes; ring holes are usually cut in the place with the maximum spurious oscillation field (and, naturally, in the minimum intensity of the operating type). Usually, as in quasi-optical waveguides, the following conditions are satisfied for dimensions of open resonators: $ka \gg 1$, $kl \gg 1$. The mirror curvature radius $r_0 (kr_0 \gg 1)$ is, naturally, a large parameter. Focusing properties of mirrors are accepted to characterize by a ratio $\nu = 2l/r_0$. In the case $\nu = 1$, the open resonator is confocal (its mirrors have a common focus), i.e., $f = l$, and the curvature center of each mirror lies on the second mirror ($r_0 = 2l$). If the curvature centers of both mirrors coincide, this resonator is the open resonator with *concentric mirrors* ($\nu = 2$, $f = l/2$, $r_0 = l$). For planar mirrors $\nu = 0$.

The resonance condition has the form:

$$2kl = \pi q + (2m - 1)\alpha + 2\pi p, \quad m = 1, 2, \ldots, \quad \alpha = \arcsin\sqrt{l/r_0}, \tag{11.6}$$

and the p parameter ($|p| \ll 1$) defining the *emission losses* is

$$p = \frac{\pi m_2}{4(M + \beta + i\beta)^2}, \quad \beta = 0.824, \quad M \approx \sqrt{2ka^2/l} \gg 1. \tag{11.7}$$

In written relationships, q defines a number of half-waves keeping within the space between the mirrors (in quasi-optical open resonators usually $q \gg 1$), m is a number of half-waves along the x axis. In the simplest case $m = 1$. The oscillation field outside the caustic surfaces decreases according to exponential law; between caustics, the field has Gaussian-like dependences defined by *Hermite polynomials* (in the two-dimensional case, the cylindrical open resonator; $\partial/\partial y \equiv 0$) or *Laguerre* polynomials (in the three-dimensional case, the spherical open resonator).

The simplest type of oscillations of the confocal resonator is $E_{00q}^{(x,y)}$ oscillations (they are sometimes referred to as oscillation of T_{00q} type). The approximate expression for the resonance wavelength of these oscillations, as it follows from Eq. (11.6), is $\lambda_{\text{res}} \approx 4l/q$.

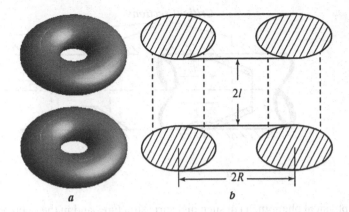

Fig. 11.10 (a) The system of two parallel tori and (b) its section. (Taken from Nefyodov and Kliuev [73], p. 365)

11.3.2 Open Toroidal Resonator

Earlier, we already became acquainted with some types of closed toroidal resonators (see the klystron toroidal resonator in Fig. 10.5a, b). In these resonators, oscillations of the internal region are studied. At the same time, there is the interesting system of two (or many) parallel tori, in which high-Q oscillations are possible as an example of the open resonator with "nonfocusing" mirrors (Fig. 11.3f–h).

The transition to the system of two coupled tori can be easily represented to take the circuit of the tuning coaxial cylindrical resonator shown in Fig. 11.3h as the initial position. Increasing the φ aperture angle of this open resonator, we come to the limit (at $\varphi \to \pi$) of the resonance structure above the ideal plane. Then, mirroring the picture obtained in the plane, we come to the required open resonator from two coupled tori (Fig. 11.10). We recommend that the reader carries out these exercises independently (see problem 2.4).

It is interesting to compare this system with the ring open resonator (Fig. 11.7h), in which oscillations are kept owing to reflection from the open edge. In the one case of absence of the azimuth field dependence (on φ), the reflection occurs as in the planar semi-infinite waveguide (Fig. 6.1b; right). If such a dependence on φ exists, we are dealing with oscillations of the *whispering-gallery* type. Oscillations in the system of two coupled tori are kept owing to wave reflection from the critical section (in Fig. 11.9b they are shown with vertical dotted lines). Emission beyond the area outside the dotted lines is small; the field beyond the oscillations area decreases according to the exponential law. Internal oscillations of the toroid are shown conditionally in Fig. 10.4c.

In the system of coupled tori (Fig. 11.10), one of them can be hollow, in which an energy is entered by the coaxial line and through the system of holes; oscillations are excited in the space between the well-conducting tori [23].

11.3.3 Open Coaxial Cylindrical Resonator with Dielectric Inserts

In many areas of physics and engineering, electrodynamic structures of the type shown in Fig. 2.3c, but with the insert from magnetic–dielectric, plasma, etc., are used. For instance, the dielectric (plasma) tube with the varying cross-section radius along the waveguide resonator (Fig. 11.11) can be used as this insert.

The lines of critical sections, from which the incident wave i in Fig. 11.10a is almost fully reflected (r), are conditionally shown by the shaded lines in the structures of Fig. 11.11. This allows the open resonator to be implemented with different locations of dielectric inserts (Fig. 11.11b, c).

Structures similar to those shown in Fig. 11.11 represent the interest from various points of view. First of all, because on their basis we can implement the open resonators with a unique rare spectrum of fundamental frequencies (see Nefyodov [23]). In particular, we can construct the open resonator with one operation frequency in the very wide bandwidth (more than an octave). We can use them in different measuring systems for determination of solid dielectric parameters, the plasma or electron flows, etc. Finally, after some variations of the cross-section shape, such structures are used in the optical range, in the quasi-optical dielectric 3D-IC of millimeter-waves [23].

Fig. 11.11 Coaxial open structures. (**a**) the key structure – the circular waveguide with semi-infinite insert. (**b, c**) Open resonators based on tube dielectric inserts of the variable sections in the circular waveguide. (Taken from Nefyodov and Kliuev [73], p. 366)

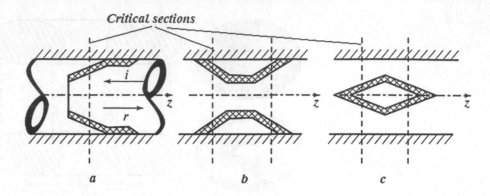

By the way, many sides of physical phenomena in such dielectric structures and in the open resonators with "nonfocusing" mirrors are awaiting researchers and engineers/designers.

11.3.4 Dielectric Resonators and Filters

Dielectric resonators are based on the open cavity structures. In the case of the disk-shaped dielectric resonator in the shield, the main oscillation type of the resonator is the H_{01m} oscillation [14].

The advantages of the dielectric resonator are small dimensions and small losses at adequate values of permittivity ($\varepsilon \gg 1$). The drawbacks are the necessity of resonator shielding, which increases the weight–dimension parameters and sensitivity to variations in environmental temperatures. The influence of temperature factors can be eliminated to a certain extent by means of the combined application of materials with additions having different temperature expansion factors.

For cylindrical dielectric resonators with face plate surfaces covered in metal, the approximate expression for the resonance frequency (in Hz) has the form

$$f_0 = \left(366/D\sqrt{\varepsilon}\right)\left(1 + 0.17(D/h)^2\right)^{1/2}. \tag{11.8}$$

where D is the cylinder diameter (in millimeters).

The resonance frequency varies when metal layers are eliminated:

$$f_0 = 24/\pi D\sqrt{\varepsilon}. \tag{11.9}$$

Expressions (11.8) and (11.9) are obtained at fulfillment of smallness conditions of the dielectric washer height: $d \ll \lambda_0/\sqrt{\varepsilon}$.

Dielectric resonators in 3D-IC of the microwave range are realized (in Russia) by means of material choice or insertion of the dielectric resonator into the necessary floor of the circuit. Thus, the preference is given to multi-layer constructions of the combined type. We know many methods of dielectric resonator excitation, which are made in the form of the cylinder or the washer, among them utilization of the asymmetric stripline located near the dielectric resonator (Fig. 11.12a) [14] or along the perimeter of the washer end (Fig. 11.12b) is, probably, the more convenient for 3D-IC of the microwave range. The strip-slot line cut, for instance, in the washer end covered by the metal is used as the exciting transmission line (Fig. 11.12c).

We must say that the highest Q-factors in open resonators were obtained precisely in disk-shaped dielectric resonators using oscillations of the *whispering-gallery* type [23].

11.4 Summary

The present chapter is devoted to the remarkable circle of issues and systems, open resonators, which present to the world a series of outstanding discoveries in the field of quantum electronics and other areas of science and technology. The idea of the open system utilization belongs to Prokhorov. Great investigations into the open systems were performed by the famous scientists Marie Paul Auguste Charles Fabry and Alfred Perot. In this chapter, we draw readers' attention to an explanation of

Fig. 11.12 Methods of dielectric resonator excitation. (**a**)With the help of a symmetric strip line. (**b**) The asymmetric stripline located near the resonator perimeter. (**c**) The strip-slot line directed along the radius of the resonator. (**d**) Dielectric resonator in the regular waveguide. (**e**) Dielectric resonators in the segment of the below-cutoff waveguide

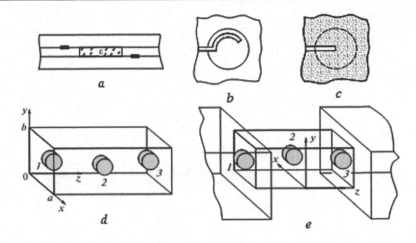

the fundamental ideas and properties of the open cavity resonators. The main property of the open resonance system consists in the fact that its spectrum of rarefaction is proportional to the first power of its linear size, whereas the rarefaction of the close resonator is proportional to the third power.

This property of the open resonator allows the creation, for instance, of a laser and other devices of quantum electronics owing to the possibility of concentrating the energy of the active substance on the single frequency.

We explain the operation principles of open resonators. In short, all of them are based on the phenomenon of effective wave reflection from the open end of the system under definite conditions. Such conditions are, say, a closeness of the wave frequency, which falls to the waveguide edge, to the critical wave for a given type (the Weinstein effect). Another model of effective reflection is the so-called critical section in the weakly irregular waveguide. And, finally, in the optical range, the high-Q oscillation is kept owing to caustic surfaces.

Sometimes, in high-power electronics, the unstable low-Q open resonators are used, in which the Sovetov formula is useful for Q-factor estimation (Sect. 10.2.2).

Section 11.2.3 is devoted to systems of diffractionally coupled open resonators, which are widely used in practice.

Applied programs in the MathCAD medium facilitate the deep assimilation of the described materials.

Checking Questions

1. The disk-shaped open resonator is shown in Fig. 11.4b. Its dimensions: a is the radius of the disks, $2l$ is the distance between them. We represent the field inside it as:

$$E_z(r,\varphi,z) = \frac{2w}{k} C J_m(wr) \cos{(m\varphi)} \left\{ \begin{array}{c} \cos \\ -\sin \end{array} \right\} \left(\frac{\pi q}{2l}z\right),$$

$$H_z(r,\varphi,z) = i\frac{2w}{k} D J_m(wr) \sin{(m\varphi)} \left\{ \begin{array}{c} \sin \\ \cos \end{array} \right\} \left(\frac{\pi q}{2l}z\right), w = k\sqrt{1-(\pi q/2kl)^2}.$$

Making use of the impedance boundary conditions of the resonance type (11.22)
, obtain the dispersion for the resonance frequencies and find out its solutions: (a) at $2kl \gg 1$ and (b) at $q = 1, 2, 3, \ldots$.

2. The plane electromagnetic wave falls under some angle α onto the planar boundary air–dielectric ($z = 0$). The permittivity is a function of z: $\varepsilon(z) = \varepsilon(0) \exp{\{0.01z/\lambda\}}$. Draw the path of the geometric–optical beams.

3. Develop the scheme of transition from the resonator with nonfocusing mirrors (Fig. 11.3h) through the scheme of a tuning resonator with nonfocusing mirrors (Fig. 11.3h) to the system from two parallel toroidal open resonators (Fig. 11.10).

4. Similar to the previous task, provide the transition from the open resonator based on the tube dielectric insert of variable sections in the circular waveguide (Fig. 11.11b) to the system from two parallel toroidal tube open resonators.

Forced Waves and Oscillations, Excitation of Waveguides and Resonators

The very important Chap. 12 about forced waves and oscillations and methods of its excitations in waveguides and resonators is the last chapter in our textbook. Up to now, we have placed great emphasis on fundamental waves in waveguides and fundamental oscillations in resonators. Now, it is time to remember the golden rule of electrodynamics (Sect. 2.2). Its sense consists in desirable and necessary continuous coupling between the field and outside sources, which do not depend upon the field under investigation.

The Kisun'ko–Feld theory turned out to be the most successful excitation theory of electrodynamic structures. The knowledge of fundamental waves in waveguides and fundamental oscillations in resonators constitutes the fundamentals of this theory. Its main difference from many other previous and less successful theories consists in the fact that the so-called head wave (i.e., that from the fundamental waves of the structure) was chosen as the Green function, whereas in usual cases the Green function of the point source can be chosen as the elementary source. This choice is not "natural" for this type of the waveguide (resonator), which leads to complicated calculation schemes (Sect. 12.1).

In microwave and millimeter-wave engineering, there is a great number of various excitation devices. The pins (metallic or magnetic–dielectric), loops with a current, holes (or hole systems and many other radiators of small dimensions (less than a wavelength) close to elementary radiators in their properties are the traditional excitation elements for transmission lines and resonators. Several types of exciters are given as examples (Sect. 12.2).

Several excitation circuits typical of three-dimensional (3D) microwave and millimeter-wave structures are examined. We would like to note that a precise multi-layer 3D structure of constriction (so-called 3D IC) allows usage of all module volumes with the most natural manner for the optimal arrangement of constitutive basing elements; thus, according to the principle of constructive conformity (the Nefyodov–Gvozdev principle), these basing elements can be performed based on any known types of guiding structures in addition to new (typical for 3D IC) types of transmission lines (Sect. 12.3).

Exciters can be divided into several classes according to the character of their application. First of all, they are ultra-wideband exciters, i.e., junctions with a direct galvanic contact (Sect. 12.3.2). This class of exciters has gained special importance after the development of the 2.5D technology. We give many examples of exciters and their parameters. Another exciter class is wideband stub junctions (Sects. 12.3.3). It is necessary to note that mounted bridges are absent in many junctions of this type, which makes them promising for high-frequency integrated circuits.

The narrowband resonance exciters of the slot type are interesting for practice (Sect. 12.3.4). They ensure, in particular, inter-layer (inter-flour) connections in microwave and millimeter-wave 3D IC. In addition, the slot junction provides the leakproofness of the connection in 3D IC, in addition to the DC isolation with external circuits, which protects the bias-circuits against external electromagnetic interference.

12.1 Introduction, General Considerations, Definitions

12.1.1 The Fundamental Eigenwaves in Waveguides and Fundamental Oscillations in Resonators

Up to now, we have considered eigenwaves of waveguides and *fundamental oscillations* of resonators. In other words, those waves and oscillations that, in principle, may exist in each structure independently on everything, i.e., we did not discuss the sources of these wave processes. Now, we must answer the question: which processes (waves, oscillations) have occurred and

© Springer International Publishing AG, part of Springer Nature 2019
E. I. Nefyodov, S. M. Smolskiy, *Electromagnetic Fields and Waves*, Textbooks in Telecommunication Engineering,
https://doi.org/10.1007/978-3-319-90847-2_12

Fig. 12.1 (a) The scheme of waveguide excitation and (b) the resonator by outside sources, which are concentrated in areas V_j with surfaces S_j

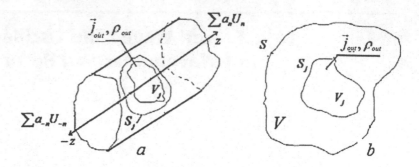

with what intensity in any system under quite definite conditions (at *excitation*, for example, by *outside sources* $\vec{j}_{\text{out}}$, ρ_{out} located in some, generally speaking, the arbitrary area V_j, which is restricted by the surface S_j (Fig. 12.1).

The fundament of the answer to this question is the property of *eigenwaves (oscillations) orthogonality*. Earlier, in Sect. 6.3, we already mentioned that this property means the *independence of eigenwaves (oscillations)* of the waveguide (resonator) with each other under the condition of the waveguide regularity, i.e., independence, say, of the shape and dimensions of its cross-section on the longitudinal coordinate z, on the uniform filling along it.

Beginning a conversation about the excitation of transmission lines and resonance structures by outside sources, it is necessary to note that the main contribution in the theory of electrodynamics structure excitation was made by Soviet and Russian scientists. The most important results belong to G.V. Kisun'ko, whose achievements surpassed the results of excellent scientists L. de Broglie, I. Mandelstam, A. Samarsky, A. Sommerfeld, A. Tikhonov, Y. Feld, and many others.

This theory gained its most complete form in the results of G. Kisun'ko and Y. Feld. Therefore, it is named Kisun'ko–Feld theory.

12.1.2 Forced Waves in Waveguides and Forced Oscillations in Resonators

It is assumed that the eigenfunction system of the considered waveguide is known, and the total waveguide field can be written (in any transverse section $S_\perp$) in the form of expansion of the rectangular waveguide (RW) type

$$\Pi_{\text{total}}^e = \sum_{m=-\infty}^{m=\infty} \sum_{n=-\infty}^{n=\infty} D_{mn} \sin\frac{m\pi}{a}x \sin\frac{n\pi}{b}y\, e^{ih_{mn}z},$$

given in (6.19). Let field sources $\vec{j}_{\text{out}}$, ρ_{out} be located in area V_j along the z-axis $z \in (z_1, z_2)$; the transverse section at $z = z_1$ is designated $S_{\perp 1} \equiv S_1$ (Fig. 12.1a). Then, for the right area ($z > z_2$; at $z = z_2$ the transverse section is S_2) from V_j, the field can be represented in the form of the following expansions over waves aspiring to $z \to +\infty$:

$$\vec{E} = \sum C_n \vec{E}_n, \qquad \vec{H} = \sum C_n \vec{H}_n \qquad (z > z_2). \tag{12.1}$$

On the left of the V_j area occupied by sources $\vec{j}_{\text{out}}$, ρ_{out}, fields can be written in the form of Eq. (12.1), but with replacement $n \to -n$, where the sign "-" means that waves from the source area go to the left (in Fig. 12.1a; $\sum C_{-n} U_{-n}$). Complex amplitudes C_n of the "direct" waves (propagating toward the positive z) and backward waves C_{-n} are obtained from the *Lorentz lemma* regarding the coupling of two fields and are determined as:

$$C_n = \frac{1}{N_n} \int\limits_{V_j} \left(\vec{j}^e \vec{E}_{-n} - \vec{j}^m \vec{H}_{-n}\right) dV, \quad C_{-n} = \frac{1}{N_n} \int\limits_{V_j} \left(\vec{j}^e \vec{E}_n - \vec{j}^m \vec{H}_n\right) dV. \tag{12.2}$$

In the latter expression, as N_n, we designate the norm of fundamental oscillation with an index n, i.e., $N_n = \mp\frac{c}{2\pi} \int\limits_{S_\perp} \left[\vec{E}_n \vec{H}_n\right] d\vec{S}$, which, in essence, is a power carried by the wave with an index n through the transverse section $S_\perp$.

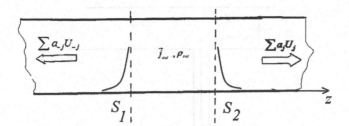

Fig. 12.2 The scheme of waveguide excitation by distributed sources in the area between planes S_1 and S_2

At the derivation and writing of the main equations in the theory of waveguide excitation (12.2), at least the following three circumstances happen to be significant. First, integration in Eq. (12.2) is carried out over the V_j area only, which is occupied by sources $\vec{j}_{\text{out}}$, ρ_{out}. Thus, as the V_j area, we can choose the volume with sources $\vec{j}_{\text{out}}$, ρ_{out}; sources can be both real (for instance, the electric current $\vec{j}^{\,e}$) and imaginary (equivalent, virtual, etc.; for example, the magnetic current $\vec{j}^{\,m}$). the plane with distributed currents and charges may serve as the excitation area, then the integral in Eq. (12.2) is transformed into the surface integral. If the thin wire acts as the exciting wire, the integral becomes the contour integral over the conducting line with a current.

Second, as the *Green function* (compare Sect. 3.3), we can choose the "oncoming" wave: for waves going to the right (Fig. 12.2a), this is the wave with index "$-n$," and if the wave goes toward negative z, the index is "$+n$." Utilization as the Green function of its representations (sometimes they are referred to as *source-wise*) in the form of the plane, cylindrical or spherical waves in the tasks of waveguide (and resonator) excitation leads to non-elegant and nonsimple expressions for determination of $C_{\pm n}$ amplitudes of Eq. (12.2) type, but also to the *system of linear algebraic equations*, which can be practically always solved by numerical methods only. The idea of "oncoming" wave application such as the Green function is attributed, apparently, to Russian scientists G. Kisun'ko and Y. Feld.

And, finally, the third circumstance related to a choice of the desired effective shape of excitation current. For example, let in the first formula of (12.2) $\vec{j}^{\,m} \equiv 0$. Then, an amplitude is $C_n = (1/N_n) \displaystyle\int_{V_j} \vec{j}^{\,e}\vec{E}_{-n}\, dV$, hence, the C_n amplitude is maximal if both integrands $\vec{j}^{\,e} \equiv \vec{j}$ and $\vec{E}_{-n}$ happen to be the "same," coinciding in the shape. We know that the scalar product of two vectors $\vec{j}$ and $\vec{E}_{-n}$ is maximal if these vectors are parallel (collinear) to each other.

Expressions (12.2) define the field outside V_j area occupied by $\vec{j}_{\text{out}}$, ρ_{out} sources. Appropriate generalizations of formulas (12.2) for determination of fields in the V_j area (Fig. 12.2) occupied by sources can be easily done; we refer the reader to Abraham and Becker, and Sommerfeld [131–132].

12.2 Some Excitation Devices

The main excitation elements for transmission lines and resonators are *pins* (metallic or magnetic–dielectric), *loops* with a current, *holes* (or *hole systems*), and many other small radiators (less than a wavelength) closed to *elementary radiators* in their properties. In the wider sense, the excitation devices represent *junctions* between different transmission lines. So, for instance, the joining of coaxial and plain (rectangular or circular) waveguides is shown in Fig. 12.3.

The junction role is especially important in 3D-ICs of microwave and millimeter-wave ranges, which contain many types of transmission lines in the one basing element and, moreover, in one functional unit. Figure 12.4a shows the excitation circuits of infinite length waveguides; hence, excitation has a "two-sided" character. In practice, excitation usually has a "one-sided" character (Fig. 12.3), and for its provision, for example, on the left of the exciting pin, the short-circuiting piston is placed. The distance from the pin to the piston is selected so that the wave (i) goes directly to the right and the wave ($-i$) reflected from the piston (r) in the waveguide would add an in-phase.

Some excitation devices are shown in Figs. 12.4 and 12.5. As examples, "traditional" circuits of waveguide excitation are presented in Fig. 12.4a–f.

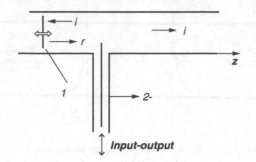

Fig. 12.3 The scheme of "one-sided" excitation of the waveguide by the coaxial transmission line: *1* a piston, *2* the coaxial line

Fig. 12.4 Methods of transmission line excitation. € rectangular waveguide (**a**) by a pin, (**b**) by a loop, (**c**) via a hole in a wall, (**d**) of the stripline by the coaxial waveguide. Compare Fig. 13.1a), (**e**) the Goubau line, (**f**) the waveguide by a pin and the moving piston. (**g, h**) Compare Fig. 13.2; excitation circuits in the space $z > 0$ of the electromagnetic field with explicitly clear component E_z of the electric field. (Taken from Nefyodov and Kliuev [73], p. 372)

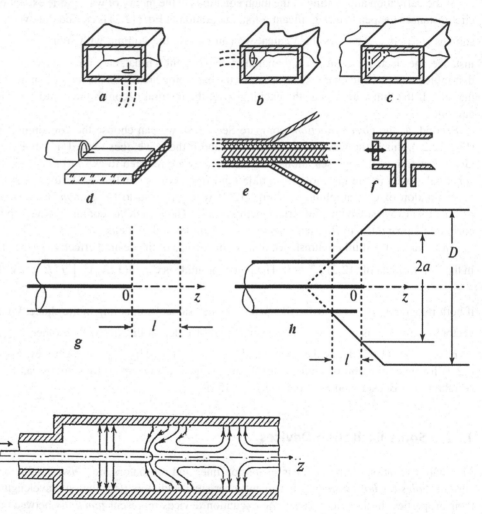

Fig. 12.5   Joining of coaxial and circular waveguides

For some applications (medicine, biology, ecology, etc.), it is interesting to study the electromagnetic field structure in the near zone of the antenna, when the R distance from the aperture with D size (Fig. 12.4h) satisfies the condition $R < D/4 + (D/2) (D/\lambda)^{1/3}$ [6, 66]. At transformation of the *TEM* wave of the coaxial waveguide (CW; Fig. 3.5; $z < 0$) into the E_{01} wave of the circular waveguide (Fig. 12.5; $z > 0$), the modulus of the reflection coefficient from irregularity is defined as $|R| = (k - h)/(k + h)$ [125]. Here h is the longitudinal wave number of the E_{01} wave.

The permanent condition of formula correctness for $|R|$ in the circuit in Fig. 12.5 is a requirement that the *TEM* wave can only propagate in the coaxial waveguide, whereas the E_{01} wave can propagate in the circular waveguide. This wave has a longitudinal component E_z of the electric field; it is kept in the near field of the conical horn (Fig. 12.4h), which is necessary for the applications under consideration [23]. We note that the expression for $|R|$ reminds us of the form of the known formula for the reflection coefficient, when two transmission lines are joined with the same wave type, namely: $|R| = (w_1 - w_2)/(w_1 + w_2)$, and wave impedances $w_{1,2}$.

A choice of the aperture angle of the conical horn and its length are defined by the necessary pattern width ($\vartheta \approx \lambda_0/D$) and are matched with angles $36°$, $76°$ or $108°$ related to angles of the golden triangle [62]. The rod length l from the cone opening expands into the region of the critical section for the E_{01} wave: $\lambda_0/2a = 1.31$.

Application of various types of inserts (dielectric or impedance) on the end of the coaxial internal conductor (Fig. 12.4), in as addition to impedance (for example, the frequently-periodic) structures 2 on the antenna mirror allows significant expansion of the area of uniform illumination (in Fig. 12.4 the almost uniform area of illumination is shaded; compare Fig. 12.4h). We can use this circumstance (and the circuit) during experiments on the interaction of the electromagnetic field with various media [27, 46].

Any adapter units (impedance *transformers*, admittance *transformers*, *converters* of wave types or oscillation types, etc.) are the specific, peculiar excitation devices.

12.3 Excitation Circuits Typical of Three-Dimensional Structures in Microwave and Millimeter-Wave Ranges

12.3.1 Main Idea

In the previous chapter, we considered in brief the main classes of directing (waveguiding) structures and some basing elements (for instance, resonators, quasi-optical devices, etc.) based on its use in microwave and millimeter-wave modules of radio electronic equipment, including on 3D-ICs. The main distinctive feature of 3D-ICs (compared with microwave planar ICs) consists in their multi-layer property. We have already mentioned above that precisely the multi-layer (three-dimensional) property of 3D-IC construction permits in the most natural manner the use of the whole module volume for optimal accommodation of its basing elements and components. Thus, in accordance with ideas of the *construction conformity principle*, these basing elements can be implemented on any known (from microwave engineering) types of guiding structures and on new types of transmission lines typical of 3D-IC. In this regard, at microwave 3D-IC implementation, the new, very interesting problem of microwave engineering arises. This problem concerns the realization of coupling between different types of transmission lines not only in the layer plane (a floor) of 3D-IC, but in the perpendicular ("vertical," inter-layer) direction. Thus, in essence, the idea of transition between transmission lines or between basing elements is reduced to the problem of excitation of one element of the circuit by another element. The necessity of implementation of inter-floor transitions–exciters leads to new (probably typical of the microwave and millimeter-wave 3D-IC) class of devices for coupling of the similar transmission lines located on the various "floors" of 3D-IC. 3D-IC differs from planar IC in the greatly developed system of inter-layer transitions, which does not "disfigure" the circuit itself, but also allows implementation of a series of additional functions with the help of transitions; namely, phase shifters, decoupling in DC, matching circuits, pass-band filters, etc.

To date, a large number of transitions–exciters have been offered and implemented, which can be classified, for example, according to their operating frequency band; in this case, devices constitute the following three groups:

- *Super-wide-band* (the band is up to several octaves – these are usually transitions with direct galvanic contact)
- *Wide-band* (the frequency band is up to one octave; more often, these are called stub transitions)
- *Narrow-band* (the frequency band in units of percent – this is usually a class of the resonance transitions that is related, as a rule, to the transition class of the slot type).

Of course, other classifications are possible, for instance, according to the class of joining transmission lines with the different transverse field structures, then the transition should perform the wave transformation task providing the maximum possible transfer coefficient in the required frequency band.

In this section, we consider the main classes of transitions used in microwave and millimeter-wave 3D-ICs. Here, we give results of quasi-static approximation for the scattering matrix. The analytical electrodynamics theory of most transition devices is still awaiting solutions. However, computer technology, which is at the engineer's disposal with appropriate software (for instance, based on high-frequency structure simulator), allows construction of acceptable excitation devices.

Fig. 12.6 Transitions with direct galvanic contact to (**a–c**) coaxial and (**d–g, i**) rectangular waveguides; (**h, j–l**) inter-layer transitions. Designations in figure: CW coaxial waveguide, ASSL asymmetric stripline, CL coplanar line, SSIL strip-slot line, RW rectangular waveguide, SSL symmetric stripline

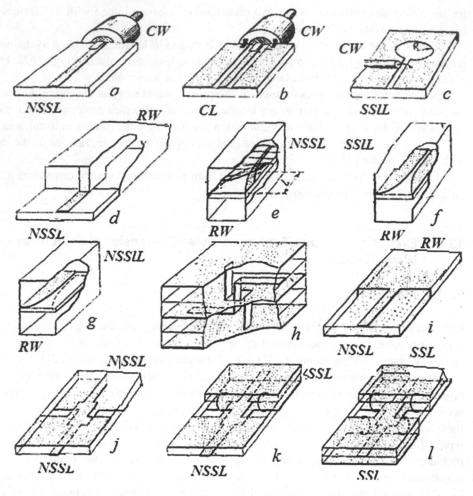

12.3.2 Super-Wide-Band Exciters: Transitions with Direct Galvanic Contacts

A large number of elements are concerned with this class of devices; some of them are shown in Fig. 12.6.[1] Conductors of joined transmission lines have a direct galvanic contact, which provides the high wide-band properties of this device class together with a similar field structure. Let us examine briefly the most typical types of these basing elements.

Transition from the coaxial waveguide to an asymmetric stripline (CW↔ASSL; Fig. 12.6a) has been known since the beginnings of microwave integrated circuitry. This transition had gained widespread application because of the coincidence of the field structure in CW and ASSL. Besides, the dispersion absence in the CW practically allows joining lines in the wide frequency band for $K_{\text{swr U}} < 1.1$ (standing wave ratio in voltage). The restriction of the maximal frequency of such an exciter is defined by the danger of higher types of waves arising in the CWs, whereas the weak dispersion of the ASSL does not have a practical influence on its characteristics. Transition at substrate thickness of the ASSL was equal to the external radius of the CW, and at a conductor width ASSL is slightly greater than the diameter of the internal conductor of the CW. The reliable electric contact with the ASSL conductor is provided by extension of the internal conductor of the CW.

Sometimes, some of the devices shown in Fig. 12.6 are considered to have lost significance for the microwave range. Nevertheless, it can seem strange that some of them seem to reappear in the millimeter-range, in which the transmission lines (in one atom in thickness) join with necessity to various components and basing structures, which are much more in dimensions. They constitute the three-dimensional structures of the element as a whole.

---

[1] A number of known exciters with galvanic contact is rather large. However, 3D-ICs is not always manufacturable for its realization in microwave; therefore in Fig. 12.6 only those types of exciters are shown, which are suitable for 3D-IC.

At realization of the exciter on the symmetric stripline (SSL), it is sufficient to deposit the dielectric layer from above on the ASSL; this dielectric is metallized from the external side, and the metal layer should be galvanically connected with the external conductor of the CW.

In the exciter of the coaxial-waveguide–coplanar line (CW↔CL (Fig. 12.6b) joining occurs in a similar manner. The galvanic contact of the external conductor of the CW with wide conductors of the CL occurs with the help of metallic bridges. Thus, they must be located as close as possible to the gap edges of the CL, the distance between them is equal to the diameter of the external conductor of the CW. The described exciters operate in the range from zero to frequencies of higher types of waves arising in the joining lines (ASSLs, CWs, and CLs). Conductance, which arises in the area of the galvanic contact of joining conductors, affects the device characteristics. Therefore, the prolonged part of the internal conductor of the CL and metallic bridges are made in the form of contact tabs.

For excitation of the symmetric slot line (SSIL), the transition with orthogonal location of the CW and symmetric slot line is successfully used (Fig. 12.6c). This is caused by the fact that the field structure in the joining lines is different (in a CW there is a T-wave, whereas in SSIL we have H-waves). In this transition, the external conductor of the coaxial resonator is connected galvanically with the edge of one side of the symmetric slot line, and the internal conductor is connected with another side. Beyond the connection area, the symmetric slot line is open-ended. The profile of the prolonged part of the internal conductor of the CW in the slot area repeats the bend of *r*adius of the electric field line. In this case, the transformation coefficient of the wave is

$$n = (\pi/2)\left|k_{tr}rH_1^{(1)}(k_{tr}r)\right|, \tag{12.3}$$

where $k_{tr} = ik(\varepsilon_{eff} - 1)^{-1/2}$, $k = 2\pi/\lambda$.

The best matching is achieved for the following condition fulfillment

$$Z_1 = Z_2 n^{-2}, \tag{12.4}$$

where Z_1 and Z_2 are wave impedances of joining lines (exciting and receiving).

The open end of the symmetric slot line (with the purpose of decreasing losses on emission) is made in the form of a circle with an R radius cut in the metallic layer.

In a similar manner, we can perform the *transition from an ASSL to a symmetric slot line*, which is often used in microwave 3D-ICs. For this, it is sufficient, instead of the CW, to insert the additional dielectric layer with the ASSL, whose conductor is connected by the metallic bridge to the opposite edge of the slot, through the hole in the dielectric.

The transition from the RW to the ASSL of the "knife" type (Fig. 12.6d) provides transformation of the H-wave into the T-wave and vice versa. Taking into consideration the necessity of a complicated transformation of wave types, the cascade connection of three types of lines is used in the transition. The RW is connected with the Π-shaped waveguide, which has a wave impedance equal to the impedance of the ASSL, with the help of the graded or stepped junction. The width of the Π-shaped waveguide ridge is equal to the width of the ASSL conductor, and the air gap is equal to the substrate thickness of the ASSL. The ridge edge is connected galvanically by the metallic bridge with the conductor of the ASSL.

Considered transitions were used at the experimental working-off of 3D-IC modules in the microwave range in connection with the necessity of their connection to standard measuring waveguide-coaxial sections.

In microwave 3D-ICs, the partially or fully shielded lines, for which the RW serves as a suitable model, find a very wide application. Some examples of a RW joining with different types of the planar transmission lines are given in Fig. 12.6e–g. The graded Chebyshev transition from the RW to the ASSL or to the symmetric slot line forms the basis of these transitions. The most complicated construction transition is from the RW to the shielded ASSL (Fig. 12.6e). Edges of the conductor and the metallic layer form the profile of the R radius. The transition area has the L length, which is approximately equal to the wavelength in the ASSL. The frequency response of the direct losses of this transition in the millimeter-wave range is shown in Fig. 12.7.

Let us pass to examination of *inter-layer transitions*, which are typical of microwave 3D-ICs. The inter-layer connections due to metallic bridges through holes in the dielectric layers (Fig. 12.6i) were one of the first transitions of this class. At assembly of multi-layer circuits on SSLs, transitions between SSLs, in which conductors and metal layers of adjacent floors are connected galvanically by the band conductors, are used. Thus, the standing wave coefficient happens to be less than 1.15 in the centimeter wavelength range (Fig. 12.8).

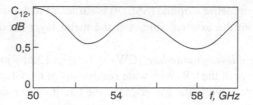

Fig. 12.7 Frequency response of the transfer coefficient in the transition ASSL ↔ RW (Fig. 12.6e)

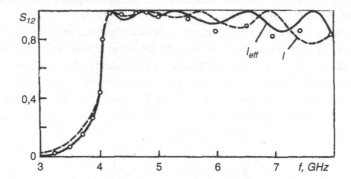

Fig. 12.8 Frequency response of the transfer coefficient in the transition ASSL↔RW filled by the dielectric (Fig. 12.6i); the solid curve is for transition taking into account the edge capacity of the open end of the waveguide; dotted line is experiment

The transition ASSL↔RW (rectangular waveguide) filled by the dielectric (Fig. 12.6i) is original. First, it opens up wide possibilities of rectangular waveguide application (or their semi-open models) in microwave 3D-IC, and second, it allows implementation of frequency-selected elements and units. The exciter ASSL↔RW is made in the dielectric layer, which has the common metallic shield from one side, whereas from the other side there is the conductor of the ASSL, which transits by a jump in the metallic wall of the rectangular waveguide. The side walls of the rectangular waveguide are metallized. In the case of a large difference in wave impedances between the ASSL and the rectangular waveguide, the grading of stepped transition is connected on the ASSL. The cut-off frequency is the main electrical parameter of the transition:

$$f_c = c/2a\sqrt{\varepsilon}, \tag{12.5}$$

The remarkable property of microwave 3D-ICs, as we have already mentioned, is the layer-wise accommodation of ASSLs and SSLs. Large dimensions, which are connected to the necessity of the graded variation of the conductor width, are the drawback of this transition. Figure 12.6j shows the simpler transition construction between two ASSLs, whose current-carrying conductors are located in different sides of the dielectric layer. In this construction, the area of graded conductor variation is changed by the SSL segments with conductors of restricted width. The minimal length of the SSL segment is selected from conditions of excitation absence of the transverse H-wave arising between edges of metal layers. In the case of a large difference in the wave impedances of joining ASSLs, the segment of the SSL with conductors of restricted width (together with the functions of the coupling element) is the $\lambda/4$-transformer. Experimental investigations into the transition between ASSLs with equal wave impedance showed that in the frequency band 1–10 GHz the transition has the reflection coefficient $K_{\text{srw U}} \leq 1.2$, and direct losses are less than 0.2 dB.

If we need to connect the ASSL with the SSL, it is sufficient to deposit the dielectric layer, which is metallized from the external side (Fig. 12.6k), on one of the conductors of the asymmetric line. To equalize potentials in the SSL formed and to suppress the volumetric waves in it, which arise in the transverse direction, the metallic layers are connected galvanically by metal bridges through holes in dielectric layers, which are simultaneously fastening connections in microwave 3D-ICs.

In a similar manner, the transition between layer-wise located SSLs (Fig. 12.6l) is formed.

12.3.3 Wide-Band Stub Transitions

Less wide-band (frequency band up to one octave), so-called stub transitions, in which potential equalizing in conductors of joining lines is provided with the help of $\lambda/4$ stubs, are used in microwave 3D-ICs. Open-ended or short-circuited stubs can

Table 12.1 Topologies, equivalent circuits, and scattering matrices of transitions

N.№	Transition name	Sketch of topology	Equivalent circuit	Scattering matrix
1	ASSL⇔ SSIL		*1:n*	$S_{11} = -S_{22} = (n^2 - 1)/(n^2 + 1)$, $S_{12} = S_{21} = 1/(n^2 + 1)$, $n = \cos p - ctg\, q \cdot \sin p$, $p = 2\pi(d/\lambda_0)\left(\varepsilon - \varepsilon_{\text{эф}}^{\text{ш}}\right)^{1/2}$, $q = p + arctg\left[\left(\varepsilon - \varepsilon_{\text{эф}}^{\text{ш}}\right)^{1/2} / \left(\varepsilon_{\text{эф}}^{\text{ш}} - 1\right)^{1/2}\right]$
2	CL⇔CL		Z_{strip}, Z_{line}, l_{strip}	$S_{11} = S_{21} = i\left(1 - C_1^2 - C_2^2\right)/R$, $S_{12} = S_{21} = -2C_2/R$, $R = i\left(1 + C_1^2 + C_2^2\right) - 2C_1 C_2$, $C_1 = ch\, \gamma_{\text{line}} l_{\text{line}} + (Z_1 + Z_{\text{line}}) sh\, \gamma_{\text{line}} l_{\text{line}}$, $C_2 = \left[\begin{array}{c}(Z_1/Z_{\text{line}} - Z_{\text{line}}) sh\, \gamma_{\text{line}} l_{\text{line}} \\ -2Z_1 ch\, \gamma_{\text{line}} l_{\text{line}}\end{array}\right]/\rho$, $Z_1 = Z_{\text{strip}} cth\, \gamma_{\text{strip}} l_{\text{strip}} + 2\pi f L$, $\gamma_{\text{strip}} = \alpha_{\text{strip}} + i\beta_{\text{strip}}, \gamma_{\text{line}} = \alpha_{\text{line}} + i\beta_{\text{line}}$
3	CL⇔CL		Z_{slot}, l_{slot}	$S_{11} = S_{22} = Z/(Z + 1)$, $S_{21} = S_{12} = 1/(Z + 1)$, $Z = Z_{\text{slot}} \cot \gamma_{\text{slot}} l_{\text{slot}}/\rho$, $\gamma_{\text{slot}} = \alpha_{\text{slot}} + i\beta_{\text{slot}}$
4	CCiL⇔ CCiL		l_{slot}, Z_{slot}	See p. 3 $Z = Z_{\text{slot}} \cot \gamma_{\text{slot}} l_{\text{slot}}/\rho$
5	CL⇔ASSL		l_{slot}, Z_{slot}	$S_{11} = S_{22} = -Z/(Z + 1)$, $S_{12} = S_{21} = 1/(Z + 1)$, $Z = Z_{\text{slot}} \tan \gamma_{\text{slot}} l_{\text{slot}}/\rho$
6	ASSL⇔ ASSL		l_{slot}, Z_{slot}	See p. 6 $Z = Z_{\text{slot}} \coth \gamma_{\text{slot}} l_{\text{slot}}/\rho$
7	CL⇔SSll		Z, l_{slot}	See p. 3 $Z = 2Z_{\text{slot}} \tanh \gamma_{\text{slot}} l_{\text{slot}}/\rho$

ASSL asymmetric stripline, *SSIL* strip-slot line, *CL* coplanar line

connect in the transition in series and/or in parallel depending on the types of joining lines. Below, we consider a series of specific examples of stub-type transitions.

The inter-layer transition ASSL↔SSIL uses a coupling in the magnetic field. This transition represents two mutually perpendicular transmission lines intersecting in loops of magnetic fields. Beyond an intercepting point, the symmetric slot line is ended by the $\lambda/4$ short-circuited stub, and the ASSL is ended by the $\lambda/4$ open-ended stub (Table 12.1, p.1). Transition frequency responses $K_{\text{swr U}}$ are shown in Fig. 12.9. Losses in the transition are less than 0.2 dB [14].

With use of the same principle, the inter-layer transition (CL ↔ CL) is built (Table 12.1, p.2; [14]). Narrow conductors of the CL transit to conductors of the SSLs of the same width. In the area of the break of metal layers of the CL, conductors of the

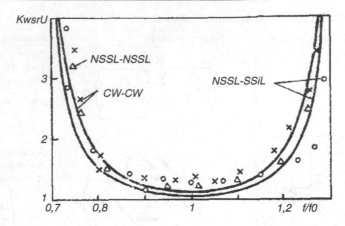

Fig. 12.9 Frequency response of the standing-wave ratio of stub transitions: solid curves – calculation; o, ×, Δ – experiment

SSL are ended by $\lambda/4$ open-ended stubs on ASSLs, which equalize the potentials of joining lines and simultaneously invert the signal phase by $180°$. The segment length of the SSL is chosen from the condition of the prevention on metal edges of emitted waves of H-type. The feature of CL $\leftrightarrow$ CL transition is the decoupling of joining lines on DC.

In the transition under consideration, the asymmetric slot line formed by the metal edges should not be excited. In principle, the asymmetric slot line can be used as the matching $\lambda/4$ stub. This type of transition is examined in [14], where transitions CL$\leftrightarrow$CL and ASSL$\leftrightarrow$ASSL (Table 12.1, pp.3 and 4) on the $\lambda/4$ open-ended stubs are analyzed. Thus, in transitions presented in pp. 3–5, stubs are connected in a series, and in p.6 they are connected in parallel. Frequency responses of transitions CL$\leftrightarrow$CL and ASSL$\leftrightarrow$ASSL (Table 12.1, pp. 2 and 6) are shown in Fig. 12.9.

Parallel connection of the $\lambda/4$ short-circuited stub on the symmetric slot line was used in the transition CL$\leftrightarrow$SSIL (Table 12.1, p.7). It should be noted that in all the transitions under consideration, the hinged bridges are absent (besides the transition CL$\leftrightarrow$CL; Table 12.1, p.3) and at adjustment we need not assemble. The manufacturability and construction simplicity of stub transitions give the possibility of applying them up to millimeter range. In the right column of Table 12.1, elements of scattering matrices are given for considered transitions of the stub type. They are obtained on the basis of equivalent schemes presented in Table 12.1. In quasi-static approximation, calculation results are in good agreement with experimental data.

In conclusion, we give a summary of designations applied in Table 12.1: S_{mn} are coefficients of the wave scattering matrix; d is the substrate thickness; ε is the substrate permittivity; ε_{eff}^{slot} is the effective permittivity of the symmetric slot line; Z_{strip}, Z_{slot}, and Z_{line} are wave impedances normalized to the impedance of joining lines; ρ, α_{strip}, α_{slot}, α_{line}, β_{strip}, β_{slot}, β_{line} are damping constants and phase constants in lines; l_{strip}, l_{slot}, l_{line} are geometric lengths of stubs on the ASSL and the symmetric slot lines; L is the inductance of joining segments; and f is the current frequency.

12.3.4 Narrow-Band Resonance Exciters of the Slot Type

The necessity for inter-layer (inter-floors) coupling arranged in microwave 3D-ICs leads to the wide application of slot transitions, when two or several transmission lines are coupled through the resonance slot. The slot transitions provide leakproofness of the microwave 3D-IC connection, in addition to DC decoupling with external circuits, which protect the bias circuits and semiconductor devices against external electric interference.

A series of specific examples is presented in Table 12.2. Slot transitions represent two transmission lines connected by the $\lambda/2$ slot resonator, which is cut in the common metallic layer and located in loops of the magnetic field of coupled lines. Transmission line coupling is performed owing to components of magnetic fields directed along the slot resonator. Conductors of symmetric and ASSLs lay across the slot symmetrically with regard to its ends and finish by the contactless short-circuit device. Usually, the $\lambda/4$ open-ended stub serves as this device.

In transitions ASSL$\leftrightarrow$RW and SSL$\leftrightarrow$RW (Table 12.2, pp. 1 and 2) the slot resonator is located on the wide wall near the short-circuited edge of the waveguide or in the middle of the edge of the waveguide between its wide walls.

In transitions ASSL$\leftrightarrow$CW and SSL$\leftrightarrow$CW (Table 12.2, pp. 3 and 4), the resonance slot is folded into the half-ring and located in the short-circuited edge of the CW symmetrically between the internal and external conductors. The simplest

Table 12.2 Sketch of the construction and loaded Q-factor of the transition

#№.	Transition name	Sketch of construction	Loaded Q-factor of the transition
1	ASSL⇔RW		$Q_{\text{load}}^{-1} = \alpha\lambda_{\text{slot}}/\pi + Z_{\text{slot}}l_{\text{res}}^2\big[1 - (\lambda_0/2a)^2\big]^{1/2}\cos^2(\pi l_{\text{res}}/2a)/15\pi^3 ab \times \Big[1 - (l_{\text{res}}/a)^2\Big]^2 + (Z_{\text{slot}}/\pi Z_3)\exp(-2\pi d/l_{\text{res}})$
2	SSL⇔RW		$Q_{\text{load}}^{-1} = \alpha\lambda_{\text{slot}}/\pi + Z_{\text{slot}}l_{\text{res}}^2\big[1 - (l_{\text{res}}/2a)^2\big]^{1/2}\cos^2(\pi l_{\text{res}}/2a)/15\pi^3 ab \times \Big[1 - (l_{\text{res}}/a)^2\Big]^2 + (4Z_{\text{slot}}/5\pi Z_{43})\exp(-2\pi d/l_{\text{res}})$
3	ASSL⇔CW		$Q_{\text{load}}^{-1} = \alpha\lambda_{\text{slot}}/\pi + Z_{\text{slot}}l_{\text{res}}^2/\big[30\pi^4 R^2 \times \ln(D/d)\big] + (Z_{\text{slot}}/\pi Z_3)\exp(-2\pi d/l_{\text{res}})$
4	SSL⇔CW		$Q_{\text{load}}^{-1} = \alpha\lambda_{\text{slot}}/\pi + Z_{\text{slot}}l_{\text{res}}^2/\big[30\pi^4 R^2 \times \ln(D/d)\big] + (4Z_{\text{slot}}/5\pi Z_3)\exp(-2\pi d/l_{\text{res}})$
5	ASSL⇔SSlL		$Q_{\text{load}}^{-1} = \alpha\lambda_{\text{slot}}/\pi + (1/Z_3 + 4/5Z_4) \times (Z_{\text{slot}}/\pi)\exp(-2\pi d/l_{\text{res}})$
6	ASSL⇔NSSL		$Q_{\text{load}}^{-1} = \alpha\lambda_{\text{slot}}/\pi + (2Z_{\text{slot}}/\pi Z_3)\exp(-2\pi d/l_{\text{res}})$
7	SSL⇔SSL		$Q_{\text{load}}^{-1} = \alpha\lambda_{\text{slot}}/\pi + (8Z_{\text{slot}}/5\pi Z_{43})\exp(-2\pi d/l_{\text{res}})$

CW coaxial waveguide, *RW* rectangular waveguide

constructions have inter-layer transitions ASSL↔SSL, ASSL↔ASSL and SSL↔SSL (Table 12.2, pp.5–7), in which conductors of joining lines are located mirror-like with respect to the slot resonator.

12.4 Summary

Chapter 12 is devoted to the study process of forced wave and oscillation implementation in the system examined earlier.

Thus, we fulfill the recommendation of the golden rule of electrodynamics (Sect. 2.2): at given outside sources $\vec{j}_{\text{out}}$, ρ_{out} and given their location, we must determine the field in the whole structure. In general, it is assumed that this problem is also solved in the area occupied by sources. The most complete form of this theory was achieved in the investigations by Kisun'ko and Feld. Therefore, it is by rights called the Kisun'ko–Feld theory.

It is essential that the eigenfunction system of the waveguide under investigation is known and the waveguide (resonator) total field can be written (for any cross-section) in the form of expansion over this system. An essence of the Kisun'ko–Feld theory consists in the choice of the emission source. Frequently, the source-wise variant of the Green function in the form of a delta-function was chosen as such a source – the Green function. Such an approach severely complicated the problem and it was reduced to the infinite system of linear algebraic equations. Thus, a problem of excitation was reduced to the numerical solution of this equation system.

Instead, in the Kisun'ko–Feld theory, one of the eigenwaves of the given system was chosen as the Green function. It was called the head wave and the whole excitation problem was reduced at the end to quadratures.

A large part of Chap. 12 is devoted to the representation of excitation devices. Generally, any junction between homogeneous or heterogeneous transmission lines can be considered to be the excitation device, which excites the field of one line by the field in another line. In 3D ICs of microwave and millimeter-wave ranges (up to optical ranges) in particular, there especially many junctions of this type.

We have described and discussed various examples of excitation systems.

Checking Questions

1. What will the pictures of field lines of electric and magnetic fields of main waves in transmission lines shown in Fig. 12.6 (regardless of excitation devices) look like?
2. What will the pictures of field lines of electric and magnetic fields of the excitation devices presented in Fig. 12.4 (with an account of the geometry of transmission lines) look like?
3. What is the sense of the Kisun'ko–Feld formula (Eq. 12.2) and what advantage does it have over other theories of electrodynamic structure excitation?
4. Why can we obtain the more uniform "light" spot in the circuit of Fig. 12.4?
5. Draw the distribution of field lines in the exciters in Figs. 12.6j–l.

Conclusion

At the end of our long and entirely nonsimple discussions, it remains for us to relate that having studied this lecture course on modern electrodynamics in the area of electromagnetic fields and waves and appropriate devices and systems (both lumped and distributed and integrated) in depth, the reader will be ready to assimilate lecture courses on the special training of the modern engineer in radio electronics, radio physics, and many other areas of modern science. The coverage of university specialties in our lecture course is not limited to radio electronics. It is much wider, and we discussed this point in our Introduction and, moreover, repeatedly mentioned it throughout the lecture course. The accurate reader had time to understand that our lecture course (compared with standard lecture courses in electromagnetics) has three distinctive features.

First, this course contains rather simple mathematics (without losing strictness) and includes many physical interpretations that are oriented to students at Bachelor's level. This is important because our long-term pedagogical experience shows that it is rather difficult for students to study mathematics for a long time (its branches can be studied at any period of an engineering career), but physical interpretation can be remembered for a long period after the student years.

Second, this lecture course contains many original educational and engineering programs prepared in the MathCAD medium (see Appendix B – http://extras.springer.com/2019/978-3-319-90847-2). In other words, this lecture course is a computerized course and it gives the reader (and engineers/designers) a wide range of possibilities to observe the processes of geometrical and physical parameter variations and initial and boundary conditions for different frequencies and transmission lines, and for various media in which waves propagate.

The third important point, the concept of the scalar magnetic field is for the first time included into a student's lecture course. This concept to a great extent widens our representations of electromagnetics as a whole (it is extremely important for future students). This approach allows elimination from the classic theory of Maxwell's equations some known restrictions (calibrations), which permit the release of generalized theory from a series of paradoxes of classic electrodynamics.

Thus, we would like to emphasize that the development of a clear representation of the *physics of phenomena and processes* in each transmission line and in the basing element, which is constructed based on this transmission line or (as we see in practice) on a combination of different transmission lines, is particularly important for modern students. This is precisely why, having finished our discussions of some specific (of course, not all) problems, electrodynamics structures, and some important methods and approaches to the analysis, we would like to emphasize again and again the persistent necessity to become acquainted with the *physics of electrodynamics*. We call upon the readers to study the remarkable results of many compatriots from different countries of the world, including Russia, who made a great contribution to revealing and explaining the physical aspects of devices and processes in classic electrodynamics. In Appendix A, the reader can find brief information about leading worldwide experts in electromagnetic and electrodynamics: this Appendix we hope will be especially interesting for foreign experts, because we think that the western scientific community has truncated information about the Russian scientific school of electrodynamics.

The development of a clear physical picture of fields in any guiding or resonance structures assumes the utilization of approximate physical and mathematical models of devices and processes. This issue is determinative in all training systems for the designing of modern equipment. Here, too, of course, it is impossible to manage (if we are not talking about the design according to a given program of some element with well-studied properties) without "mutually profitable" cooperation of the electronic engineers and the mathematician programmers. Our long-term practice shows that each of these experts cannot separately manage the whole electrodynamics problem, i.e., developing a model and selecting a convenient method of analysis, and as a result the method of parametric synthesis. Exclusions from this common rule are rare.

The manifold approximate expressions presented in our lecture course for determination of any element parameters may serve as "zero" approximation before construction of some consecutive iteration process for specification of this result. Precisely such an approach happens to be quite effective at solving the problems in the design of transmission lines and basing

© Springer International Publishing AG, part of Springer Nature 2019
E. I. Nefyodov, S. M. Smolskiy, *Electromagnetic Fields and Waves*, Textbooks in Telecommunication Engineering,
https://doi.org/10.1007/978-3-319-90847-2

elements in microwave and millimeter-wave ranges. This approach, based on understanding of physical features of the element under design or investigation, allows the operation speed of the software for analysis to be increased by several (2–4) orders, and, therefore, the possibility of obtaining the optimal solution, i.e., to solve the problem of parametric synthesis of devices according to characteristics given in advance.

Knowledge of the fundamentals of computing electrodynamics allows the radio engineer, first, to be oriented freely enough in manifold and surprisingly dissimilar (at the level of form and content) literature on computing mathematics and physics. Frequently, in publications of this profile, under a supposedly ostentatious approach, the physical "emptiness" is concealed, the wrong interpretation of the physical process sense, etc. Usually, the boundary problem is reduced to infinite systems of integral–differential equations, whose coefficients are the multi-dimensional integrals of many parameters. The solution of such a system requires powerful (huge) computers and frequently leads to convergence on a nonphysical result. Possibly, in this connection, the most demonstrative are the various approaches to the theory of slot transmission lines of almost complete inverse operator, which solved this problem to a great extent. We emphasize once more that only a deep understanding of its operation principles allowed an acceptable approximate solution to be obtained with estimation of its error. Evidently, there are other examples of unnecessary "mathematization" of engineering electrodynamics.

On the other hand, assimilation of the fundamentals of computing electrodynamics will permit the electronic engineer to discuss (as equals) problems of a numerical search for an acceptable solution with experts in computing mathematics. Based on this "one in two" approach to the solution of practical problems of design and investigation, success will come.

Let us finish this lecture course and wish our students and readers success in the future assimilation of one of the most magnificent sciences – electrodynamics.

Information About the Authors

Eugene I. Nefyodov was born on 19 August 1932 in Taganrog-city (Russia). In 1951, he graduated from secondary school with the gold medal, and was drafted to Leningrad. He graduated from the Radio Engineering Faculty of the Army and was directed to the Military Academy of Communication in 1956. In 1964, he graduated from a Ph.D. course at the Institute of Radio Engineering and Electronics (IREE) of the Russian Academy of Sciences. Now, he works in the Friazino Branch of IREE as the Chief Researcher and as a professor of the Moscow State University of Radio Engineering, Electronics, and Automatics. During this period he participated in a series of significant scientific and applied fields. He took part in the creation of a CAD system for integrated resources, in the development of mathematical simulation methods and affective algorithms for a CAD system of ultra-fast information processing on microwave and millimeter-wave frequencies (EHF) on the basis of three-dimensional integrated circuit (3D-IC). He made many efforts in medical–biological research into development mathematical models of bio-information technologies, medicine, and other areas of microwave and millimeter-wave ranges, methods of mathematical simulation, and effective algorithms for analysis and development of bio-energy information signals and fields in various media and organisms.

E.I. Nefyodov has an academic degree of Doctor of Sciences (Physics and Mathematics). He was elected as academician of several Russian and International Academies: the Academy of Engineering Sciences, the International Academy of Informatization, the Academy of Military Sciences, the Academy of Medical-Engineering Sciences, the London Scientific Association, etc.

He is Chief Editor of the journal *Electrodynamics and Engineering of UHF, EHF, and Optical Frequencies*, a member of the editorial board of journals *Bulletin of New Medical Technologies*, *Physics of Wave Processes and Radio Engineering Systems*, and *Automation and Modern Technologies*.

Being a famous scientist in the field of microelectronics, electrodynamics, and microwave engineering, information engineering, and medical instrumentation, E.I. Nefyodov is actively and fruitfully working in the scientific schools of thought supervised by him that have worldwide priority: the development of theory and practical implementation of systems for ultra-fast information processing in 3D-IC (the "topological computer," the computer in field coupling); the development of the theory of the unified information field of the Earth noosphere; fundamental research in the field of medical informatics and medical radio engineering.

He is the author of more than 250 scientific publications, including more than 30 books, textbooks, and about 50 Author's Certificates and patents. He was supervisor of 12 Dr. Sci. and 14 Ph.D. He makes reports at conferences and delivers lectures at foreign universities (in particular, in Germany, Greece, Kazakhstan, Poland, USA, Czech Republic, France, etc.). Many of his pupils became Heads of departments in Russian universities; some pupils are working in foreign universities and companies.

Professor E. Nefyodov has made a considerable contribution to hold regular international and All-Russia conferences and seminars in current problems of the theory and design of radio engineering and medical–biological systems. He is an Honorary Member of the Popov Scientific-Engineering Association on radio engineering, electronics, and communication

© Springer International Publishing AG, part of Springer Nature 2019

E. I. Nefyodov, S. M. Smolskiy, *Electromagnetic Fields and Waves*, Textbooks in Telecommunication Engineering, https://doi.org/10.1007/978-3-319-90847-2

Head of the Section Electrodynamics and Microwave Engineering. He is a scientific supervisor and a head of the organizing committees of many scientific conferences and schools.

He is an Honorary Radio Engineer of the USSR (1988) and the Russian Federation (2001), and was awarded the Popov Gold Medal. He is the Honorary Professor of the Europe–Asian National University of Kazakhstan. He is the adjunct-professor of the Durban University of Technology (RSA).

Sergey M. Smolskiy was born on 2 January 1946 in Moscow (Russia). In 1964, he graduated from secondary school with a medal and entered the Moscow Power Engineering Institute (Technical University) (MPEI). He graduated from the Radio Engineering Faculty of MPEI in the specialty Radio-Physics and Electronics in 1970 and was recommended for the Ph.D. course. During the Ph.D. course at MPEI he received a scholarship named after Lenin and he defended his Ph.D. thesis in 1974 on the theory of high- frequency transistor oscillators. After finishing the Ph.D. course he worked at the Radio Transmitter Department of MPEI as the Senior Researcher, the Head of Research Lab, the Head of Research Division, and afterward he moved to the position of MPEI Deputy Vice-Rector on research activity and then to the position of MPEI Vice-Rector on international relations. In 1993, he defended wrote his Doctor of Sciences thesis on radar and radio navigation systems, in 1994 became a full professor, and in 1995 was elected as Head of the Radio Receiver Department. Since 2006, he has worked as a professor in this department, the Deputy Director of the MPEI Institute of Radio Engineering and Electronics.

During this time, he worked on a series of significant scientific and applied problems. In particular, he participated in the development of the theory of autonomous and synchronized transistor self-oscillation systems, methods of mathematical modeling, and effective algorithms for analysis of self-oscillating systems. He actively developed the theory and implementation of autodyne (self-oscillating mixer) systems for short-range radar, research in the field of medical electronics, radio monitoring systems, self-similar processes in telecommunications, and investigations into chaotic oscillations and signals. Professor S.M. Smolskiy is a recognized expert in these (and other) areas.

S.M. Smolskiy has an academic degree a Doctor of Engineering Sciences. He was elected as academician of several Russian and International Academies: International Academy of Electrical Sciences, International Academy of Informatization, International Academy of Higher Educational Institutions, a member of IEEE, and Honorary Doctor of three foreign universities.

Being a famous scientist in the field of nonlinear oscillating systems and radar devices, information science, and medical instrumentation, during the past few years, S.M. Smolskiy has been actively and fruitfully working in the scientific school of thoughts created with his participation: development of the theory and practical implementation of short radar systems and self-similar systems; improvement of radio monitoring and chaotic signal theory; research into the field of medical instrumentation, etc.

He is author of more than 330 scientific publications including 18 books, textbooks, including those published by Artech House Publishers, John Wiley and Sons, Springer US, and Scientific Research Publishers.

Under his scientific supervision, two Doctor of Sciences and 11 Ph.D. theses have been prepared. He takes part in conferences with reports and reviews and with lecture courses at foreign universities. Some his pupils have become Heads of departments in various Russian universities and different companies.

Professor S.M. Smolskiy made a serious contribution to holding All-Russia and international conferences and symposiums on the current problems of the theory and implementation of oscillating and radar systems.

He is a member of the Popov Scientific-Engineering Association of Radio Engineering, Electronics, and Communication and Head of the Section "Radio Engineering Methods in

Power Engineering"; he is a scientific supervisor and a member of organizing committees of international scientific conferences.

He is an Honorary Radio Engineer of Russia, he was awarded the Gold Medal of Popov "For services in the development of radio electronics and communication." He is an Honorary Worker of Russian Higher Education, was awarded the Polish order, the medal or Riazansky from the Academy of Space. He is adjunct-professor of Durban University of Technology (RSA).

Eminent Scientists Who Made an Important Contribution to the Development of Theory and Practice of Electrical Engineering, Electrodynamics, Antennas systems, and Radio Wave Propagation

Alfvén, Hannes Olof Gösta (1908–1995) was a Swedish electrical engineer, plasma physicist, and winner of the 1970 Nobel Prize in Physics for his work on magnetohydrodynamics (MHD).

He described the class of MHD waves now known as Alfvén waves. He was originally trained as an electrical power engineer and later moved to research and teaching in the fields of plasma physics and electrical engineering. Alfvén made many contributions to plasma physics, including theories describing the behavior of aurorae, the Van Allen radiation belts, the effect of magnetic storms on the Earth's magnetic field, the terrestrial magnetosphere, and the dynamics of plasmas in the Milky Way galaxy.

Ampere, Andre Marie (1775–1836) is the famous French physicist, mathematician, and chemist, professor of the Norman School, which was the main higher education institution in Paris.

He discovered the magnetic interaction of currents, developed the law of this interaction (*Ampere law*) and came to the conclusion that all magnetic phenomena can be reduced to purely electrical effects. According to the Ampere hypothesis, a great number of circular electric currents circulate inside any magnet and its action causes the magnetic forces. He discovered the magnetic effect of the coil with current, referred to as a solenoid.

Andreev, Nikolay N (1880–1970) is Russian physicist; expert in acoustics; founder of the scientific school of acoustics.

He was an academician of the USSR Academy of Sciences (1953), an honored Russian worker in science and technology (1960), Hero of the USSR (1970).

His main research was in the field of acoustics (hydro-acoustics, architecture, biologic, and nonlinear acoustics). He developed the strict theory of sound distribution in moving media. He performed research on the theory of sound distribution along absorbing surfaces, the theory of acoustic filters, and sound waves of finite amplitudes. A series of his projects related to the study of the damping oscillation spectrum, with oscillation investigation of crystal and anisotropic media, problems of sound reverberation, and acoustic insulation. Under his supervision, in Russia, research into nonlinear acoustics, sound propagation in laminar media, and electromechanical active materials was begun. He created a scientific school in physical and engineering acoustics.

© Springer International Publishing AG, part of Springer Nature 2019
E. I. Nefyodov, S. M. Smolskiy, *Electromagnetic Fields and Waves*, Textbooks in Telecommunication Engineering,
https://doi.org/10.1007/978-3-319-90847-2

Armstrong, Edvin Howard (1890–1954).

American engineer-electrician, the inventor of frequency modulation. Graduated from Columbia University in the specialty "Electrical Engineering" in 1913. Worked as an assistant-professor of the Electrical Department during 1913–1914, then he moved to the university research laboratory named after M. Hartley and from 1914 to 1935 conducted investigations and experiments in the field of radio engineering. In 1934, he became professor of Columbia University. In 1912, having tried to understand how an audion (vacuum tube invented in 1905 by De Forest) operates, E. Armstrong offered the structure called "regenerative" (circuit with positive feedback) with the help of which he obtained 1,000-times amplification of high-frequency signals. In 1913, he suggested the configuration of the regenerative radio receiver, and in 1918 he developed a super-heterodyne receiver with high sensitivity and flat gain in the whole band of received frequencies. In 1911, he built the super-regenerator with a larger gain, which immediately found application in mobile communication. In 1925, E. Armstrong began investigations into the elimination of interference, which were finished by the discovery of frequency modulation (FM) – a slow variation of electromagnetic oscillation frequency in accordance with some time law. This approach is used now for transmission of an audio signal on TV and FM radio broadcasting. In 1937, he established a private FM radio station in New Jersey, and in March claimed the multiplex method, allowing the broadcasting of several programs via one FM channel. He is a laureate of the Nobel Prize for physics.

Bohr, Niels Henrik David (1885–1962) was a Danish physicist-theoretic, one of the founders of modern physics.

He was a laureate of the Nobel Prize for physics in 1922, a member of the Danish Royal Society (1917) and its President from 1939. He was a member of more than 20 Academies of Science.

N. Bohr was famous as a creator of the first quantum theory of the atom and an active participant of the development of the first fundamentals of quantum mechanics. He also made a great contribution to the theory of atomic nuclear, nuclear reactions, and processes of interaction of the elementary particles with the media.

Bose, Sir Jagadish Chandra (1858–1937), was a Bengali polymath, physicist, biologist, biophysicist, botanist, and archaeologist, and an early writer of science fiction.

Living in British India, he pioneered the investigation into radio and microwave optics, made significant contributions to plant science, and laid the foundations of experimental science in the Indian subcontinent. IEEE named him one of the fathers of radio science. Bose is considered the father of Bengali science fiction, and also invented the crescograph, a device for measuring the growth of plants. A crater on the moon has been named in his honor.

He conducted his research with the Nobel Laureate Lord Rayleigh at Cambridge and returned to India. He joined the Presidency College of the University of Calcutta as a professor of physics. There, despite racial discrimination and a lack of funding and equipment, Bose carried on his scientific research. He made remarkable progress in his research into remote wireless signaling and was the first to use semiconductor junctions to detect radio signals. However, instead of trying to gain commercial benefit from this invention, Bose made his inventions public to allow others to further develop his research.

Bose subsequently made a number of pioneering discoveries in plant physiology. He used his own invention, the crescograph, to measure plant response to various stimuli, and thereby scientifically proved parallelism between animal and plant tissues. Although Bose filed for a patent for one of his inventions because of peer pressure, his objections to any form of patenting were well known. To facilitate his research, he constructed automatic recorders capable of registering extremely slight movements; these instruments produced some striking results, such as the quivering of injured plants, which Bose interpreted as a power of feeling in plants. His books include *Response in the Living and Non-Living* (1902) and *The Nervous Mechanism of Plants* (1926).

Brekhovskikh, Leonid M (1917–2005) was a Russian/Soviet scientist known for his work in acoustic and physical oceanography.

During the Second World War, L.M. Brekhovskikh began to work in the team of academician N.N. Andreev (see above) on the development of methods and devices of ship protection against German acoustic mines. In 1947, he defended the Dr.Sci. thesis on the theory of sound and electromagnetic waves propagation in the laminar media. In 1951, he was awarded the State Prize for the discovery of the underwater sound channel.

In 1953, Leonid Brekhovskikh was confirmed as a professor of Lomonosov Moscow State University and was elected as corresponding member of the USSR Academy of Sciences. During 1954—1980, he worked in the Acoustic Institute of USSR Academy of Sciences.

Brewster, Sir David (1781–1868) was a British physicist.

He described the chromatic polarization phenomenon in single-axis and double-axis crystals. He discovered the circular polarization of light and the double refraction phenomena in environments with artificial anisotropy.

Brillouin, Léon Nicolas (1889–1969) was a French physicist. He made contributions to quantum mechanics, radio wave propagation in the atmosphere, solid-state physics, and information theory.

Brillouin was a founder of modern solid-state physics for which he discovered, among other things, Brillouin zones. He applied information theory to physics and the design of computers, and coined the concept of negentropy to demonstrate the similarity between entropy and information.

Brillouin offered a solution to the problem of Maxwell's demon. In his book, *Relativity Reexamined*, he called for a "painful and complete re-appraisal" of relativity theory which "is now absolutely necessary."

Carlson, Chester (1906–1968) was an American physicist, the inventor of xerography (1938). His biography is typical of inventors of the last century, when the genius should fight his way from the very low layers of society and then with great efforts push those inventions that will be in the future integral features of human civilization. He received a Bachelor's degree in physics at the California Institute of Technology in 1930.

Carlson often thought on the copying method. All free time he spent in the New York Public Library and after several months began to understand photo-copying. Having been a physicist in education, Carlson assumed that at light incidence on such materials the electrical conductivity on the surface varies depending on the illumination of the specific part. In other words, the conductivity on brightly illuminated parts should be greater than on non-illuminated parts. Hence, the distribution of the surface conductivity of the material repeats the image projected. It is necessary only to transfer it onto paper.

Having examined the theoretical points, Carlson proceeded to the practical implementation of his idea. As often happened in the life of a technical genius, his own kitchen in his New York flat became scene for his first experiment. Precisely there, Carlson carried out his first experiments that laid down the basic principles of what he called "electrostatic photography." Carlson obtained his first patent in 1937. A historic event occurred on 22 October 1938 during experiments over a zinc plate, which was thoroughly covered by finely divided sulfur. This type of material, being dielectric under usual conditions, began to slightly conduct an electrical current under great light illumination. In other words, the photo conductivity effect occurred. On 22 October 1948, the first public demonstration of the new copying apparatus took place, and the first copiers were offered onto the market in 1949. With the help of its invention, Chester Carlson earned $150 million and donated approximately $100 million to philanthropic causes.

Coulomb, Charles-Augustin (1736–1806) is a famous French physicist and military engineer.

In 1784, he discovered electric balance, which is based on the property of metal filaments having a reaction force at torsion that is proportional to the torsion angle, and applied this balance for measuring the repulsion force of charged balls. In 1785, he discovered the fundamental law in electricity named after him.

Simultaneously with the investigation of electric forces, he studied the interaction of the magnet needles and derived that the force of magnetic attraction is inverse to the square of distance between the magnetic molecules.

D'Alembert, Jean-Baptiste le Rond (1717–1783) was a French mathematician, mechanician, physicist, philosopher ,and music theorist.

Until 1759, he was also co-editor with Denis Diderot of the *Encyclopédie*. D'Alembert's formula for obtaining solutions to the wave equation is named after him. The wave equation is sometimes referred to as d'Alembert's equation.

De Forest, Li (1873–1961) was an American inventor.

He graduated from Yale University in 1896, with a Ph.D. under the scientific supervision of famous physicist G. Gibbs. From 1900, he began to deal with wireless telegraphy. Having tried to attract attention to the electrolytic receiver (developed by him), which was called a "responder", he created a wireless telegraph for information transmission from an international competition of yachtsmen. De Forest mounted the first radio stations in five largest US Navy bases; he developed horizontal receiving, loop, and finding antennas and developed the duplex mode of transmission and reception. In 1910, he made the first musical broadcast from the theatre the "Metropolitan Operahouse." He worked on improving his electronic tube, whose first variant was made in 1905, when he looked for a way to perfect radio wave reception. In 1907, he had obtained a patent for the "audion" – a three-electrode tube with a controlling grid-like electrode (a triode). In 1913, for the first time, he demonstrated an acting oscillator on an audion. In 1915, De Forest executed a radio communication session between Arlington (Virginia) and Paris. Thus, a receiver was mounted on top of the Eiffel tower. In 1916, he arranged the first news broadcast on a radio channel.

In 1920, De Forest began to develop sound films, which were called "phono-film." He was the first to find the method of the photographing of audio waves onto the same film as images. He participated in the development of TV technology and the technique of high-speed photo-telegraph transmission of images.

Eisler, Paul was an illegal immigrant from Vienna and he created in the UK in 1936 the first printed board. He mounted it in a radio receiver of his own design, which he was going to sell. He had neither the permission to work nor money to live. However, his invention did not interest anybody: the work of the young women manually assembling the parts of electrical equipment was much cheaper than the manufacture of printed boards.

In 1941, he had obtained residence and a permit, and sold the rights to all his inventions for one pound sterling to the company "Hender-Son and Spulling," which dealt with the printing of music (this company became famous for the first edition of the complete collected works of Beethoven). He had obtained in return the position of director and the only employee of a department of musical instruments with a salary of eight pounds per week. He wanted to mount a printed board inside an electrical machine for printing notes, but during the war period such production was not purchased. The company tried to sell the Eisler patent to the British Military complex, which manufactured military electronic equipment, but the British manufacturers did not support it. The Americans showed a much more far-sighted view. Eisler's idea was estimated with dignity by the representatives of the National Bureau of Standards in Washington. The license was purchased by the company "Globe Union Inc.," who applied the printed boards in the manufacture of electrical detonators for anti-aircraft proximity fuses. These shells were transported to the UK by seaway and applied in the struggle with V-1 missiles, which were used by Germany for bombing London in 1944. Eisler can be proud of himself: the shells in which his printed board worked destroyed approximately 4000 V-1 missiles. In 1957, Lord Heilsham called the printed board one of the six main scientific achievements in the twentieth century. Meanwhile, the patent was transited to the Governmental National Corporation for Research and Development, where printed boards produced a real revolution in electronics, which took the first steps. All this was practically without payment.

Faraday, Mickle (1791–1867) was a great British physicist, chemist, and founder of a doctrine of the electromagnetic field.

He discovered the rotation of a magnet around a conductor with a current and the rotation of a conductor with a current around the magnet (1821). He discovered the phenomenon of electromagnetic induction, which underlies electrical engineering. He proved the identity of the different kinds of electricity: obtained from the friction, from "living animals," from "magnetism," etc. He discovered the laws of electrolysis, which were strong arguments in favor of the discreteness of substance and electricity. He introduced the concept of lines of force. He discovered the phenomena of para- and diamagnetism (1845), observed a rotation of the light polarization plane in a magnetic field (Faraday effect), which was the first illustration of a relation between magnetic and optical phenomena that gave further confirmation of Maxwell's electromagnetic theory of light.

Fleming, John Ambrose (1849–1945) was a Britain physicist, a member of the London Royal Society (from 1892).

He graduated from London University in 1870. During 1877–1881, he worked in the Cavendish Laboratory, during 1881–1882 he was a professor at Nottingham University, and 1885–1926 professor at London University. His results were devoted to radio telegraphy, radio telephony, and radio and electric engineering. He devised the rule of the right hand for determination of the induction current direction in a conductor (Fleming's rule). He investigated the "Edison effect" and the discovery of Thompson. On this basis, he invented and patented in 1904 the "rectifier," or the two-electrode tube called the "Fleming diode."

Fock, Vladimir A (1898–1974) was a Soviet theoretical physicist and academician. He is famous for his fundamental results on quantum mechanics and electrodynamics, and on general relativity.

He developed a method of the self-congruent field for the multi-electron system (Hartree–Fock method), the method of secondary quantization, results on radio wave propagation, mathematical physics, and philosophical problems in physics. His key results were devoted to gravity theory, quantum field theory, the theory of multi-electron systems, functional methods in field theory and in mathematical statistics, radio wave propagation, the theory of diffraction, and mathematical physics.

Fraunhofer, Joseff (1787–1826) was a famous German physicist, professor of Munich University.

His key results were devoted to physical optics. He improved the manufacturing technology of the large achromatic lens, and discovered the ocular microscope and the heliometer. He observed and for the first time explained the presence of the absorption lines in the solar spectrum (Fraunhofer lines, 1814). He used for the first time diffraction gratings for the spectrum investigations (1821). He offered and observation method for light diffraction in parallel beams.

Fresnel, Auguste Jean (1788–1824) is a famous French physicist, member of London Royal Society, creator of the wave theory of light.

His key results are the creation of the complete theory of diffraction based on the application of Huygens principle in the original formulation of Fresnel; experimental investigations into the influence of polarization on the interference and substantiation of the transversal character of light waves (1821); explanation of the problem of the color of crystal plates; discovering the circular and elliptic polarization and description of these phenomena from the point of view of the wave theory (1822); description of the phenomenon of polarization plane rotation as the double refraction of the circular polarized light (1822); discovering the laws of reflection and refraction (Fresnel formulas); discovering elliptic polarization at complete internal reflection; creation of the double refraction theory and the substantiation of crystal optics.

Gauss, Karl Friedrich (1777–1855) was a famous German mathematician, astronomer, geophysicist.

He worked at Gettingent University. He made a fundamental contribution to theoretical and applied mathematics, astronomy, and geodesy. His book entitled *General theory of terrestrial magnetism* (1838) is well-known. He derived the more simple (compared with M.V. Ostrogradskiy) and less general formula for conversion of the volumetric integral into surface ones, which does not use the divergence operation (1841). (This was strictly proved by Russian academician A.N. Tikhonov).

Heaviside, Oliver (1850–1925) was an English self-taught electrical engineer, mathematician, and physicist who adapted complex numbers to the study of electrical circuits, invented mathematical techniques for the solution of differential equations (equivalent to Laplace transforms), reformulated Maxwell's field equations in terms of electric and magnetic forces and energy flux, and independently co-formulated vector analysis. Although at odds with the scientific establishment for most of his life, Heaviside changed the face of telecommunications, mathematics, and science for years to come.

Helmholtz, Herman Ludwig Ferdinand (1821–1894) was a German physicist, mathematician, physiologist, and psychologist.

He worked as a professor at the universities of Königsberg, Bonn, Heidelberg, and Berlin. He gave the mathematical substantiation of the law of energy conservation and having analyzed the physical phenomena most well-known at the time, he showed this law of universality, in particular, that the processes existing in living organisms obey the law of energy conservation. This was the strongest argument against the concept of the peculiar "living force," which appeared to control the organisms. He proved for the first time the adaptability of the principle of least action to the thermal, electromagnetic and optical phenomena and discovered this principal connection with the Second Thermodynamic Basis (the law of energy degradation). His results on electromagnetism, optics, and acoustics were mostly connected to his physiological investigations. He discovered the phenomenon of the oscillating discharge of the Leyden jar and this fact had played an essential role in the development of the theory of electromagnetism. He tried to measure the propagation speed of the electromagnetic disturbances, but failed. On the basis of his suggestion, though, H. Hertz carried out experiments with electromagnetic waves. Helmholtz developed the theory of irregular dispersion. In 1881, he offered the idea of the corpuscular structure of electricity.

Henry, Joseff (1797–1878) was an American physicist.

He created powerful electrical magnets and an electrical motor. He discovered (independently from M. Faraday) the self-induction phenomenon (1832) and found out the oscillating character of the condenser discharge (1842).

Hertz, Heinrich Rudolf (1857–1894) was a great German physicist, one of the founders of electrodynamics.

He was an assistant of H. Helmholtz and worked as a professor of the Hochschule in Karlsruhe and Bonn University. His main results are connected with electrodynamics. Based on Maxwell's equations, Hertz experimentally proved in 1886–1890 the existence of electromagnetic waves and investigated their properties (the reflection from the mirrors, refraction inside the prisms, etc.). Hertz obtained electromagnetic waves with the help of a vibrator discovered by him (the Hertz vibrator). He confirmed the conclusions of Maxwell's theory that the speed of electromagnetic wave propagation in the air is equal to the light speed. He proved the identity of the key properties of the electromagnetic waves and the light waves. Hertz also studied the propagation of the electromagnetic waves inside the conductor and offered a method for measuring its propagation speed. Having developed Maxwell's theory, Hertz derived equations of electrodynamics in symmetric form, which can correctly find out the complete interconnection between electric and magnetic phenomena. He developed the electrodynamics of moving objects based on the hypothesis that the ether is carried by moving objects. However, his electrodynamics appeared to be in contradiction with the experiments and later it gave way to the electronic theory of Lorenz. The Hertz results in electrodynamics played the greatest role in the development of science and technology and led to the appearance of wireless telegraphy, radio communication, TV technologies, radar technology, etc. He observed and described for the first time the external photo-effect, developed the theory of the tuned resonance circuits, studied the properties of cathode rays, investigated the influence of ultraviolet rays upon the electric discharge. He offered the theory of the elastic ball impact and calculation of the time of collision, etc. The unit of oscillation frequency was named after Hertz.

Kamerlingh-Onnes, Heike (1853–1926).

He graduated from Groningen University and studied in Heidelberg under the supervision of Kirchhoff. From 1882, he worked as a professor at Leyden University. From 1894, he was the founder and director of the Leyden Cryogenic Laboratory. He is famous for his experimental research in the field of low-temperature physics and super-conductivity. He developed and implemented the liquefier unit. In 1906, he obtained liquid hydrogen. In 1908, for the first time, he obtained liquid helium and could achieve the record-breaking low temperature of 0.9 K. This was rewarded with the Nobel Prize for physics in 1913. In 1911, he observed for the first time the sharp decrease in electrical resistance of mercury at temperatures lower than 4.1 K. This phenomenon was called super-conductivity. In 1913, he discovered the destruction of super-conductivity by strong magnetic fields and currents.

Kapitza, Piotr L (1894–1984) was a great Russian physicist, professor of Moscow Physical Technology Institute, academician, Nobel Prize laureate.

He suggested (together with N.N. Semionov) the method for the determination of magnetic moments of atoms inside the atomic beam, offered and proved the pulse method of obtaining super-strong magnetic fields, discovered in the strong magnetic fields the linear electrical resistance dependence of some metals upon the field intensity (Kapitza's law). He developed the method and created the installations for helium and air liquation (1934–1939). He discovered the superfluidity of liquid helium (1938) and created ultra-high-frequency (UHF) oscillators of super-high power. He discovered the stable plasma filament at UHF discharge in solid gas.

Katselenenbaum, Boris Z (1919–2014*)*, radio-physicist. Dr. Sci. (Phys.-Math.), Professor.

After graduation from the Lomonosov Moscow State University in 1941, he was heavily wounded in the war. After demobilization, he worked in the R&D Radio Engineering Institute and as a teacher. From 1954, he worked in the Institute of Radio Engineering and Electronics of the USSR Academy of Sciences. His main scientific publications were in theoretical radio physics and mathematical physics. He suggested a classification for excitation types in dielectric waveguides and investigated outgoing waves. In the theory of irregular waveguides, he developed a method of cross-sections, devoted to the system of differential equations and had obtained adiabatic solutions. He developed the asymptotic theory of waveguide breaks and the generalized method of proper oscillations in diffraction theory. He performed a constructive synthesis of antennas with semi-transparent surfaces, the so-called resonator antennas. He proved an imperfection of the field system, created by currents on some surfaces and debated the consequences of this imperfection (non-approximability).

Kisun'ko, Georgiy V (1918–1998) was one of the founders of the Soviet system of anti-missile defense. The general lieutenant engineer (1967), Doctor of Sciences (Eng.) (1951), Professor (1956), corresponding member of the USSR Academy of Sciences (from 1958), The Russian Hero, The Lenin Prize Laureate.

He worked for the Department of Theoretical Radar Technology at the Military Engineering Academy of Communication (1944–1950). During his work at this Academy, he wrote the famous book on the electrodynamics of hollow systems, which was way ahead of its time. In 1961, G. Kisun'ko was appointed head of the newly-established R&D Bureau and the general designer of the anti-missile system, A-35. In March 1966, this R&D Bureau was transformed into the R&D Bureau VYMPEL and G. Kisun'ko began to work as its Director and scientific supervisor.

In August 1949, he was appointed as scientific consultant in the R&D Institute of the USSR Ministry of Defense.

Kohlrausch, Rudolf Hermann Arndt (1809–1858) was a German physicist.

In 1854, Kohlrausch introduced the relaxation phenomena, and used the stretched exponential function to explain relaxation effects of a discharging Leyden jar (capacitor). In 1856, with Wilhelm Weber (1804–1891), he demonstrated that the ratio of electrostatic to electromagnetic units produced a number that matched the value of the then known speed of light. This finding was instrumental towards Maxwell's conjecture that light is an electromagnetic wave. Also, the first usage of the letter "c" to denote the speed of light was published in an 1856 paper by Kohlrausch and Weber.

Kotelnikov, Vladimir A (1908–2003) is a Soviet scientist in the field of radio engineering, professor of Moscow Power Engineering Institute, academician, Director of the Institute of Radio Engineering and Electronics of the Academy of Sciences of the USSR.

His main research was devoted to the problems of improving the reception methods, investigating reception interference and development of the methods of fighting with them. His results in the field of the theory for potential noise immunity was of great importance. He was a supervisor of many projects on the radar investigation of Mars, Venus, and Mercury. He was awarded the State Award (1943, 1946), Lenin's Award, 4 Lenin Orders, Russian Orders, and many Russian and foreign medals.

Khvorostenko, Nikolay Piotr (born in 1933) graduated from the Military Engineering Academy of Communication (1956). A famous theorist in the field of communication, he made a great contribution to the theory of potential noise immunity. He obtained important results on the reception of PSK signals and orthogonal and simplex signals with redundancy. He developed the theory of the optimal diversity reception of signals.

He achieved significant success in the creation of generalized electrodynamics, developing the microscopic theory of longitudinal electromagnetic waves.

Langmuir, Irving (1881–1957) was an American chemist and physicist.

His most noted publication was the famous 1919 article "The Arrangement of Electrons in Atoms and Molecules" in which, building on Gilbert N. Lewis's cubical atom theory and Walther Kossel's chemical bonding theory, he outlined his "concentric theory of atomic structure".[4] Langmuir became embroiled in a priority dispute with Lewis over this work; Langmuir's presentation skills were largely responsible for the popularization of the theory, although the credit for the theory itself belongs mostly to Lewis. While at General Electric from 1909–1950, Langmuir advanced several basic fields of physics and chemistry, invented the gas-filled incandescent lamp and the hydrogen welding technique, and was awarded the 1932 Nobel Prize for Chemistry for his work in surface chemistry. The Langmuir Laboratory for Atmospheric Research near Socorro, New Mexico, was named in his honor, as was the American Chemical Society journal for surface science, called *Langmuir*.

Lebedev, Piotr N (1866–1912) was a great Russian physicist–experimentalist who for the first time experimentally confirmed Maxwell's conclusion about the presence of light pressure.

He was a founder of the first Russian scientific physical school of thought, a professor of Moscow University. In 1899, Lebedev, using masterly experiments, confirmed the theoretical Maxwell's prediction about light pressure on solid bodies, and in 1907 – on gases (discovery of the light pressure effect). This research was an important step in the science of electromagnetic phenomena. One famous physicist of those times (William Thomson) said: "All my life I struggled with Maxwell not recognizing his light pressure, and Lebedev forced me to refuse from my opinion due to his experiments».

Piotr Lebedev also investigated problems of electromagnetic wave influence on resonators and carried out deep investigations concerning molecule interactions, investigating acoustic problems, in particular, hydro-acoustics.

Leontovich, Mikhail A (1903–1981) was a Soviet physicist, academician; author of publications on plasma physics. The laureate of the Lenin prize (1958), the Gold Medal of A.S. Popov (1952).

He performed a whole series of large investigations, including research into the general theory of thin wire antennas, and fundamental investigations into radio wave propagation along the Earth's surface. He suggested an idea regarding the introduction of fluctuation outside the electromagnetic field in electrodynamics equations. This cycle of Leontovich investigations became the basis of the Soviet theoretical school of radio physics.

Lomonosov, Mikhail V (1711–1765) was the greatest Russian scientist/naturalist of universal importance; poet, linguist, literature critic, painter, historian, advocate of the development of native education, science, and economy, and academician.

In 1748, he created the first chemical laboratory in Russia. Under his initiative, the Moscow University was created (1755). He developed the atomic–molecular ideas for substance structure. In the period of previous heat theory domination, he stated that the warming is caused by corpuscle (molecules) motion. He formulated the principle of conservation of substance and motion. He excluded the phlogiston from the list of chemical agent. He created the fundamentals of physical chemistry. He investigated the atmospheric electricity and the gravity force, offered the study of color and created some optical devices. He discovered the atmosphere of Venus and described the Earth's structure, explaining the origin of much mineral wealth. He published a manual on metallurgy and emphasized the importance of the development of the North Sea Route and Siberia.

Lorentz, Hendrik Antoon (1853–1928), a physicist–theorist from the Netherlands, laureate of the Nobel Prize for Physics (1902) and other awards, a member of the Netherlands Royal Academy of Sciences (1881) and various foreign academies. H. Lorentz is mainly known for his results in electrodynamics and optics. Integrating the concept of the continuous electromagnetic field with the representation of discrete electric charges included in the matter's structure, he created classic electronic theory and applied it for the solution of a number of specific tasks. He obtained the expression for the force acting on the moving charge from the electromagnetic field (the Lorentz force), he devoted a formula connecting the refraction index of the matter with its density (the Lorentz–Lorenz formula), developed the theory of light dispersion, explained some magneto-optical phenomena (in particular, the Zeeman effect) and some properties of the metals. Based on electronic theory, he developed electrodynamics of the moving media, including a hypothesis on body reduction in the direction of its movement (the Lorentz–FitzGerald contraction), introduced the concept of "local time," obtained the relativistic expression for mass dependence on the speed, devoted relations between coordinates and time in moving inertial reference systems (Lorentz transforms). Lorentz's investigations facilitated the formulation and development of ideas of special relativity theory and quantum physics. In addition, he obtained a series of significant results in thermodynamics, the kinetic theory of gases, the general relativity theory, and the theory of heat emission.

Makeeva, Galina S, Doctor of Sciences (Phys.-Math.), Professor of Penza State University (Russia). After graduation from Penza Polytechnic University in 1968, she was educated on a Ph.D. course at the Leningrad Electrical Engineering Institute. In 1980–1982, she joined the Institute of Radio Engineering and Electronics of the USSR Academy of Science for preparation of her Dr.Sci. thesis. In 1997, she defended her Dr. Sci. thesis on the theme "Electrodynamics of integrated waveguiding structures with thin-film semiconductor and ferrite layers and inserts." She is known in Russia and in foreign countries as one of the leading specialists in the field of radio physics and computing electrodynamics, in particular, in the field of nano-electronics and the theory of graphene electrodynamics.

She has published more than 350 scientific works, including in the USA, Japan, Germany, Hungary, Poland, Bulgaria, France, Norway, the Netherlands, Italy, Spain, Czech Republic, Morocco, and Malta, including two books.

Maliuzhinets, Georgiy D (1910–1969). After graduation from Moscow State University in theoretical physics, he studied the diffraction theory of acoustic and electromagnetic waves. He developed the theory of heterogeneous absorbing media, which was the basis for the highly effective layered absorbent covering for the dome of the Great Concert Hall in Moscow. To calculate the layered covering, he offered an original method on the impedance locus. After the Second World War, he developed the cycle of investigations devoted to the general Sommerfeld integral theory, the mathematical formulation of stationary diffraction problems in the arbitrary areas, in addition to new mathematical methods in diffraction theory.

For the last 15 years of his life, he worked in the Acoustic Institute. His contribution to the development of approximate methods of diffraction theory is inestimable. He used the method of transverse diffusion to the problems of radio wave propagation in heterogeneous media. His attention was particularly drawn to the development of calculation methods in the theory of stationary diffraction. The results of his fundamental investigations in the field of wave propagation in heterogeneous media are the basis for methods of effective noise and vibration damping in conformity with shipbuilding and other areas.

Mandelstam, Leonid I (1879–1944) was a Soviet physicist, one of the founders of the Russian scientific schools of thought in radio-physics, and academician.

In 1928, he discovered (together with G.S. Landsberg) the combinative dispersion of light. Together with N.D. Papaleksi, he executed fundamental research into nonlinear oscillations, developed the method of parametric excitement of electric oscillations, and offered the radio interferential method. He had great results on light dispersion.

Maxwell, James Clerk (1831–1879) was the greatest British physicist, creator of classic electrodynamics, and one of the founders of statistical physics.

He worked as a professor at Marischal College in Aberdeen and also at London and Cambridge Universities. He was a founder of the first specially equipped physical laboratory in the United Kingdom, called the Cavendish Laboratory. Maxwell's research activity included the problems of electromagnetism, the kinetic theory of gases, optics, the theory of elasticity, etc. His first research was carried out when he was 15 years old and jis first investigations related to physiology, physics of color vision, and colorimetry. He conducted a theoretical investigation into the stability of Saturn's rings and proved that they can be stable only if they consist of solid particles not related each other.

In his research, Maxwell theoretically developed the Faraday's ideas about the role of the intermediate medium in electrical and magnetic interconnections. He tried (after Faraday) to interpret this medium as a fully penetrating universal ether; however, these attempts failed. The further development of physics showed that the electromagnetic field is a bearer of electromagnetic interactions. Maxwell created the theory of this field. In this theory, he generalized all the facts on macroscopic electrodynamics that were known then, and for the first time offered current displacement, which gave rise to the magnetic field like the usual electric current (conductive current of moving electrical charges). Maxwell expressed the laws of the electromagnetic field in the form of system consisting of four differential equations in partial derivatives. The general and irrefragable character of these equations became apparent in the fact that this analysis allows the prediction of many unknown phenomena and regularities. Thus, the existence of electromagnetic waves, which was discovered further by Hertz's experiments, followed Maxwell's equations. Having investigated these equations, Maxwell concluded the electromagnetic nature of light and proved that the speed of any other electromagnetic waves in a vacuum is equal to the speed of light. It resulted from Maxwell's theory that the electromagnetic waves produce pressure, which was proved experimentally by the Russian scientist P.N. Lebedev.

Minkowski, Hermann (1864–1909) was a mathematician, and professor at Königsberg, Zürich, and Göttingen universities. He created and developed the geometry of numbers and used geometric methods to solve problems in number theory, mathematical physics, and the theory of relativity.

Minkowski is perhaps best known for his work in relativity, in which he showed in 1907 that his former student Albert Einstein's special theory of relativity (1905) could be understood geometrically as a theory of four-dimensional space–time, since known as "Minkowski spacetime".

Moog, Robert (1934–2005) became famous by the creation of a musical instrument, which in the 1960s was used by "The Beatles" and "The Doors" in their songs.

When he was 14 years old, R. Moog assembled his first musical instrument, the "Termenvox," which was invented by Russian physicist Lev Termen. In 1970, he offered to the market his first compact synthesizer the "Mini-Moog."

At first, R. Moog and his invention attracted the attention of musicians in 1968, when the album of interpretations of I.S. Bach music appeared, which was executed on the analog synthesizer by Wendy Carlos. This disk was highly successful, and he received a Grammy. In 2004, the first festival "MoogFest" was held in New York.

Moore, Gordon (born in 1929).

He graduated from Berkley University and received a Bachelor's degree in chemistry (1950). After that, he received a Ph.D. in chemistry and physics (after graduation from the Technology University of California).

G. Moore became more popular still in April 1965, when his publication *Law of Moore* appeared, which touched on the semiconductor area. After 30 years, G. Moore introduced serious corrections to his law, which soon became the fundamental law of semiconductors.

Newton, Isaac (1643–1727) was a great British physicist and mathematician.

He laid the theoretical foundations of mechanics and astronomy, and discovered the law of gravity. He developed (along with G. Leibniz) the differential and integral calculus, and was inventor of the mirror telescope and carried out important experiments in optics.

Nikolaev, Gennady V (1935–2008). He was born in Uzbekistan. In 1967, he graduated from Tomsk Polytechnical Institute. In 1967, he presented in a scientific seminar his first book *Laws of mechanics and electrodynamics in the real near-Earth space*. In 1970, he presented his first Ph.D. thesis, but it was not accepted for defense due a lack of argument regarding the problem. In 1973, he theoretically proved the existence of the scalar magnetic field and the longitudinal force of magnetic interaction. In 1989, he prepared his second Ph.D. thesis, but it was also not accepted for defense.

He actively participated in scientific projects, actively investigated problems of noospherical interactions and enormous phenomena in the USSR. His field of activity was the discovery of a new type of magnetic field and the new Nikolaev theory of electromagnetism was interesting for Austrian physicist Stefan Marinov, who reproduced some of the experiments of Nikolaev, and based on descriptions built the operating paradoxial electric motor, which should not operate in accordance with known laws of electrodynamics. This motor was named in honor of Nikolaev.

Ohm, Georg Simon (1787–1854) was a German physicist, professor at Munich University. His main results were related to electricity, optics, crystal optics, and acoustics. In 1826, he discovered the law named after him.

Ørsted, Hans Christian (1777–1851) was a Danish physicist and chemist who discovered that electric currents create magnetic fields, which was the first connection found between electricity and magnetism.

He is still known today for Ørsted's Law. He shaped post-Kantian philosophy and advances in science throughout the late nineteenth century. In 1824, Ørsted founded *Selskabet for Naturlærens Udbredelse* (SNU), a society for disseminating knowledge of the natural sciences. He was also the founder of predecessor organizations that would eventually the Danish Meteorological Institute and the Danish Patent and Trademark Office. Ørsted was the first modern thinker to explicitly describe and name the thought experiment.

A leader of the so-called Danish Golden Age, Ørsted was a close friend of Hans Christian Andersen and the brother of politician and jurist Anders Sandøe Ørsted, who eventually served as Danish prime minister (1853–54). The oersted (Oe), the centimetre–gram–second unit of magnetic H-field strength, is named after him.

Ostrogradskiy, Mikhail V (1801–1861) was a great Russian scientist, mathematician, professor of officers in the Navy Corp, Main Pedagogic Institute, Railway Engineers Institute, and Main Artillery Military School.

His main results were related to mathematical analysis, theoretical mechanics, mathematical physics, etc. He derived and proved the general formula of the volumetric integral conversion into surface ones, which includes the divergence expression (1835). He solved the important task regarding wave propagation on the liquid surface in the pool, which the form of a circular cylinder. His results on the theory of elasticity, celestial mechanics, theory of differential equations, variation principles, etc., are well-known.

Papaleksi, Nikolai D (1880–1947) was a Soviet physicist, professor of Odessa Polytechnic Institute, and academician.

From 1923 to 1935, together with L.I. Mandelstam, he was head of the scientific division of the Central Radio Laboratory in Leningrad. Beginning from 1935, he worked in Moscow in the Physical and Power Engineering Institutes. He was the head of the Soviet Scientific Council on Radio Physics and Radio Engineering of the Russian Academy of Sciences. From 1914 to 1916, he conducted research into guided radio telegraphy, radio communications, and the remote control of submarines, additionally developing the first Russian radio tubes. Together with L.I. Mandelstam, he executed fundamental research into nonlinear and parametric oscillations; they discovered and studied the n-type resonance, combination and parametrical resonance, developed the method of parametric excitation of the electric oscillations. With the help of the offered interferential approach, they investigated radio wave propagation above the Earth's surface and fulfilled the precise measurement of its speed.

Planck, Max Karl Ernst Ludwig (1858–1947) was an outstanding German physicist. As a founder of the quantum theory, he predetermined the key direction of physics development from the beginning of the twentieth century.

He graduated from high school in Munich, where together with high endowments for many disciplines he demonstrated great diligence and efficiency. The decision to become a physicist was not very simple because music and philosophy attracted his attention in addition to the natural sciences, especially as one of his teachers declared that nothing new can be discovered in physics.

After defense of a Ph.D. thesis, he taught from 1885 to 1889 in Kiel and after that from 1889 to 1926 in Berlin. From 1930 to 1937 Planck was a head of the Kaiser William Society (from 1948 it was reorganized as the Max Planck Society). His investigations were mainly devoted to thermodynamics. He became famous after explanation in 1900 of the spectrum of the so-called "absolute black body," which was laid as a fundament of quantum physics. As opposed to the physical representations regarding the continuity of all the processes, which was the basis of a physical picture of the world built by Newton and Leibniz, Planck introduced representation of the quantum nature of radiation. In accordance with his theory, radiation can be emitted and absorbed by the quantum with the energy of each quantum. As a result, having applied more or less standard theoretical approaches to statistical physics, Planck had found out the correct formula for the spectral density of radiation of the absolute black body heated to temperature T (which can be reduced to Planck's distribution in modern terminology).

Poisson, Siméon Denis Poisson (1781–1840) was a French mathematician, mechanician, and physicist.

The number of his scientific publications exceeds 300. They are concerned with different areas of pure mathematics, mathematical physics, theoretical mechanics, and celestial mechanics.

In mathematical physics, the most important publications were papers on electrostatics and magnetism, especially the latter which was devoted to the basis of the theory of temporal magnetization.

Popov, Aleksandr S (1859–1906) was a great Russian physicist and electrical engineer, the inventor of wireless electrical communication (radio communication).

He worked as a teacher of physics and electrical engineer of Mine Officers and Navy Technical Military School in Kronstadt and as a professor in physics of the St. Petersburg Electrical Institute.

First, scientific researchers of Popov were devoted to the analysis of the most profitable action of electrical dynamo and to Hughes' inductive balance. After publication of Hertz's results on electrodynamics, Popov began to study electromagnetic phenomena. Having tried to find a method to demonstrate effectively Hertz's experiments in front of a large audience, he began to design a more obvious indicator of the electromagnetic waves radiated by the Hertz vibrator. Having clearly understood the necessity for Russia of the means for wireless signaling, he stated a task behind himself in the 1890s to apply electromagnetic waves to signaling. The searches for this task solution were fulfilled in two stages: searching the electromagnetic wave indicator, which was sensitive enough; development of the device, which was able to reliably register the electromagnetic waves radiated by the Hertz vibrator. Popov chose a radio-conductor (coherer) as this indicator. By spring 1895, he constructed the sensitive and reliably operated receiver adaptable for wireless signaling (radio communication). This receiver could detect the radiation of radio signals sent by the transmitter at a distance of up to 60m. When conducting his experiments, Popov observed that the connection to the coherer, the vertical metal wire (an antenna) leaded to an increased in the distance of reliable reception.

On 25 April (7 May according to the new calendar) 1895, during the session on the physical division of the Russian Physical–Chemical Society, Popov made a scientific report about his invention of a communication system without wires and demonstrated this system. In spring 1897, in experiments in Kronstadt harbor, the 600-m distance was achieved, and in summer 1897, when testing the ships achieved a distance of 5000 m. He found out at that time that the influence of metal ships on the electromagnetic wave propagation and offered a method of determination of the direction to the operated transmitter (radar technology). During the experiments in 1897, he used the electromagnetic waves that lay on the boundary between the decimeter and meter wavelength ranges. The results of Popov's X-ray investigations may be concerned with that period of his activity; he made the first photographs in Russia of objects and human extremities on X-rays.

Poynting, John Henry (1852–1914) was an English physicist.

He was a professor of physics at Mason Science College, from 1880 to 1900, and then the successor institution, the University of Birmingham, until his death.

He was the developer and eponym of the Poynting vector, which describes the direction and magnitude of electromagnetic energy flow and is used in the Poynting theorem, a statement about energy conservation for electric and magnetic fields. This work was first published in 1884. He performed a measurement of Newton's gravitational constant by innovative means during 1893. In 1903, he was the first to realise that the Sun's radiation can draw in small particles toward it [5]: this was later named the Poynting–Robertson effect.

He discovered the torsion–extension coupling in finite strain elasticity. This is now known as the (positive) Poynting effect in torsion.

Poynting and the Nobel prizewinner J.J. Thomson co-authored a multi-volume undergraduate physics textbook, which was in print for about 50 years and was in widespread use during the first third of the twentieth century [6]. Poynting wrote most of it [7].

Prokhorov, Alexander M (1916–2002). His scientific research was devoted to radio physics, accelerator physics, radio spectroscopy, quantum electronics and its applications, and nonlinear optics. In the first papers he investigated the radio waves propagation along the Earth's surface and in the ionosphere. After the Second World War, he began to develop of methods of oscillator frequency stabilization, which was based on his Ph.D. thesis. He suggested a new method of millimeter-wave generation in the synchrotron, discovered their coherent character, and prepared his Dr.Sci. thesis for this investigation (1951).

Developing quantum frequency standards together with N.G. Basov, he formulated the main principles of the quantum amplification and generation (1953), which was realized at the development of the first quantum oscillator (microwave amplification by stimulation emission of radiation [a maser]) on ammonia (1954). In 1955, they suggested a three-level diagram of inverse population creation, which found wide application in masers and lasers. The next few years were devoted to investigations of parametric amplifiers in microwaves, in which he suggested using a series of active crystals, such as a ruby, the detailed investigation of which happened to be extremely useful for creation of the ruby laser. In 1958, Prokhorov suggested using the open resonator at the development of quantum oscillators. For fundamental research in the field of quantum electronics, which led to creation on laser and maser. A. Prokhorov and N. Basov were awarded the Lenin Prize in 1959, and in 1964, together with C. Townes, the Nobel Prize for physics.

Schukin, Alexander N (1900–1990) was a Soviet scientist in the field of radio engineering and radio physics, the author of works in theory and calculation methods of the long-range, short-wave communications, the founder of the theory of the underwater reception of radio signals. He was a USSR academician, general lieutenant engineer, and twice USSR hero. He was also laureate of the Lenin and State Prizes. He introduced for the first time the impedance boundary conditions on a well-conducting surface. This simplified to a great extent the solution of many problems of diffraction and scattering of electromagnetic waves.

Shatrov, Alexander D (1944–2017).

He graduated from the Moscow Physical-Technical Institute, worked for his whole his life at the Friazino Branch of the Institute of Radio Engineering and Electronics of the Russian Academy of Sciences, named after V.A. Kotelnikov (see above), heading the Department of Electrodynamics after B.Z. Katsenelenbaum (see above). His research covers a huge frequency range from meter waves to the optical range. He obtained fundamental results for electrodynamics of the chiral structures and the theoretical fundamentals of nano-engineering.

Shawlow, Arthur (1921–1999) was an American physicist. He chose the specialty of radio engineering at Toronto University.

Before 1944, he worked as a teacher of military personnel at Toronto University, after which he joined the team developing a microwave antenna for radar equipment. In 1945, he returned to Toronto University and prepared his Ph.D. thesis in 1949 on optical spectroscopy under the supervision of Malcolm F. Crowford. After that, he worked at California University together with C. Townes on problems of microwave spectroscopy and the superconductivity discovered by Dutch physicist Heike Kamerlingh-Onnes. In 1957–1958, Townes and Shawlow were involved in searching for the methods of obtaining of the maser effect in a visible light range, and in December 1958 they published in *Physical Review* the paper "Infrared and Optical Masers," in which they explained how it can be done. In 1960, the American physicist from "Hughes Aircraft," Theodor Meiman, had demonstrated the first acting laser. Precisely at this time, Russian physicists N. Basov and A. Prokhorov created masers and lasers independently of their American physicists. A. Shawlow received the Nobel Prize in 1981 for "the contribution in development of laser spectroscopy.

Shiller, Nikolay N (1848–1910) was a Russian physicist, Professor of Khar'kov University.

Having worked at H. Helmholtz's laboratory, he measured the substance permittivity by a method offered by him. He proved in his experiments the principle of current insularity (1900). As a consequence of the law of energy degradation, he formulated the principle of the impossibility of the continuous decreasing or increasing of temperature by means of closed adiabatic processes (a similar formulation was suggested by German scientist C. Caratheodory in 1909). In 1874, he used for the first time the electrical oscillation method to determine permittivity and checked the equation.

Shockley, William (1910–1989) entered California University at Los Angeles and after 1 year moved to the California Technological Institute, from which he graduated in 1932 with a Bachelor's degree.

After that, he graduated from the Ph.D. course at the Massachusetts Institute of Technology (MIT) and in 1936 defended his Ph.D. thesis on the theme "Calculations of Wave Functions for Electrons in Sodium Chloride Crystals." Solid-body physics was studied by Shockley at MIT, and his results on crystals became a strong basis for further research activity.

During the Second World War, Shockley worked on military projects, at first on electronic equipment for the ground-based radar of Bell Company. From 1942 to 1944, he executed the obligations oft he Director of Research of the group, dealing with investigations into anti-submarine operations, at Columbia University in New York. In 1947, J. Bardeen and

W. Brattain achieved the first successful development of the semiconductor amplifier or transistor. Shockley, Bardeen, and Brattain were awarded the Nobel Prize in physics for "investigations of semiconductors and discovery of transistor effect" in 1956. Shockley was an invited professor of the California Institute of Technologies and a supervisor of a group for the estimation of the weapon systems of the US Ministry of Defense.

Shockley was appointed First Professor of Engineering and Applied Sciences at Stanford University, where he worked until retirement in 1975.

Shuleikin, Mikhail V (1884–1939) was one of the first-rate Soviet radio physicists and radio engineers, professor of some universities, and academician.

He developed the theory of short-wave dispersion in the uniform ionized medium, derived the formula for determination of the equivalent permittivity of the ionosphere, the wave propagation factor, and the wave absorption factor as a function of frequency. He showed how it is necessary to determine the phase velocity value. He introduced the concept of the critical frequency at which the refraction factor of the ionosphere is equal to zero, and he proved that the propagation of waves, which are longer (in wavelength) than the critical ones, takes place without refraction from the atmosphere and this wave comes away to beyond atmosphere space. He developed the theory of long-wave antennas and conducted research into loop aerials.

Snell, Willebrord (1580–1626) was a Dutch astronomer and mathematician, and professor at Leyden University.

He discovered the law of light refraction that is named after him. His results were devoted to plane and spherical trigonometry.

Sommerfeld, Arnold Johannes Wilhelm (1868–1951) was a German theoretical physicist who pioneered developments in atomic and quantum physics, and also educated and mentored a large number of students for the new era of theoretical physics.

He served as Ph.D. supervisor for many Nobel Prize winners in physics and chemistry (only J.J. Thomson's record of mentorship is comparable with his).

He introduced the second quantum number (azimuthal quantum number) and the fourth quantum number (spin quantum number). He also introduced the fine-structure constant and pioneered X-ray wave theory.

Sommerfeld was a great theoretician, and besides his invaluable contributions to quantum theory, he worked in other fields of physics, such as the classic theory of electromagnetism. For example, he proposed a solution to the problem of a radiating hertzian dipole over a conducting earth, which over the years has led to many applications. His Sommerfeld identity and Sommerfeld integrals are still to the present day the most common way of solving this kind of problem. Also, as a mark of the prowess of Sommerfeld's school of theoretical physics and the rise of theoretical physics in the early 1900s, as of 1928, nearly one-third of the ordinarius professors of theoretical physics in the German-speaking world had been students of Sommerfeld [42].

Tamm, Igor E (1895–1971) was a Soviet physicist and academician.

After graduating (1918) from Moscow University he worked as a teacher at various universities. In 1924–1941 and from 1954, he worked as a teacher at Moscow University. From 1934, he worked at the Physical Institute. His main results were devoted to quantum mechanics and its application, the theory of radiation, the theory of cosmic rays and interactions of nuclear particles. He created the quantum theory of light dispersion in solid bodies (1930) and the theory of light dispersion by electrons (1930). In the field of quantum theory of metals, he created together with S.P. Shubin the theory of the photo-effect in metals (1931). He showed theoretically the opportunity of existence of the specific conditions of electrons on the crystal surface (so-called Tamm levels, 1932), which was further based on the explanation of the various effects on crystals. In 1934, he suggested and mathematically developed the quantitative theory of nuclear forces based on this approach, which was used by the modern meson theory of nuclear forces. In 1937, together with I.M. Frank, he created the radiation theory of fast moving electrons. In 1945, he developed the approximate method of explaining the interaction of nuclear elementary particles. In 1950, he offered to use the warm plasma in the magnetic field to obtain a controlled thermonuclear reaction. He created the schools of thought of theoretical physicists, and was a Nobel laureate (1958).

Tesla, Nikola (1856–1943) was a famous Serbian inventor in the field of electrical and radio engineering.

From 1884, he lived in the USA. In 1888, he described (independently from the Italian physicist) the phenomenon of the rotating magnetic field. He developed multi-phase electrical motors and the distribution schemes for the multi-phase currents. He is a pioneer of high-frequency technologies (generators, transformers etc.; 1889–1891). He investigated the possibility of transferring the electrical signals and electrical power without wires (according to the results of Popov).

Tomilin, Alexander K (born in 1955).

He is a Russian scientist, Dr.Sci. (Phys.-Math.), Professor of Tomsk State University (Russia). An honorary educational worker, he has a degree from the "International Teacher of Engineering University".

The scientific interests of Professor Tomilin are related to problems of mechanical system oscillations and the general problems of electrodynamics.

Townes, Charles (born in 1915) was a famous American physicist and one of the founders of quantum electronics.

He entered Furman University in Greenwill and graduated in 1925 with double honors. He had obtained the Bachelor of Physics and Bachelor of Arts in the field of modern languages. Although he chose physics as his main subject. attracted by its logic and the elegance of its structure, in the real life he perfectly read books in the French, German, Spain, Italian, and Russian languages. After a post-graduate course at Duke University, he received a Master's degree in physics in 1936 and after that a Ph.D. degree in 1939 at the Technological Institute of California. His doctoral thesis was named "The Separation of Isotopes and the Determination of the Spin of the Nucleus of Carbon 13." In 1954, he constructed the first maser, and is a laureate of the Nobel Prize for physics in 1964. He proved (together with A. Shawlow) the possibility of laser creation.

Umov, Nikolay A (1846–1915) was a great Russian scientist, engineer, and professor at Moscow University.

He is famous for many notable results, the first of which were devoted to the oscillation processes in elastic media. He extended this theory to thermal mechanical phenomena in these media (1870). He entered the concept of energy flow density (1870; 10 years earlier than British scientist Poynting), offered a general solution for the task on the distribution of the electric currents on the conducting surfaces of a general type (1875). He discovered the phenomenon of light chromatic depolarization, which falls on the suffused surface. He executed a number of brilliant investigations on the theory of terrestrial magnetism.

Vavilov, Sergey I (1891–1951), Soviet physicist, the founder of the scientific school of physical optics in the USSR, academician (1932), and the President of the Academy of Sciences (1945–1951).

The main scientific direction for S.I. Vavilov were investigations in the field of physical optics, in particular, the luminescence phenomenon. In 1925, he fulfilled a series of experiments, during which he discovered the reduction of the absorption coefficient of the uranium glass at highlight intensity. The effect observed was laid base on nonlinear optics. He introduced the concept of a luminescence quantum yield and investigated the dependence of this parameter upon the wavelength of exciting light (Vavilov's law). He investigated the luminescence phenomenon polarization and became a founder of new scientific direction – micro-optics. He contributed a great deal to the development of nonlinear optics. Together with his Ph.D. student P.A. Cherenkov, in 1934, he discovered the Vavilov–Cherenkov effect; for this discovery, P. Cherenkov in 1958 (after Vavilov's death) won the Nobel Prize. S.I. Vavilov himself was twice elected for the Nobel Prize (in 1957 and 1958).

Volta, Alessandro (1745–1827) was an Italian physicist and physiologist, and one of the founders of the electricity doctrine.

He created the first chemical current source, the so-called Volta column (1800). He discovered the contact potential difference.

Von Neumann, Johann (1903–1957) was a Hungarian–German mathematician who contributed to quantum physics, quantum logic, functional analysis, set theory, information science, economics, and other scientific fields.

He is most famous as a forefather of modern computer architecture (the so-called von Neumann architecture), the application of operator theory to quantum mechanics (see von Neumann algebra), and a participant in the Manhattan project and creator of the game theory and concept of cellular automata. He made a serious contribution to the development of quantum mechanics, which is a corner-stone of nuclear physics, and developed the game theory, which is a method of analysis of interconnections between people that has found wide application in various areas, from economics to military strategy.

The interest of von Neumann in computers was related in some degree to his participation in the top-secret Manhattan project for creation of an A-bomb, which was developed in Los Alamos, New Mexico. Von Neumann proved the practicability of the explosive manner of A-bomb detonation. Then, he thought about a much more powerful weapon such as the H-bomb, the development of which required very complicated calculations.

Nevertheless, von Neumann understood that a computer is nothing more than a simple calculator, and that is why (at least, potentially) it represents a universal tool for scientific research. In July 1954 (less than 1 year after joining the Mawchley and Ekkert's group), von Neumann prepared a 101-page report in which he generalized the plans of work on the powerful Electronic Discrete Variable Automatic Computer. Von Neumann's preliminary report became the first report on digital electronic computers, which was studied by a wide circle of t he scientific community.

In the future, Von Neumann worked at the Princeton Institute of Advanced Research, participated in the development of the several computers of newest configurations. There was (among the others), in particular, the computer that was used for the solution of the problems related to creation of the H-bomb. Von Neumann wittily called it MANIAC, the abbreviation from Mathematical Analyzer, Numerator, Integrator, and Computer. Von Neumann was also a member of the Commission on Atomic Energy and the Head of the Consulting Commission of the US Air Force on ballistic missiles.

Vvedenskiy, Boris A (1893–1969) was a Soviet physicist, professor of the All-Soviet Research and Development Electrical Engineering Institute and of Lomonosov Moscow State University, and academician of the Soviet Academy of Sciences.

He was born into the family of the Moscow Religious Academy. He worked in the Laboratories of N.N. Andreev, V.K. Arkad'ev, M.V. Shuleikin. His key publications cover the area of radiophysics and radio engineering; fundamental research into methods of radio wave generation and reception, especially within the UHF range. He proved that propagation features of the UHF waves in straight visibility can be caused by interference of the incident rays and the rays reflected from the Earth's surface. He found for the first time a solution of the diffraction propagation of the UHF waves in the case of spherical Earth and UHF wave propagation around the globe, taking into account not only the diffraction, but also the refraction in the atmosphere. He was the Main Editor of the *Great Soviet Encyclopedia*.

Weber, Wilhelm Eduard (1804–1891) was a German physicist and, together with Carl Friedrich Gauss, inventor of the first electromagnetic telegraph.

In 1831, on the recommendation of Carl Friedrich Gauss, he was hired by the University of Göttingen as professor of physics, at the age of 27. His lectures were interesting, instructive, and suggestive. Weber thought that, to thoroughly understand physics and apply it to daily life, mere lectures, although illustrated by experiments, were insufficient, and he encouraged his students to experiment themselves, free of charge, in the college laboratory. As a student for 20 years, he, with his brother, Ernst Heinrich Weber, Professor of Anatomy at Leipzig, had written a book on *Wave Theory and Fluidity,* which brought its authors a considerable reputation. Acoustics was a favourite science of his, and he published numerous papers upon it in *Poggendorffs Annalen,* Schweigger's *Jahrbücher für Chemie und Physik,* and the musical journal *Carcilia.* The "mechanism of walking in mankind" was another study, undertaken in conjunction with his younger brother, Eduard Weber. These important investigations were published between the years 1825 and 1838. Gauss and Weber constructed the first electromagnetic telegraph in 1833, which connected the observatory with the institute of physics in Göttingen.

Weinstein, Lev A (1920–1989), Soviet scientist–radio physicist, corresponding member of the USSR Academy of Science, and laureate of the State Prize.

The scientific results of L.A. Weinstein concerned the diffraction theory, the theory of open resonators, to wave propagation in the near-Earth space, to UHF electronics (he developed the nonlinear theory of the traveling-wave tube), to the theory of signal transmission. Theoretical methods developed by L.A. Weinstein were of great importance for the design of radar systems, microwave engineering devices, and lasers. The universal function he introduced in 1953, through which the solution of a series of quasi-optical problems, further received the name of Weinstein's diffraction $U(s,p)$-function. During his last years, he investigated the statistics of photo-samples, the theory of flicker noise, the theory of cooperative emission, and the theory of a laser on free electrons.

Whittaker, Edmund Taylor (1873–1956) was an English mathematician who contributed widely to applied mathematics, mathematical physics, and the theory of special functions.

He had a particular interest in numerical analysis, but also worked on celestial mechanics and the history of physics. Near the end of his career he received the Copley Medal, the most prestigious honorary award in British science. The School of Mathematics of the University of Edinburgh holds The Whittaker Colloquium, a yearly lecture in his honour.

In the theory of partial differential equations, Whittaker developed a general solution to the Laplace equation in three dimensions and the solution to the wave equation. He developed the electrical potential field as a bi-directional flow of energy (sometimes referred to as alternating currents). Whittaker's pair of papers in 1903 and 1904 indicated that any potential can be analyzed by a Fourier-like series of waves, such as a planet's gravitational field point-charge. The superposition of inward and outward wave pairs produces the "static" fields (or scalar potential). These were harmonically related. By this concept, the structure of electric potential is created from two opposite, though balanced, parts. Whittaker suggested that gravity possessed a wavelike "undulatory" character.

Yashin, Alexey A was born in 1948.

Russian Doctor of Sciences (Eng) and Doctor of Sciences (Biol) and professor, he made a great contribution to the electrodynamics of engineering and live systems at both a macroscopic and a microscopic level. Under his supervision and direct participation, devices were developed based on his offered theory of interaction of physical fields with biological objects.

He has a prominent literary talent and is author of many belletristic books, which have been awarded many prestigious prizes.

References[1]

1. *Nefyodov E., Kliuev B.* Transmission lines of microwave and mm-wave ranges: computerized lecture course - Textbook. LAP LAMBERT Academic Publishing GmbH & Co., Saarbrücken, Germany, 2016.-544 p.
2. *Nefyodov E. I., Smolskiy S. M.* (2012), Understanding of Electrodynamics, Radio Wave Propagation and Antennas: Lecture course for students and engineers. Scientific Research Publishers, USA, 2012. ISBN 978-1-61896-041-2.-426 pp.
3. *Kaganov V.I.* Computer calculations in EXCEL and MathCad (in Russian) // Moscow, Telekom-Hot Line Publ., 2003.-328 p.
4. *Prakash Bhartia, Protap Pramanick.* Modern RF and microwave filter design. Artech House, Boston-London. 402 p. 2016. ISBN-13: 978-1-63081-157-0.
5. *Khvorostenko N.P.* Longitudinal electromagnetic waves (in Russian)// Izvestia vuzov. Physics, 1992, № 3, p.24–29.
6. *Nefyodov E.I., Subbotina T.I., Yashin A.A.* Modern bioinformatics (in Russian).- Moscow: Telecom-Hot Line Publ., 2005.-272 p.
7. *Tomilin A.K., Misiucenko L., Vikulin V.S.* Relationships between Electromagnetic and Mechanical Characteristics of Electron // American Journal of Modern Physics and Application 2016; 3(1): pp.1–10.
8. *Kasterin N.P.* About wave propagation in heterogeneous medium. Part 1. Sound waves (in Russian). Moscow: University Publ., 1903.-165 p.
9. *Misiuchenko I.L.* The last God mystery (electric aether) (in Russian). Sankt-Peterburg, 2009.-267 p.
10. *Erokhin V.V.* Fundamentals of constructive electrodynamics, part 1.-2002 / Magnetic field in non-relativistic approximation (in Russian).
11. *Nikolaev G.I.* Secrets of electromagnetism and free energy: new conceptions of physical world (in Russian).- Tomsk: 2002.- 150 p.
12. *Tomilin A.K.* The Fundamentals of Generalized Electrodynamics. http://arxiv.org/pdf/0807.2172; *Tomilin A.K.* The Potential-Vortex Theory of the Electromagnetic Field (in English). http://arxiv4.library.cornell.edu/ftp/arxiv/papers/1008/1008.3994.pdf
13. *Gvozdev V.I., Kuzaev G.A., Nefyodov E.I., Yashin A.A.* Physical bases of VIC SHF and mm-range modeling (in Russian)// Uspekhi fisicheskikh nauk, 1992, Vol.162, № 3, p. 129–160.
14. *Gvozdev V.I., Nefyodov E.I.* Volumetric integrated circuits in microwaves (in Russian)- Moscow: Nauka Publ., 1985.-256 p.
15. *Gvozdev V.I., Nefyodov E.I.* Volumetric integrated microwave circuits – the elements base of analog and digital radio-electronics (in Russian).- Moscow: Nauka Publ., 1987.-111 p.
16. *Gridin N.V., Nefyodov E.I., Chernikova T.Yu.* Electrodynamics of structures in extra high frequencies (in Russian) / Under edition of ⬚.M. Belotserkovsky.- Moscow.: Nauka, 2002.-360 p.
17. *Neganov V.A., Nefyodov E.I., Yarovoy G.P.* Strip-slot structures of microwave and mm-wave ranges (in Russian). - Moscow: Nauka Publ., 1996.-304 p.
18. *Neganov V.A., Nefyodov E.I., Yarovoy G.P.* Modern methods of transmission line and resonator design (in Russian): Textbook. - Moscow: Pedagogika Press Publ., 1998.-328 p.
19. *Neganov V.A., Nefyodov E.I., Yarovoy G.P.* Electrodynamics methods for design of SHF devices and antennas (in Russian).-Moscow: Radio I Sviaz Publ., 2002.-416 p.
20. *Nefyodov E.I., Kozlovsky V.V., Zgursky A.V.* Microstrip emitting and resonance devices (in Russian).- Kiev: Tekhnika Publ., 1990.-160 p.
21. *Nefyodov E.I., Saidov A.S., Tegilaev A.R.* Wideband microstrip control microwave devices (in Russian).- Moscow: Radio I sviaz Publ., 1994. - 168 p.
22. *Henning F. Harmuth,* Nonsinusoidal Waves For Radar And Radio Communication (Advances In Electronics & Electron Physics Supplement). Academic Pr. 1981. 396 p. ISBN-13: ISBN 9780120145751
23. *Nefyodov E.I.* Open coaxial resonance structures (in Russian).- Moscow: Nauka, 1982.-220 p.
24. *Nefyodov E.I.* Electrodynamics, people, life. – Tsaritsyn, Stalingrad-Volgograd: (in Russian). Blank Publ. 2002.-504 p.
25. *Nefyodov E.I.* Radio electronics of our days (in Russian).- Moscow: Nauka Publ., 1986.-192 p.
26. *Nefyodov E.I.* Engineering electrodynamics: textbook for universities (in Russian).- Moscow: Academia Publisher, 2008.-416 p.
27. *Nefyodov E.I.,* Microwave devices and antennas: textbook for universities (in Russian).- Moscow: Academia Publ., 2009.-384 p.
28. *Nefyodov E.I.* Antenna-feeder devices and radio wave propagation: textbook for colleges (in Russian) / 2nd edition.- Moscow: Academia Publ., 2009.-320 p.
29. *Julius Adams Stratton.* Electromagnetic theory. Wiley-Interscience. 2007. ISBN-13 978.0-470-13153-4.
30. *B. Rama Rao, W. Kunysz, R.Fante, K.McDonald.* Antennas for GPS/GNSS . Artech House Publisher, Boston-London. 2013. ISBN-13: 978-1-59693-150-3.
31. *Nefyodov E.I., Ermolaev Yu.M., Smelov M.V.* Experimental research of excitation and propagation of knotted electromagnetic waves in various media (in Russian).// Radiotekhnika, 2014, № 2, p. 31–35.

---

[1] This is the list of literature, which is recommended for interested students and specialists.

© Springer International Publishing AG, part of Springer Nature 2019
E. I. Nefyodov, S. M. Smolskiy, *Electromagnetic Fields and Waves*, Textbooks in Telecommunication Engineering,
https://doi.org/10.1007/978-3-319-90847-2

32. *Perunov Yu.M., Matsukevich V.V., Vasiliev A.A.* Foreign radio electronic means (in Russian)./ Under edition of Yu.M. Perunov. Book 3. Antennas. -Moscow.: Radiotekhnika, 2010.-304 p.

33. *Smelov M.V.* Excitation and propagation of knotted electromagnetic waves. Germany. LAP LAMBERT Academic Publishing. 2015. P. 333. ISBN 978-3-659-77846-9

34. *Tesla N.* Lections (in Russian).- Samara: «AGNI» Publ., 2010.- 312 p.

35. *Sacco B., Tomilin A.* The Study of Electromagnetic Processes in the Experiments of Tesla. http://viXra.org/abs/1210.0158

36. *G.K. Southworth.* Principles and applications of wave-guide transmission, New York, Van Nostrand [1950], xi, 689 p. illus. 24 cm. Bell Telephone Laboratories series. 1950.

37. *Bhat B., Koul S.K.* Analysis, Design and Application of Fine Lines.- Artech House Publ, 1987.-475 p.

38. *Nefyodov E.I.* Antenna-feeder devices and radio wave propagation: text-book for college (in Russian).; 2nd edition.- Moscow.: Publ. Center «Academia», 2009.-320 p.

39. *Paul C.R., Whites K.W., Nazar S.A.* The Mathcad Electronic Book for Introduction to Electromagnetic Fields, third edition / Visual Electromagnetics for Mathcad / Ed. Whites K.W. // The McGraw-Hill Companies, Inc., Cambridge, Massachusetts, 2000.

40. *Hoffmann R.K.* Handbook of Microwave Integrated Circuits.- Artech House Publ., 1987.-527p

41. *E.T. Whitteker.* "A History of the Theories of Aether and Electricity" (From the Age of Descartes to the Close of the Nineteenth Century). LOGMANS, GREEN and Co., 39 PATERNOSTER ROW, LONDON, 1910. 502 P.

42. Faraday's Dairy on Experimental Investigations. HR Direct Publ., 2008, USA. 560 p. ISBN 9780981908328.

43. *Nikolaev G.V.* Electromagnetics Secrets and the free energy: new concepts of the physical world (in Russian). Tomsk. 2002.- 150 p.

44. *Nikolaev G.V.* Modern electrodynamics and reasons of its paradoxicality: prospect of consistent electrodynamics development (In Russian). Book 1.- Tomsk: Tverdynya Publ. 2003. -149 p.

45. *Pursell E.M., D.J. Morin.* Electricity and Magnetism. 3rd Edition, Cambridge University Press, 2013, 853 pages, ISBN: 1107014026.

46. *Born M., Wolf E.* Principles of Optics.- (4th.ed.) Pergamon Press 1970. -790p.

47. *Smyth W.R.* Static and Dynamic electricity, Third Edition, Revised Printing. Taylor & Francis: 1989, 623 pages.

48. *Luzin B.A.* The aether and the universe or the end of relativism (in Russian).- Perm: 2003.-424p.

49. *Nefyodov E.I.* Radio wave propagation and antenna-feeder devices: textbook for universities (in Russian).- Moscow: Academia Publ., 2010.-320 p.

50. *Katsenelenbaum B.Z., Korshunova E.N., Sivov A.N., Shatrov A.D.* Chiral electrodynamics objects (in Russian)// Uspekhi fisicheskikh nauk, 1997, vol. 167, № 11, p. 1201.

51. *Neganov V.A., Osipov O.V.* Reflecting, waveguiding and emitting structures with chiral elements (in Russian).- Moscow: Radio I Sviaz Publ., 2006.- 280 p.

52. *Nefyodov E.I.* Electromagnetic wave diffraction on dielectric structures (in Russian).- Moscow: Nauka Publ., 1979.-272 p.

53. *Nikolsky V.V., Nikolskaya T.B.* Electrodynamics and radio wave propagation: Textbook for universities (in Russian).- Moscow: Nauka Publ., 1989. – 543 p.

54. *Tamm I.E.* Fundamentals of electricity theory: textbook for university (in Russian). – Moscow: Nauka Publ., 1989. -504 p.

55. *Zhilin P.A.* Reality and mechanics (in Russian) // Proceedings of XXIII school-seminar «Analysis and synthesis of non-linear mechanical oscillating systems», Sankt-Peterburg, 1-10 July, 1995. Institute of Machinery Problems Publ. 1996. Pp. 6–49.

56. *Kostrov B.V.* Mechanics of tectonic earthquake center (in Russian). — Moscow: Nauka Publ., 1975. — 173 p.

57. *Kurushin E.H., Nefyodov E.I.* Electrodynamics of anisotropic waveguiding structures (in Russian).- Moscow: Nauka Publ., 1983.-223 p.

58. *Nefyodov E.I., Fialkovsky A.T.* Strip transmission lines: theory and calculation of typical irregularities. (in Russian).- Moscow: Nauka Publ., 1974.-128 p.

59. *Shchukin A.N.* Radio wave propagation: textbook for universities (in Russian).-Moscow: Sviazizdat Publ., 1940. — 399 p.

60. *Nefyodov E.I., Sivov A.N.* Electrodynamics of periodic structures (in Russian). - Moscow: Nauka Publ., 1977. –208p.

61. *Kuzelev M.V., Rukhadze A.A.* Electrodynamics of Dense Electron Beams in Plasma, English completed edition Plasma Free Electron Lasers Edition Frontier Paris. 1995.

62. *Yasinsky S.A.* Applied "gold" mathematics and its applications in communications (in Russian).- MDoscow: Hot Line-Telecom Publ., 2004.-239 p.

63. *Edward M. Purcell.* Electricity and magnetism. Berkeley Physics Course. Vol.2. 853 p. Cambridge Academ., 1975. ISBN-13(EAN): 9781107014022

64. *Feynman R.F., Leighton R.B., Sands M.* The Feynman lectures on physics. Vol.5. Electricity and magnetism. Vol.6. Electrodynamics - ADDISON-WESLEY Publishing Company. 1963–64.

65. *Vaganov R.B., Katsenelenbasum B.Z.* Fundamentals of diffraction theory (in Russian). - Moscow: Nauka Publ., 1982.-272 p.

66. *Katsenelenbasum B.Z.* High-frequency electrodynamics: fundamentals of the mathematical apparatus (in Russian). –Moscow: Nauka Publ., 1966. -240 p.

67. *Marcuvitz N.* Waveguide Handbook, New York: Dover Publications, 1998. — 428 p. — ISBN-10: 0863410588; ISBN-13: 978–0863410581

68. *Maliuzhinets G.D.,* Development of representations on wave diffraction phenomena (in Russian)// UFN, 1959, vol. 69, №2, p.321–334

69. *Weinstein L.A.* Diffraction theory and factorization method (in Russian). –Moscow: Sovetskoe Radio, 1966.-432p.

70. Computational Electromagnetics/under edition of Ray Mittra. Springer Science+Business Media, New York, 2014, 704 pages, ISBN: 1461443814. http://www.twirpx.com/file/1469452/

71. *Ufimtsev P.Ya.* Fundamentals of the Physical Theory of Diffraction, John Wiley & Sons, Inc., Hoboken, New Jersey, 2007

72. *L.D. Landau & E.M. Lifshitz* Electrodynamics of Continuous Media (Volume 8 of A Course of Theoretical Physics) Pergamon Press 1960

73. *Nefyodov E.I., Kliuev S.B.* Electrodynamics and radio wave propagation (in Russian): textbook for university students.- Moscow: KURS Publ., 2016.-344 p.

74. *Delorme B. Antennas.* Site engineering for mobile radio networks. Artech House, Boston-London

75. *Honl H., Maue Au.W., Westpahl K.* Theorie der Beugung. Springer Verlag, 1961.-320 c.

76. *Grinberg G.A.* Selected issues of mathematical theory of electric and magnetic phenomena (in Russian).- Moscow-Leningrad: AN of USSR, 1948. -728 p. www.twirpx.com/file/21425/

77. *Makeeva G.S., Golovanov O.A.* Mathematical modeling of electronically-controlled devides of terahertz range on the base of a graphen and carbonic nano-tubes. (in Russian) – Penza: Penza State University Publ., 2018. 304 p.

78. *Kosterin N.P.* Generalization of main equations of aerodynamics and electrodynamics (in Russian).- Moscow: AS USSR Publ., 1937.-21 p.

79. *Mitkevich V.F.* Physical action at a distance. Proceedings of the Russian Academy of Science. Series VII. Division of mathematical and natural science, 1391–1409 (1933).

80. *Kiryako A.G.*: Theories of origin and generation of mass. http://electricaleather.com/d/358095/d/massorigin.pdf

81. *van Vlaenderen K. J., Waser A.* Generalization of classical electrodynamics to admit a scalar field and longitudinal waves// Hadronic Journal 24, 609–628 (2001).

82. *Woodside D.A.* Three-vector and scalar field identities and uniqueness theorems in Euclidean and Minkowski spaces// Am. J. Phys.,Vol.77, № 5, pp.438–446, 2009.

83. *Arbab A. I., Satti Z. A.* On the Generalized Maxwell Equations and Their Prediction of Electroscalar Wave// Progress in physics, 2009, v.2. – s. 8–13.

84. Podgainy D.V., Zaimidoroga O.A. Nonrelativistic theory of electroscalar field and Maxwell electrodynamics// http://arxiv.org/pdf/1005.3130.pdf

85. *Stokes G. G.* On some cases of fluid motion. Internet Archive. Transactions of the Cambridge Philosophical Society 8(1): 105–137(1843).

86. *Fock V.A.* Theory of space, time and gravitation., Pergamon Press Ltd. 1964.

87. *Feynman R., Layton R, Sands M.* Feynman Lectures on Physics. Volume 6: Electrodynamics. Addison Wesley Publishing Company. Sixth printing, 1977. - 515 p..

88. *Helmholtz H.* About integrals of hydrodynamic equations, which correspond to the vortex motion. Crelles J. 55, 25 (1858).

89. *Rohrlich F.* The dynamics of a charged sphere and the electron. American Journal of Physics 65 (11): pp.1051–1056 (1997). 1997, https://doi.org/10.1119/1.18719

90. *Schwinger J.* Electromagnetic mass revisited// Foundations of Physics, 13 (3): pp. 373–383, (1983). https://doi.org/10.1007/BF01906185

91. *Fedosin S. G.* The Integral Energy-Momentum 4-Vector and Analysis of 4/3 Problem Based on the Pressure Field and Acceleration Field. American Journal of Modern Physics. Vol. 3, №. 4, pp. 152–167 (2014).

92. *Lorentz G.A.* The theory of electrons and its application to the phenomena of light and heat radiation. Dover Publications, 2011, 352 p.

93. *Kozyrev N.A.* Selected works (in Russian). Leningrad University Publ., 1991.- 438 p.

94. *Sovetov N.M.* UHF engineering: theoretical bases (in Russian) Moscow: Vyshaya Skola Publ., 1976. 184 p.

95. *Gluschenko A.G., Zakharchenko E.P.* Beyond-cutoff waveguide structures and media with amplification (in Russian). Samara: Scientific Center of RAN, 2009.-85 p.

96. *Shevchenko V.V.* Direct and backward waves: three definitions, their mutual connection and application conditions // Uspekhi fisicheskikh nauk, 2007, Vol.177, №3, p.301–306.

97. *Chaplin A.F.* Analysis and Synthesis of antenna arrays. – Lviv, Ukraine: Lviv State University Publ., 1987.-180 p.

98. *Nefyodov E.I., Fialkovskiy A.T.* Asymptotic diffraction theory of electromagnetic waves on the finite structures (in Russian). – Moscow. Nauka Publ. 1972. –204 p.

99. *Sivukhin D.V.* General course of physics: optics (in Russian). Vol. 4 - Moscow: Nauka Publ., 1985.-752 p.; Elcctricity. Vol.3.- Moscow: Fizmatlit, 1983.

100. *Schukin A.N.* Radio wave propagation: textbook.- Moscow: Sviazizdat, 1940.-399 p.

101. *Noble B.* Methods Based on the Wiener–Hopf Technique.- Pergamon Press, London, New York, Paris, Los Angeles, 1958. 279p.

102. *Kugushev A.M., Golubeva N.S., Mitrokhin V.M.* Fundamentals of radio electronics. Electrodynamics and radio wave propagation (in Russian). Textbook.- Moscow: Bauman Technical University Publ., 2001.-368 p.

103. *W.Wood R.* Physical Optics, New York, The MACMILLAN COMPANY. 1934. 895 p.

104. *Ramo S. and Whinnery J.R.,* Fields and Waves in Communication Electronics, John Wiley and Sons, New York. 1965.

105. *Snyder A.W., Love J.D.* Optical Waveguide Theory, London-York, Chapman and Hall, 1983, 656 p.

106. *Unger H.G.* Planar Optical waveguides and Fibers, CLARENDON PRESS, Oxford, 1977. 656 p.

107. *Marcuse D.* Light Transmission Optics, N-Y, Cincinatti, Toronto-London-Melbourne, 1972. 570 p.

108. *Belov L.A., Smolskiy S.M. and Kochemasov V.N.* RF, Microwave and Millimeter Wave Components. 685 Canton Street, Norwood, Massachusetts 02062, 2012.

109. *Buduris G., Shevenie P.* Microwave Circuits / Transl. from French. Moscow: Sovradio Publ., 1979.-288 p.

110. *Gupta K.S. and Singh A.* Microwave Integrated Circuits, Halsted Press, New York. 1975.

111. *Vaĭnshteĭn L A.* Open resonators and open waveguides, Golem Press, 1985. 439 p. www.twirpx.com/file/506951/.

112. *Kisun'ko G.V.* Electrodynamics of hollow systems (in Russian). Leningrad, Communication Academy Publ., 1949. -426p.

113. *Hoffmann R.K.* Handbook of Microwave Integrated Circuits.- ARTECH HOUSE,INC, 1987.-527pp

114. *J. Fjelstad, T. DiStefano, A. Faraci.* Microelectronics International, MCB UP Ltd, 1982. ISSN: 1356-5362.

115. *Gunston M.A.R.* Microwave transmission lines impedance data. Van Nostrand Reinhold Company LTD. New York. Cincinati. Toronto. Melburn. 1972. ISBN 442 02898 9. — 125 p

116. *Grigo S.* λ/8-coupling in BIC of UHF and EHF ranges (in Russian) // Electrodynamics and UHF EHF and Optical Range Engineering, 2001, vol. 9, №1, p.43–55.;

117. *Felsen L.B., Marcuvitz N.* Radiation and Scattering of Waves.- Prentice Hall, New York, V.1, V.2.- 1973.

118. *Balanis C.A.* Advanced Engineering Electromagnetics, J.Wiley & Sos, 2012.

119. *Brekhovskikh L.M.* waves in laminate media (in Russian). — Moscow: AN USSR Publ., 1957. — 507 p.

120. *Fletcher C.A.J.* Computational Galerkin Method, Springer Verlag, N.-Y., Berlin, Heidelberg, Tokyo. 1984

121. *Stutzman W. L. and Thiele G. A.*: Antenna Theory and Design, 2nd edition, Wiley NY. 1998.

122. *Sheluhin O., Smolsky S., Osin A.* Self-Similar Processes in Telecommunications. John Wiley & Sons, The Attrium, Southern Gate, Chichester, West Sussex. 2007. 314 p

123. *Eoin Carey, Sverre Lidholm.* Millimeter-Wave Integrated Circuits. 2005 Springer Science + Business Media, Inc..268 p. ISBN 978-0-387-23665-0.

124. *Mitra R., Lee S.* Analytical methods of the waveguide theory. — New York. MacMillan: 1971. 302p.

125. *Weinstein L.A.* Electromagnetic waves (in Russian). —Moscow: Sovradio Publ., 1988. — 440 c.

126. *Paul C.R.* Introduction to Electromagnetic Compatibility, 2006, J.Wiley & Sons INC Publication, New Jersey.

127. *Panchenko B.A., Nefyodov E.I.* Microstrip antennas (in Russian). - Moscow: Radio i Svaiz Publ. 1985.-144pp.

128. The Electrical Engineering Handbook, under edition of Wai Kai Chen, Elsevier Academic Press, Amsterdam-Boston-Heidelberg, 2004, ISBN 0-12-170960-4, 1208 p.

129. *Ochkov V.A.* MathCad 14 for students and engineers: Russian version.- Sankt Peterburg. 2009.-512p.

130. *Benson F.A.* (ed), Millimeter and Submillimeter Waves, Iliffe Books, London. 1969.

131. *Abraham M., Becker R.* Klassische Electrizität und Magnetismus, Berlin. 1932.

132. *Sommerfeld A.* Theorie der Beugung. Chapter 20 in the book Differential and Integralgleichungen der Mechanik und Physik. 1935. Vol. 2, Physical Part. Editors: F.Frank and R.V.Mizes. Frie. Vieweg & Sohn, Braunschweig, Germany. American publication: New York, 1943, 1961.

133. *Tikhonov V.I, Samarskiy A.A.* Equations of mathematical physics (in Russian).-Moscow: Nauka, 1972.

Index

© Springer International Publishing AG, part of Springer Nature 2019
E. I. Nefyodov, S. M. Smolskiy, *Electromagnetic Fields and Waves*, Textbooks in Telecommunication Engineering,
https://doi.org/10.1007/978-3-319-90847-2

Printed in the United States
By Bookmasters